Environmental Challenges and Solutions

The Environmental Challenges and Solutions series aims to improve our understanding of the Earth's most important environmental challenges, and how we might more effectively solve or at least mitigate these challenges. Books in this series focus on environmental challenges and solutions in particular geographic regions ranging from small to large spatial scales. These books provide multidisciplinary (technical, socioeconomic, political, etc.) analyses of their environmental challenges and the effectiveness of past and present efforts to address them. They conclude by offering holistic recommendations for more effectively solving these challenges now and into the future.

We are welcoming proposals specifically on the following topics: -Environmental sciences and social aspects -Environmental sciences and human conflicts -Boosting de-carbonization of energy system to face climate and economic emergency -Climate emergency and health: sources, science-based analysis and global solutions -Indoor and outdoor chemicals, environmental quality and public health in a changing world -Demographic, climate and environmental change: evidence and recommendation for global sustainability -Technological, industrial and corporate change for supporting environmental and social sustainability -Nanotechnology applied to water decontamination/purification processes -Sorbents and other materials in environmental protection -Raw materials supply and circular economy -Strategies to boost clean mobility

All books are written in a concise and readable style, making them suitable for both specialists and non-specialists starting at first year graduate level. Proposals for the book series can be sent to the Publishing Editor, Éva Lörinczi eva.loerinczi@springer.com.

Frode Skarstein • Lili-Ann Wolff
Editors

Sustainability and Science Education

Exploring Roles and Challenges

Springer

Editors
Frode Skarstein
Department of Education and Sports Science, Faculty of Arts and Education
University of Stavanger
Stavanger, Norway

Lili-Ann Wolff
Faculty of Educational Sciences
University of Helsinki
Helsinki, Finland

ISSN 2214-2827 ISSN 2214-2835 (electronic)
Environmental Challenges and Solutions
ISBN 978-3-032-24407-9 ISBN 978-3-032-24408-6 (eBook)
https://doi.org/10.1007/978-3-032-24408-6

This work was supported by Karlstad University, Sweden; University of Stavanger, Norway; University of Helsinki, Finland; and University of Inland, Norway.

This Springer imprint is published by the registered company Springer Nature Switzerland AG
The registered company address is: Gewerbestrasse 11, 6330 Cham, Switzerland

Introduction: Integrating Sustainability Thinking into Science Education—A Multifaceted Challenge

The challenges of sustainability are not only ecological and technical but also cultural, political, and moral. They demand that societies negotiate diverse values, identities, and interests while at the same time remaining within the biophysical boundaries of the Earth. In this negotiation, science has a crucial, though often contested, role. It provides indispensable insights into ecosystems, climate dynamics, biodiversity, energy, and health. Yet science is only one way of knowing among many, and its authority is not always accepted without question. To serve sustainability, science needs to be taught and learned in ways that allow people to both appreciate its contributions and critically examine its assumptions and limits.

Science education is central in this effort. It has long been justified either by the need to prepare future scientists or by the ambition to transmit established facts to all citizens. Over the last decades, however, a broader vision has emerged. In this vision, science education is not only about knowledge acquisition but also about fostering democratic citizenship. Rooted in a *Bildung* tradition, this vision understands education as formative of individuals and societies, linking knowledge with identity, values, and collective decision-making. It acknowledges that science is a cultural practice as well as a body of knowledge, and that to be educated in science is to be able to inquire, critique, and engage with uncertainty. This shift makes science education particularly relevant to sustainability. To teach sustainability through science is not only to introduce new content but to reimagine the very purposes and practices of science education.

This anthology gathers a wide variety of contributions that address this task from various vantage points. Together, they show that the integration of sustainability into science education is multifaceted: it involves philosophical and normative questions about the aims of education; structural and curricular innovations across disciplines; insights into students' worldviews and ways of relating to nature; and tangible pedagogical practices that succeed or struggle in the classroom. The authors come from a variety of countries and traditions, with several contributions rooted in international collaborations such as the COST 'network PHOENIX—CA19123—Protection, Resilience, Rehabilitation of Damaged Environment (2020–2024). Like the authors, the diversity of approaches is also intentional. It demonstrates that there is no single blueprint for aligning science education with sustainability thinking, but rather a need to explore multiple pathways, each with its own tensions and opportunities.

The book is organized into four parts, moving from the broadest conceptual framings to tangible classroom practices, and ending with editorial reflections that draw together the themes of the volume.

Part I: Conceptual and Critical Perspectives sets out the philosophical and value-laden foundations of sustainability in relation to science education. *Niklas Gericke* analyzes the tensions that arise when sustainability perspectives are integrated into science education, showing how the field must negotiate between maintaining disciplinary coherence and addressing societal needs. *Kerry Shephard* examines how higher education and science education can contribute to a critical citizenry, one that is capable of thoughtful engagement with sustainability challenges. *Fiona Van Schaik* highlights the importance of uncertainty, arguing that recognizing and teaching uncertainty is essential for preparing students to deal with complex sustainability issues. *Claire Polo* and *Kristine Lund* approach vaccine hesitancy during the Covid-19 pandemic as a socio-scientific controversy, revealing how cognitive, social, and emotional dimensions are inseparably linked in such debates, and how science education could respond. Finally, *Tuula Sarvas Skarstein* and *Frode Skarstein* introduce the idea of evidence-based hope in ocean education. They argue that while fear-based narratives may paralyze learners, scientifically grounded and solution-oriented narratives of hope can mobilize engagement and agency. Taken together, these chapters provide a conceptual platform for the rest of the book, showing why science education must be rethought in light of sustainability.

Part II: Interdisciplinary Pathways turns to curriculum and professional practice, asking how sustainability can be woven into science education by crossing disciplinary boundaries. *Anne Bergliot Øyehaug*, *Jill Tove Buseth*, and *Robert Didham* discuss how teacher education programs can cultivate interdisciplinary approaches that engage with sustainable development, pointing to institutional challenges but also to the opportunities for teacher educators to prepare future professionals for this task. *Noora Jaakkola* investigates the integration of sustainability into Finnish engineering curricula, illustrating how technical training is increasingly reshaped by the imperative of sustainability. *Eija Yli-Panula*, *Isabel Banos-González*, *Eila Jeronen*, and *Sandra Bueno-Férez* focus on biology education, showing how biodiversity can be taught in ways that are not only scientifically accurate but also transformative in cultivating a sense of responsibility. *Eldri Scheie* provides a detailed example of how collaboration across subjects can engage students with wicked problems, using pollinators as a case that reveals the interdependence of ecological, social, and economic dimensions. These chapters illustrate the structural and curricular changes needed for science education to contribute meaningfully to sustainability thinking in education.

Part III: Students, Nature, and Worldviews shifts attention to learners' own perspectives. S*laven Gasparovic*, *Rawad Chaker*, and *Maria Impedovo* examine how children's engagement with the virtual world of Minecraft shapes their conceptions of plants and ecosystems. They show both the dangers of plant blindness and the potential of digital environments for fostering ecological sense-making. *Gregor Torkar*, *Iris Zajc*, and *Bruce Johnson* explore primary school students' knowledge about water, their environmental worldviews, and their reported behaviors, shedding light on how scientific understanding, values, and everyday practices intersect in the minds of young learners. These contributions underline the importance of understanding students' perspectives and worldviews when linking science education to sustainability.

Part IV: Pedagogical Practices and Challenges turn to classrooms and teaching practices. *Anna Uitto* develops the notion of action competence in science education, asking how science teaching can empower students not only to understand sustainability but also to act. *Lili-Ann Wolff*, *Sari Vuorenpää*, and *Pia Sjöblom* describe a project in which students cultivated bacteria in collaboration with a scientist, demonstrating how hands-on inquiry can foster health literacy and include sustainability-related issues in science education. These chapters provide vivid examples of the promises and obstacles of teaching sustainability in diverse contexts.

In the *Concluding Reflections*, we bring the volume to a close. Here *Frode Skarstein*, *Lili-Ann Wolff*, and *Kerry Shephard* attempt to identify and gather threads that run through the book. Across the three earlier parts, a common insight emerges: sustainability cannot be integrated into science education simply by adding new topics such as climate or biodiversity. Rather, it requires a re-examination of the aims and practices of science education itself. It involves rethinking how science is taught, how it relates to other disciplines, how it connects with students' worldviews, and how it fosters the capacity to act.

In conclusion, this anthology demonstrates that when dealing with sustainability issues, science education must be both critical and constructive. It should foster deep understanding of scientific knowledge while also cultivating openness, interdisciplinarity, and the courage to act. By combining conceptual analyses, curriculum studies, learner perspectives, and classroom practices, the chapters gathered here provide a rich resource for educators, researchers, and policymakers who seek to align science education with the urgent demands of a sustainable future.

Stavanger, Norway — Frode Skarstein
Helsinki, Finland — Lili-Ann Wolff

Contents

Part I
Conceptual and Critical Perspectives

Chapter 1
The Challenges of Including a Sustainability Perspective in Science Education

Niklas Gericke

Abstract In the curricula of many countries, sustainability education is the responsibility of teachers of all subjects, including science teachers. This chapter will show that teaching science from a sustainability perspective may prove challenging for science teachers because such a perspective draws on knowledge structures and teaching traditions that lie outside of the realm of science. Based on an analysis of the didactical questions of "the why," "the what," and "the how" these challenges will be outlined, which include the normative perspective on education, that is, including values and emotions in addition to scientific concepts, and the inclusion of an anthropocentric perspective. Also, the complexity of sustainability issues calls for multi- or interdisciplinary perspectives, while the "wicked" nature of sustainable development (as opposed to the "tame" issues science commonly deals with) makes it difficult to apply predefined solutions, which is a common approach in science education. A possible way to overcome these challenges is applying a cross-curricular approach to the organization of sustainability education in which teachers representing separate subjects or disciplines collaborate. It can be argued that this complementary perspective could widely benefit science teachers, who could focus on contributing to sustainability education based on their specific disciplinary competencies rather than reorienting their teaching in a way to accommodate the challenges associated with providing a comprehensive view of sustainability.

Keywords Cross-curricular teaching · Education for sustainable development · Interdisciplinarity · Knowledge structures · Multidisciplinary · Science education

---

N. Gericke (✉)
Karlstad University, Karlstad, Sweden
e-mail: niklas.gericke@kau.se

F. Skarstein, L.-A. Wolff (eds.), *Sustainability and Science Education*, Environmental Challenges and Solutions,
https://doi.org/10.1007/978-3-032-24408-6_1

1.1 Introduction

This chapter will address the possible challenges that science teachers may face when including a sustainability perspective in their teaching. It is now well-established that sustainability is important for human beings' survival as a species. Consequently, the importance of addressing sustainability in schools and teaching is pivotal to future attempts at harmonious survival on planet Earth, and education is now a part of Sustainable Development Goal 4 in the UN policy framework (Sachs et al. 2017). This has motivated numerous countries, including Norway, Sweden, and Finland, to draft steering documents concerning sustainability education and/or education focusing on sustainable development. However, these terminologies sometimes encompass different ideas of what education is or what it should be. As a result, these coexisting, and potentially oxymoronic, concepts can make it difficult for science teachers to know how to include sustainability in their teaching and what aspects to teach students.

This chapter will draw a didactic perspective of teaching in which it is central to differentiate between *the why*, *the what,* and *the how* of teaching. The why refers to the purpose or objectives of teaching. The what refers to the content that the teacher has selected to teach. The how refers to the approaches or methods used when teaching. I will show that the decision to include a sustainability perspective in science education requires embracing multiple aspects of different knowledge cultures, each of which influences the why, the what, and the how of teaching. The purposes, contents, and methods that are relevant for sustainability education are often grounded in disciplinary and teaching traditions that are distinct from those related to science education; this makes it inherently difficult to align science education with the approaches used to teach sustainability. Consequently, science teachers—who are accustomed to viewing the curriculum and pedagogy based on a science tradition—might experience challenges in aligning their teaching with the sustainability perspective.

The end of the chapter suggests how to overcome these possible challenges by applying a cross-curricular teaching approach, which include teachers from multiple school subjects, and thereby draws from knowledge structures and teaching traditions of different disciplines. Such a teaching approach provides the opportunity for science teachers to make use of their own specialized knowledge, while at the same time they create a more holistic and interconnected learning experience for students. Hopefully this chapter will shed some light on the inherent differences between science education and sustainability education so that science teachers manage to overcome them.

1.2 The Characteristics of Sustainability Issues and the Associated Consequences for Science Curricula: The Why and the What

The content within science education is normally delimited to the natural world, without any connection to humans. Knowledge is seen as objective, and the application of knowledge is often not considered to be the responsibility of the scientist or the learner. For example, biology teachers educate students about ecosystems, but the knowledge does not always describe the impact on humans. In chemistry, the teacher explains organic chemistry without addressing the implications for humans or ecosystems. In geoscience, the teacher describes sedimentary rocks, but will not generally cover the importance of geology for human life. In physics, the teacher discusses various forms of energy without necessarily addressing the relationships to the environment. Of course, this is a simplification, because teachers address how science affects humans and the environment in more normative ways when discussing socioscientific issues (which are included in science education) or when addressing the technological applications of science (Hodson 2014). Nevertheless, these aspects are most often not at the center of teaching, but rather extensions of the underlying scientific concepts. Hence, teaching science from a sustainability perspective will inevitably lead to a change in the teaching approach in which the teacher must discuss more normative aspects.

Sustainability can be described as a social goal in which humans successfully coexist with each other and the surrounding nature on Earth over long periods of time (Purvis et al. 2019). There is currently no unanimous definition of sustainability, with the descriptions varying in literature and based on context, as well as over time (Giddings et al. 2002). The *Oxford Dictionary of English* defines sustainability as "the property of being environmentally sustainable; the degree to which a process or enterprise is able to be maintained or continued while avoiding the long-term depletion of natural resources" (Sustainability 2024). Hence, this definition emphasizes how sustainability over time is connected to the goal of maintaining natural resources. This reveals the first, and perhaps most problematic, obstacle to overcome when incorporating a sustainability perspective into science education—dealing with a *normative* concept. As such, sustainability is not a scientific concept that science teachers, whether teaching biology, chemistry, geoscience, or physics, are used to teach. The scientific concepts taught in educational institutions normally describe a phenomenon in the natural world, whether it is an object or a process. In other words, the concept has no direction and does not prescribe what the learners should do with the knowledge once they have understood the concept; it is precisely the opposite when a sustainability perspective is included in science education.

The inclusion of a sustainability perspective in teaching requires shifting to an overarching perspective of the long-term survival of humankind and nature, which—in the scientific field—could be more appropriately described as maintaining biodiversity; this shift involves including certain *values* in teaching. Teachers cannot be totally objective and only teach scientific concepts when including sustainability in the curriculum because sustainability issues are both *complex* and *wicked* (see the next paragraph). Here, the term complex refers to the multifaceted nature of sustainability issues, which require various kinds of knowledge and concepts to address, that is, a *multidisciplinary* perspective. As such, one must include ideas and concepts from diverse knowledge domains to be able to understand a sustainability issue. For example, many problems linked to biodiversity are related to invasive species that need to be understood from a biological or ecological perspective, that is, the invasive species lacks competitors or predators in a new ecosystem and can therefore take over an ecological niche and outcompete native species. However, to understand how to prevent the spread of invasive species (the normative part of the issue), which is largely the result of various human activities, one must have sufficient knowledge about human actions in relation to economic, psychological, and social incentives (Luke et al. 2023). Hence, this complexity requires science education to include knowledge elements or concepts from various disciplines, often from both science and social science. This multidisciplinary perspective is an approach that school science teachers seldom are accustomed to (Sund and Gericke 2020, 2022).

In addition to complexity, sustainability issues are often wicked. Wicked problems are problems that are difficult or impossible to solve because of incomplete, contradictory, or changing requirements that are often difficult to distinguish. The concept of wicked problems was introduced by Rittel and Webber (1973) to describe issues that cannot be easily fixed, do not have one single solution, and show a certain resistance to resolution. The authors introduced the term as science is generally used to solve *tame* problems, which are problems with a single, absolute solution. Sustainable problems are often wicked because they are the symptom of other problems that are generally strongly engrained into human society. Perhaps the most important issue is the connections between sustainability issues and the idea of economic growth, which consequently increases consumption (a predominant contemporary indicator of the success of a society) (Costanza et al. 2013).

For example, it could be argued that it is a good idea to open a mine for rare earth metals in a poor country because the extracted materials can be used to produce electric batteries. This type of endeavor would—in the long-term—decrease CO_2 emissions because electric machines can replace the current machines that run on fossil fuels. Also, the mine could improve the standard of living (through income and education) of the community near the mine. On the other hand, the mine might have severely negative impacts on nature, leading to the extinction of local species and a loss of biodiversity. Consequently, there is no correct answer to whether the mine should be opened or not. It is a wicked problem, and decision-makers need to choose which aspect to prioritize (the standard of living of the community or nature preservation). The issue is a political issue that involves values connected to various priorities, which need to be grounded in scientific facts. In this way, most sustainability issues involve both values and facts. This is something that science teachers traditionally have more trouble coping with than—for example—social science teachers (Borg et al. 2012; Oulton et al. 2004).

A further obstacle that science teachers must overcome, related to the content aspect of sustainability issues, is their *anthropocentric* character. Teaching science from a sustainability perspective involves incorporating the human perspective, as all sustainability issues stem from human activity. This time in history is referred to as the geological era of the Anthropocene (Zalasiewicz et al. 2011). This means that humans are among the main drivers in the transformation of planet Earth. Notably, humans are transforming matter in a way that influences the natural geological cycles, for instance, the carbon, nitrogen, and phosphorus cycles (Mackenzie et al. 2002). Humans are introducing new chemicals and substances into ecosystems, and they are creating crop monocultures on such a large scale that many natural ecosystems are being deteriorated. In this way, humans are becoming an intrinsic stakeholder in what is happening to nature.

Nature was previously regarded as essentially independent of human will (Bonnett 2002). Western culture, in which science developed, includes a dichotomy between nature and culture. As such, nature has generally been regarded as being separate from the human world. This separation of humans from nature goes back hundreds of years in Western Christian culture and is reflected in the organizational structures of various disciplines all the way back to medieval universities. In medieval times, academic disciplines fell under the Seven Liberal Arts. These were divided into the quadrivium that included music, astronomy, geometry, and arithmetic, and the trivium that included grammar, rhetoric, and philosophy or logic (Holmes and McLean 2019). The former four arts—music, arithmetic, geometry, and astronomy—later evolved into the sciences, while the arts of the trivium—grammar, logic, and rhetoric—evolved into the humanities (Muller 2012). Johan Muller and others have described how this distinction between the Trivium and Quadrivium yielded two fault lines: one between the sciences and humanities; and the other between pure and applied disciplines. Consequently, sciences are traditionally disciplines that focus on purely concept-centric content; this remains the case today. Later in this chapter this issue will be explored more in depth, but for now it is fitting to conclude that the divide separating culture and nature in education is rooted in disciplinary traditions where the sciences did not include the anthropocentric aspect.

However, the inclusion of the sustainability perspective brings the human perspective back into science education. This is noticeable with the inclusion of the word *environment*, rather than *nature*, in most sustainability discourses. The term environment can have many different meanings. The *Oxford English Dictionary* defines the word environment as "the physical surroundings or conditions in which a person or other organism lives, develops, etc., or in which a thing exists" (Environment 2024). In the biological sciences, for example, the environment is mostly restricted to natural living conditions that do not consider humans. In contrast, the sustainability perspective includes the human when discussing the environment by describing how humans affect other living creatures through social activities, and how humans—as social animals—are affected by other living and nonliving factors of nature. This means that bringing the sustainability perspective into science education inevitably introduces the human, or an anthropocentric, perspective. This is often difficult for science teachers to grasp because the disciplinary tradition of the sciences does not include this perspective.

Increasing critique toward the anthropocentric focus of sustainability education can be noticed in the literature (e.g., Kopnina 2014). Along with many others, Kopnina (2014) argues that education should not favor human rights or justice over other living creatures in nature. As such, Kopnina and others argue for a more eco-centric or bio-centric perspective in sustainability education. This debate is not about disciplinary traditions, as discussed in the previous paragraph, but more tied to the issue of values in education, or rather which values to include—human values or bio-centric values. This debate is also relevant at the disciplinary level of the biological sciences due to the rising importance of the field of Biology Conservation, which evolved to a subdiscipline within biological sciences and addresses the human impact on nature as well as how to address human social aspects (Cooke et al. 2021). This debate also resonates with the issue of whether the term sustainable development or sustainability should be used, as sustainable development addresses the developmental aspects of humankind, and not other beings (this debate could be covered for many more pages, but due to space constraints, readers are referred to the work by Kopnina 2014).

To summarize this section, I have discussed the consequences of including a sustainability perspective within science education in terms of the purpose and content of teaching, that is, the why and what questions. The argument being posed is that teaching science from a sustainability perspective breaks with *the canon* of traditional science curricula concerning what to teach and why to teach. In other words, the inclusion of a sustainability perspective changes the purpose of teaching from the traditional learning goals of science education (to learn the concepts, models, and laws of science, to conduct scientific experiments, and to learn about the history and epistemology of science; Hodson 2014) to a more normative idea of learning *for sustainability*. Consequently, this type of education should influence how students think and act in the world rather than just learn about sustainability.

However, this is a disputed issue within the discussion of sustainability education, with some researchers questioning whether the teaching will become too normative and might lead to indoctrination. This is a line that sustainability education should not cross and has been discussed in detail by, for example, Jickling and Wals (2012). Still, being aware of and not crossing this line, few would argue against that the main overarching goal of sustainability education is the promotion of human well-being, and as discussed herein to include this perspective would shift the purpose of science education.

Furthermore, this section has shown that sustainability issues, presumably the content that should be taught, are often complex and require knowledge from multiple disciplines, potentially even those that lie outside the realm of sciences. Moreover, the issues are often wicked in that they are not possible to definitively solve and do not have one simple solution. Instead, the issues are value-driven; as such, in addition to factual knowledge, other values need to be addressed when prioritizing sustainability challenges. Finally, an anthropocentric perspective, or at least the human as an actor in the environment, must be part of the educational content. To conclude, the purposes and contents of science education and the ways these might influence science teachers when including a sustainability perspective have been discussed. It has also been argued that science teachers might experience the purposes and contents of sustainability education as obstacles because they are not always included in traditional science education curricula. In the next section, the how question will be addressed in the light of different teaching traditions in science and sustainability education.

1.3 The Character of Sustainability Education and How It Relates to Science Education: The How

Sustainability education has been described in many ways in the literature. From a semantic perspective, it has been termed sustainability education, education for sustainability, and education for sustainable development. These three terms have slightly different connotations, especially regarding if and how development can be connected to sustainability. In this chapter, these nuances will not be investigated further. For the interested reader, there are other publications that delve into this issue (e.g., Smyth 2008). Instead, from herein these concepts will more or less be treated as synonyms. Notably, Education for Sustainable Development (ESD) is the most commonly used term in international (e.g., UNESCO 2017) as well as many national policies (Gericke et al. 2020b). Therefore, in this section the term ESD will be used, but considered synonymous with sustainability education—the term used in other sections of this chapter.

ESD is a concept that developed—at the policy level—in response to increasing awareness about how environmental education had not slowed the accelerating environmental degradation, even if it has been a major curricular area for over 50 years in Sweden and various other countries (Gericke et al. 2020b). Traditionally, environmental education has focused on teaching students about specific environmental problems, such as climate change, ocean acidification, the eutrophication of bodies of fresh water, and ozone layer depletion. Usually, environmental and sustainability issues have been incorporated into the teaching of school subjects that are most relevant to these problems, such as biology, geography, and civics, but have also been taught as a specific subject. The main idea has been transmitting factual knowledge about how and why the problems emerge, and how to counter them, while the underlying assumption has been that a better understanding of specific environmental problems would foster a kind of civic knowledge that would encourage people to act at personal and societal levels in ways that would gradually solve the problems. This way of teaching about environmental issues has been termed as a fact-based teaching tradition in the literature (Öhman 2008) and has for decades been recognized by most science teachers as a natural part of science education, especially biology education. Here, scientific ideas are at the heart of education. Hence, what was addressed about sustainability education in the previous section—the normativity issue, the inclusion of values, and the anthropocentric perspective—are not widely emphasized in traditional environmental education teaching practices. Therefore, a science teacher often feels more accustomed to teaching according to the environmental education tradition rather than the sustainability education tradition (Sund and Gericke 2020). How can teaching practices—the how—be understood in ESD or sustainability education? This will be addressed in the next paragraph.

As research has shown, greater knowledge of environmental problems, as previously taught in environmental education, does not automatically translate into environmentally favorable behavior once students leave school and participate in society (Kollmuss and Agyeman 2002). Therefore, environmental education, as well as research into this type of education, has reoriented with an emphasis on addressing affective aspects of teaching and learning, that is, emotional and attitudinal aspects, in addition to cognitive aspects (Marcinkowski and Reid 2019). This field has drawn inspiration from psychological research related to attitudes, values, emotions, and behavior, based on the insight that all aspects of a student's consciousness must be addressed for educational initiatives to empower students to act sustainably. Consequently, teaching approaches that intend to transform learners through education, often described as more progressive, have been frequently advocated in sustainability education research (Eilam and Trop 2010). These more progressive and holistic teaching approaches, which are positioned outside of the school subject syllabi, have been collectively named ESD. ESD is evident in international policies, national curricula, and research literature (Eilam and Trop 2010) advocating progressive and transformative teaching approaches, as shown in this quote from an UNESCO educational policy:

> What ESD requires is a shift from teaching to learning. It asks for an action-oriented, transformative pedagogy, which supports self-directed learning, participation and collaboration, problem-orientation, inter- and transdisciplinarity and the linking of formal and informal learning. Only such pedagogical approaches make possible the development of the key competencies needed for promoting sustainable development. (UNESCO 2017, p. 7)

In the ESD, the content knowledge is viewed holistically, and environmental problems are regarded as sociopolitical problems because there is an incentive to resolve socioeconomic conflicts (Öhman 2004). Therefore, ESD promotes a holistic perspective on sustainable development that integrates ecological, economic, and social aspects. There is no consensus in the educational research community regarding the teaching approaches that ESD should employ. At the policy level, for instead the UNESCO quote included above, ESD is a container concept that includes various definitions—often portrayed as long lists of what ESD can be—without differentiating how these ideas about content and teaching relate to each other. However, in a research review of the ESD field, Eilam and Trop (2010) outline a framework of four essential principles of ESD pedagogy based on accumulated theory and experience in the field.

Eilam and Trop's (2010) framework was created in a bid to encompass the most prevailing teaching approaches within a set of four basic principles. The first principle, the traditional academic style of teaching and learning, exists within traditional school environments and is decontextualized from the learning objective. Eilam and Trop (2010) argue that this principle supports the development of analytical-rational modes of intelligence among students. These modes are closely related to the core of science education. The second principle is multidisciplinary teaching and learning by use of inter- and/or cross-disciplinary approaches. This approach combines knowledge from a variety of disciplines and is considered capable of supporting the acquisition of systemic thinking and the formation of linkages between cause and effect within systems (Eilam and Trop 2010). The third principle, multidimensional teaching and learning, can be achieved when academic learning (principle 1) and a multidisciplinary system (principle 2) are linked with dimensions of time and space. This approach facilitates the development of contextual ways of thinking (Hopkins and McKeown 2002) and the formation of out of the box thinking; as such, students should be able to investigate systems based on their relations to other systems, spaces, and times. According to Eilam and Trop (2010), the fourth principle denotes emotional teaching and learning, which involves the inclusion of emotions in learning activities that simultaneously activate processes of value and ethics clarification. When exposed to this mode of learning, it is argued that students are motivated to take part in activities that make them feel any type of emotion, ranging from enjoyment to distress. In other words, they are led on a path of emotional learning—to care.

When does science education enter the equation? It could be argued that science teachers can quite easily contribute to the first and third teaching and learning principles, that is, the academic style of teaching and multidimensional principles, respectively, while science teachers struggle more with incorporating multidisciplinary and emotional learning principles that aim to address the learning objectives addressed in the previous section, which are the normativity issue, the inclusion of values, the complexity and wickedness of sustainability issues, and the anthropocentric perspective. Eilam and Trop (2010) argue that it is necessary to include all these principles in ESD teaching to encourage action competence in line with sustainability among the learners (Mogensen and Schnack 2010). This leads to the following question: do science teachers feel comfortable in including all four of the presented principles in their own teaching? This may hold true at the primary education level, but studies focusing on secondary education indicate that this is not the case, and that teachers of different disciplines need to collaborate when covering all aspects of sustainability education (Sund and Gericke 2020; Sund et al. 2020). Also, studies of ESD at the primary level argue for the need for collaboration, for example, a study of 14 school projects in a Norwegian primary school showed that teachers from multiple subjects had to collaborate to cover the entirety of sustainability (Munkebye et al. 2020).

Consequently, most of the literature recommends that ESD should be taught in an interdisciplinary way (Eilam and Trop 2010) based on a holistic perspective and pluralistic teaching approaches that strive for action orientation (e.g., Sinakou et al. 2019). In the context of teaching, pluralism emphasizes a democratic approach by striving to promote various perspectives, values, and emotions when dealing with authentic questions and problems concerning the future of the world (Tryggvason et al. 2023). Moreover, pluralistic teaching has been found to affect students' self-reported pro-sustainability actions (Boeve-de Pauw et al. 2015). Hence, there is solid evidence for advocating this perspective in sustainability education. However, as argued in this chapter, this way of teaching is in stark contrast to the traditional teaching methods used in science education, which often focus on teacher-led methods recontextalizing science practices rather than applying knowledge in contextualized authentic situations. As such, it becomes clear that the incorporation of a sustainability perspective in science education requires the teacher to not only reconsider the purpose and content of the education, but also how to teach.

In line with this prognostication of science teachers' potential problems in teaching from a sustainability perspective, research has shown that science teachers tend to focus on scientific facts and concepts in their sustainability teaching, with noticeable difficulties in addressing more multidisciplinary and emotional aspects of ESD (e.g., Borg et al. 2012; Uitto and Saloranta 2017). Science teachers often favor more teacher-centered approaches in which the learning of conceptual knowledge is at the fore, rather than values and emotions. Why does one find these discrepancies in the teaching of sustainability between science teachers and teachers of social science and humanities? This—I would argue—is mainly due to different disciplinary traditions, which will be discussed in the next section.

1.4 Knowledge Structures and Knower Structures: Science Versus Other Disciplines

In this section, the knowledge structures of the sciences, and how they differ from those of other disciplines, such as social studies and humanities, will be discussed. This is important because the way knowledge is defined and understood in science differs from the learning and teaching theories that dominate the discourse surrounding sustainability education. This will provide an additional layer for understanding the differences that I have outlined in the preceding two sections regarding the why, what, and how of teaching science compared to sustainability.

As already discussed in the introduction, the premise for this chapter is that science educators who want to include a sustainability perspective in their teaching will have to move away from traditional teaching and incorporate aspects of sustainability that stem from other disciplines and various learning and teaching traditions. Consequently, a science educator might experience challenges when attempting to enact teaching in an alternative way. For teachers to overcome these challenges, and be able to change their own teaching practices, they must sufficiently understand their own disciplinary and didactical assumptions, and be able to compare them to those of teachers representing other disciplines (Sund and Gericke 2020). Consequently, this section will account for what the literature, especially within curriculum theory, has to say about these differences between disciplines and school subjects.

In a seminal paper from Biglan (1973), more than 200 university-level teachers representing 30 distinct disciplines were asked to group disciplines according to the perceived level of similarity. The analysis of the results used a multidimensional scaling technique based on their judgments and revealed two dimensions across which the disciplines differed and could be described (Fig. 1.1). The first dimension describes a horizontal, hard–soft continuum. Hard disciplines include those that have fixed and well-defined boundaries, with soft disciplines having the opposite character. The second dimension describes a vertical, pure-applied continuum (Fig. 1.1). Pure means that a discipline is oriented toward academic, conceptual-based knowledge, while applied knowledge is more practical, with a focus on solving authentic problems. As shown in Fig. 1.1, all sciences are categorized as pure and hard, with certain differences in the extent to which they are hard; biology, for instance, is perceived as a bit less hard. For comparison, a couple of other

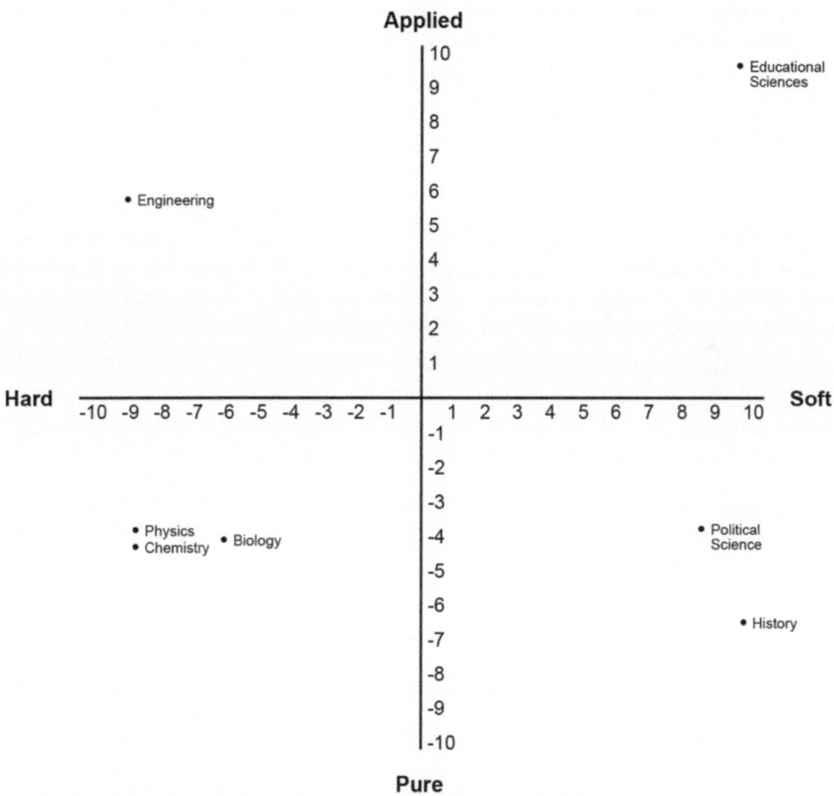

Fig. 1.1 Classification scheme of a few examples of disciplines according to the dimensions hard–soft and applied-pure. (Revised after Biglan 1973)

disciplines, such as engineering, education, history, and political sciences, are included to illustrate the differences in how disciplines are perceived (see Fig. 1.1). The reason is that a sustainability perspective provides a holistic picture that incorporates various aspects of various disciplines. What can these perceived differences be about?

One of the most influential scholars in the field concerning knowledge structures of different disciplines and their related school subjects is Basil Bernstein. He described that disciplines, and their associated school subjects, are related—to various degrees—to two main typologies of knowledge structures: vertical and horizontal discourse (Bernstein 1999). The science disciplines are categorized as vertical knowledge structures, whereas the humanities and social sciences are categorized as horizontal knowledge structures. There is some variation in the extent to which different disciplines belong to one of these two structures, but these differences will not be addressed here.

In disciplines with vertical knowledge structures, substantive knowledge, that is, the concepts and content of the discipline, takes the form of a coherent, explicit, and systematically principled structure (see the left side of Fig. 1.2). In such disciplines, the methods to conduct research are commonly agreed upon and not discussed. According to Bernstein, in hierarchical knowledge structures theory serves as both a way to connect the concepts (an example of systematic and coherent information), and to project explanations for the future through theoretical as well as empirical methodologies. In physics, if one knows the forces influencing a car that is rolling down a slope, one can then predict how long it will roll. In biology, one can predict the chance or risk of inheriting a specific gene over the generations. Hence, horizontal knowledge structures are verified from empirical data. The language used within disciplines representing this knowledge structure is fixed and the concepts used most often have an agreed meaning that does not need to be explained (Bernstein 1999). What consequences does this have for teaching and learning in school? First, it means that the content to be taught, the most central concepts to learn, as well as the disciplinary practices to portray in teaching can be easily agreed upon in the school community, and will often stay rather constant when moving the content

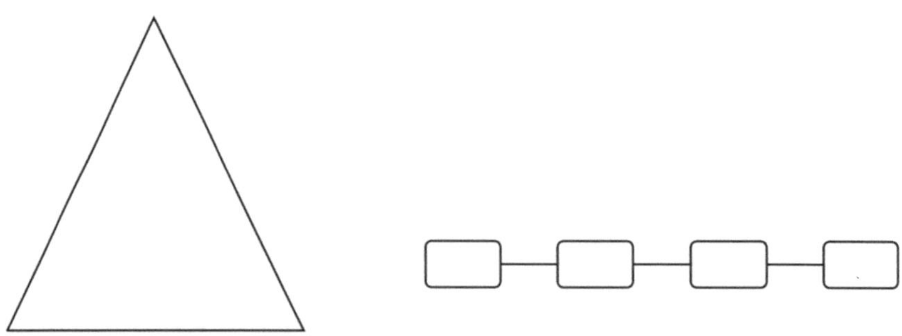

Fig. 1.2 Vertical and horizontal knowledge structures. (Adapted from Bernstein 1999)

from the university level to the school level, albeit in a more didacticized (adapted for the teaching at that specific level) manner (Gericke et al. 2018).

In Fig. 1.2, the triangle to the left represents a vertical knowledge structure, which can be found in disciplines where the content of the discipline takes the form of a coherent and systematically principled structure. The segmented picture to the right represents a horizontal knowledge structure, which can be found in disciplines where the content can be described as local theories wherein the concepts included in different theories (segments) not connected to each other.

The teacher does not need to negotiate what to teach, but *the canon* of vertically arranged school subjects often follows an international canon that is taken for granted. Hodson (2014) described this as the curricular goals of (a) learning *in* science, (b) learning *to do* science, and (c) learning *about* science in different topic domains such as genetics and evolution in school biology, acid and bases and organic chemistry in school chemistry, and electricity and astronomy in physics education. Hence, the curriculum and teaching instructions in science education traditionally follow this hierarchical structure. As Bernstein (1999, p. 164) argues, "the acquirer does not have the problem of knowing whether she/he is speaking physics or writing physics, only the problem of correct usage."

Nevertheless, this paradigm, that the content of school science subjects is in close alignment with the mother disciplines at the university level, has been questioned in the past few decades of science education in the Science, Technology, Engineering, and Mathematics (STEM) movement. This movement predominantly promotes teaching approaches that teach students how to apply scientific knowledge rather than reproduce it (Bencze et al. 2020). This movement has also led to the development of socioscientific issues (SSI) to teach science; in other words, teaching science considering related, authentic societal problems rather than according to the inherent vertical knowledge structures (Erduran 2023; Sadler 2009). Similarly, scholars have proposed that science education should be reoriented toward broader learning objectives focusing of scientific literacy promoting critical-democratic participation to transform society and culture in a sustainable way (Sjöström and Eilks 2018). The increased demand for a sustainability perspective in science education has further challenged the traditional way of reproducing science in the science classroom, as these new ways of thinking demand connections to horizontal knowledge structures, which will be addressed in more detail in the next paragraph.

Disciplines and the associated school subjects within horizontal knowledge structures, such as the humanities and social sciences, can, according to Bernstein (1999), be described as a set of segmentally ordered local theories in which the concepts included in different theories (segments) cannot necessarily be connected to each other (see right side in Fig. 1.2). Instead, the theories can be viewed as complementary in the way that they show different perspectives of a certain phenomenon, or give alternative truths about it. For example, education, which can be regarded as a social science, includes many theories about learning. In cognitive learning theory, functions of the individual brain are important such as for the notion on cognitive load (Sweller 2010), while sociocultural learning theories frame learning as an activity or practice that occurs in a social setting with diverse learners (e.g., Engeström 1999). Neither of these theories can be said to provide the true and only description of learning, and the concepts in these theories would be difficult to compare to each other because they are defined and framed differently; as such, the knowledge theories are segmented.

Also, the methods used to study and investigate education differ between these segments because knowledge and learning are understood differently and as taking place in either an individual or community setting. This is only one example and was meant to demonstrate how transforming, or to use the term presented by Bernstein (1999), recontextualizing, knowledge from the discipline to the school subject is very difficult for educators and teachers in school subjects with horizontal knowledge structures. During this process, questions that might arise include: Which one of these local theories should be chosen? What should we teach about? In horizontal knowledge discourses, these types of questions are difficult to find consensus about and are thus constantly debated.

Karl Maton (2006) has written about this, and he argues that disciplines and school subjects placed with vertical knowledge structures, such as the sciences, have horizontal *knower structures* (see Fig. 1.3). Here, knower structures relate to power relations, more specifically, the issue of who decides which knowledge is relevant, and—consequently—what should be selected and taught. In hierarchical organized disciplines, new knowledge is confirmed based on how it relates to the

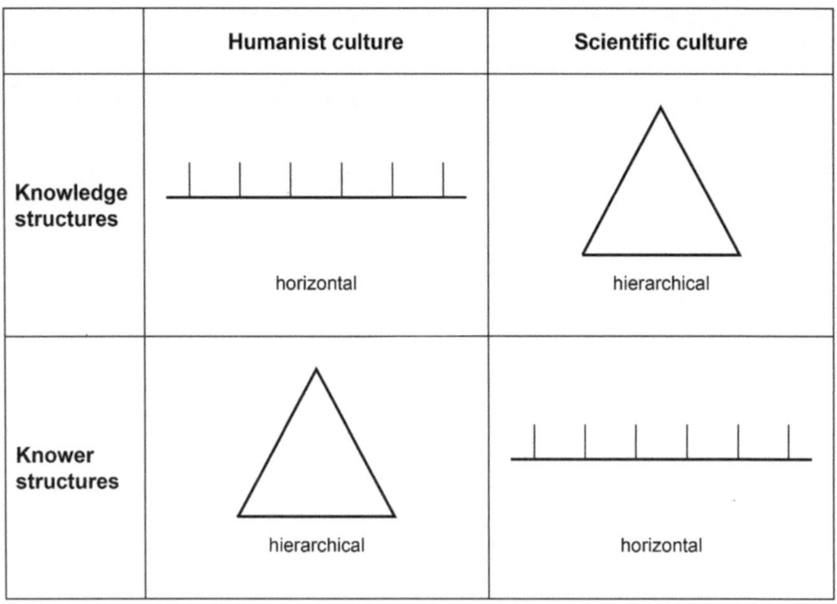

Fig. 1.3 Knowledge and knower structures in both scientific and humanistic discourses. (Adapted from Maton 2006)

existing structure, and how it can be verified by experimental and empirical studies.[1] In this way, as long as one follows the accepted methods of inquiry, everyone can contribute important new knowledge to the field. According to Maton (2006), this is not the case in disciplines that follow a horizontal knowledge structure of segmented knowledge. Here, it is more important who makes a claim, meaning that the credibility of whoever is making a new knowledge claim comes to the fore. Because these disciplines do not involve a fixed method, or a knowledge structure with common consensus on defined concepts, it is nearly impossible to have objective criteria for evaluating the knowledge claim. Instead, it is the argument that is important, and each argument can only be valid within the specific knowledge segment with which it aligns. This leads to the question about how important the position of the presenter of a claim is, and whether others find the individual to be credible or not. Due to this caveat, Maton (2006) argues that horizontally organized knowledge also has a vertical, or hierarchical, knower structure.

In Fig. 1.3, the column under humanist culture represents segmented horizontal knowledge structures, such as educational sciences, with hierarchical knower structures. In the third column under scientific culture, a hierarchical or vertical knowledge structure is represented, with horizontal knower structure. To conclude, Maton (2006) argues that knower structures relate to power relations in the discipline or subject, that is, the issue of who decides which knowledge is relevant, and what should be selected and taught. The next section will clarify how these theoretical discussions are relevant for the inclusion of sustainability in science education.

1.5 Knowledge Structures and Knower Structures: A Way to Understand Science Educators' Challenges with Including a Sustainability Perspective

As argued previously, science educators who strive to teach science with a sustainability perspective need to address social and economic aspects related to science. This can be done in a multidisciplinary fashion by collaborating with teachers from other disciplines, or in an interdisciplinary fashion by integrating various aspects into their own teaching. What is clear is that these aspects cannot be neglected; in this way, the provision of classic environmental education—which completely relies on scientific facts—must be connected to how environmental problems emerge in society and how they can subsequently be solved. The point made in this section is that science teachers—when teaching from a sustainability perspective—will have to face disciplines and school subjects with knowledge and knower structures that differ from the discipline they are accustomed to. This will require them

[1] Within the sciences, paradigmatic knowledge ruptures can occur, and these instances do not follow the pattern, as pointed out by Kuhn (1996); however, this seldom happens, so it will not be discussed further herein.

to recontextualize the education with a complete understanding of the differences in disciplinary traditions, or to collaborate with other teachers. First, an example based on economics will shed more light on this problem.

One of the most important aspects of economic sciences is understanding and measuring economic progress and prosperity, which is also important for sustainability. One of the most contested ideas in sustainability research is the idea of economic growth (Costanza et al. 2013). There are competing theories or ideas that can be described to belong to different local theories (or knowledge segments when using the terminology of Bernstein) about how to best organize society and define economic progress. One theory is the neo-classical economic concept of sustainability, which is also referred to as *Weak Sustainable Development* in the literature (Nilsen 2010). Most economists use this definition of economic progress, and it can be defined as a form of sustainability in which utility or consumption does not decline over time (Pezzey 1997). Hence, the aim of this theory is incessant economic growth. However, this way of perceiving wealth has been criticized in society and by other economists for five decades. A competing theory exists within another segment and is often presented by professionals who refer to themselves as ecological economists. They argue that natural capital should be preserved for future generations and consider zero economic growth to be a viable option. In other words, the economy and nature are seen as complementary, and both must be sustained (Nilsen 2010). The ecological economist view of sustainable development is also referred to as *Strong Sustainable Development*. This is—of course—only a brief, and simplified, version of the debate, but is nevertheless relevant to the argument concerning the question: what theories and segmented knowledge structures should be included in teaching?

The solution, often suggested in sustainable education literature, to this problem with competing theories or perspectives is teaching from a pluralistic perspective that includes various perspectives (Tryggvason et al. 2023). Notably, there is research on the specific issue of what economic perspectives should be addressed in sustainable education (Berglund and Gericke 2018, 2022). Briefly, a pluralistic approach suggests introducing multiple and conflicting perspectives on the issue, and then letting students discuss and value the different arguments, each representing a distinct theory. The aim of pluralistic teaching approaches is providing students with a framework that they can use to make informed decisions based on both scientific facts and values.

The main argument covered in this section is that it might be difficult for science teachers to promote sustainability in their teaching because they lack an understanding of the underlying knowledge structures of adjacent disciplines and school subjects that they need to involve in teaching when teaching sustainability. The vertical knowledge structure of science is often described as an objective and positivist understanding of knowledge that might be perceived as universal, and Bernstein (1999) commented that there is a risk that within a hierarchical knowledge structure the perception of "this is the only and sole pathway to 'truth'" (p. 165) might arise. Hence, it is important to be aware of these differences when considering how to integrate sustainability education into science teaching.

An English report on teachers' perspectives on sustainability education identified tensions between teachers' perceptions of sustainable education and the current, perceived curricular prioritization of teaching academic disciplinary knowledge (Glackin et al. 2018). Previous research has also revealed extensive differences in how teachers of various school subjects address different aspects of sustainability, such as social or ethical topics (e.g., Levinson and Turner 2002). In some school subjects (e.g., mathematics and psychology), sustainable topics are covered only briefly or tangentially, yet they receive a larger spotlight when compared to the case in other subjects (e.g., environmental science and geography) (Dawe et al. 2005). Levinson and Turner (2002) also found differences between science and humanities teachers' understanding of the aims of their teaching, and concluded that science teachers focus on presenting the facts of their subject, whereas humanities teachers are more confident when discussing ethical and social questions.

Interestingly, it has been reported that humanities teachers use class discussions more frequently than science teachers (Oulton et al. 2004). Summers et al. (2005) also found that science teachers use sustainability-inspired teaching methods less frequently than geography teachers, while similar patterns were also detected in a large survey of more than 3000 Swedish teachers (Borg et al. 2012). The similarities in findings of studies performed across various countries suggest that teachers of different school subjects understand and teach sustainability issues in various ways, with science teachers presenting topics in a way that makes it difficult to incorporate aspects of disciplines with horizontal knowledge structures in their teaching, as this would demand pluralistic-oriented teaching.

1.6 Conclusions: How to Align Science Education with a Sustainability Perspective

In the previous sections of this chapter, I have argued that teaching science education from a sustainability perspective might pose challenges for science teachers because the purposes, the content, as well as the teaching methods need to be modified to some degree from traditional science teaching. These changes encompass the incorporation of a more normative perspective on education, the inclusion of values and emotions other than scientific concepts, and taking into account the anthropocentric perspective; this is because all sustainability problems, and the possible solutions to these problems, arise in the social world of humans. Finally, the complexity of sustainability issues calls for multi- or interdisciplinary perspectives, and the wicked nature of sustainability problems makes it difficult to teach any predefined solutions. What further complicates the inclusion of sustainability in science education is that all these aspects of teaching are interrelated or entangled, and not easily discerned from each other. These various challenges have been discussed with the overarching goal of helping explain and clarify to teachers why it might be experienced as difficult to teach science from a sustainability perspective.

So far, this chapter paints a rather grim picture of how to transform teaching practices so that a sustainability perspective can be included in science education, and some teachers may ask themselves: should I teach science in a way that also covers sustainability? That, however, is not an option considering the mounting scientific evidence of the enormous problems that the Earth faces in the Anthropocene (Rockström et al. 2009). Consequently, it is crucial for all teachers, including science teachers, to address the sustainability perspective, an issue that has already been included in national curricula and international policy documents. However, if science teachers start to teach sustainability issues without reflecting how their teaching is influenced by a certain discipline, it is likely that they will quickly revert to teaching science based on scientific facts. The prevalence of this traditional environmental education approach has been shown many times in previous research (Borg et al. 2012, 2014; Oulton et al. 2004; Uitto and Saloranta 2017).

There is nothing wrong in teaching science from a scientific teaching tradition. Actually, this seems to have been a very proficient way of teaching science over the centuries that this teaching tradition has developed, as this method of teaching is adapted for the knowledge structure it is supposed to reproduce. However, science education that includes a sustainability perspective must involve a shift in teaching toward a more anthropocentric, normative, and value-driven direction, which is critical to promoting action competence for sustainability (Sass et al. 2020). Consequently, both the content and teaching approach need to be changed and aligned with this educational goal to be effective (Boeve-de Pauw et al. 2015). The teaching must encapsulate content from other disciplines, and sustainability issues need to be taught in a way that exposes values and evokes emotions among students. How can this transition of teaching be accomplished?

One way is that science teachers include an interdisciplinary perspective in their own teaching by including aspects of economy and society. Content related to sustainability can also be taught using more progressive teaching approaches—such as debates, group discussions, role play, or inquiry—to activate and engage students to express values and evoke emotions. This approach is demanding. First, a science teacher used to a hierarchical knowledge structure will most likely struggle to validate new content from a subject with a segmented horizontal knowledge structure (see the discussion in the preceding section about different views on economic growth). Second, it is difficult to reorient and change existing teaching traditions (Sund 2016).

Another way to promote sustainability from a multidisciplinary perspective is via cross-curricular collaboration between teachers of various school subjects. By working in this way, the science teachers can continue teaching their expertise from a more traditional content perspective, while at the same time teachers of other disciplines can contribute with their disciplinary skills. For example, social science teachers tend to incorporate more progressive teaching methods such as debates and role play in their disciplinary teaching tradition that could complement science teaching (Borg et al. 2012; Oulton et al. 2004; Sund and Gericke 2020).

Emotional learning to develop identity and action competence is a learning objective in language and literature education teaching (Maijala et al. 2023; Sund and Gericke 2020). Hence, the disciplinary teaching traditions associated with social sciences, humanities, and language are well suited for covering most of the aspects of sustainability that lie outside of the traditional realm of science teaching (Gericke 2025a, b). By utilizing cross-curricular collaboration, science teachers can receive support with the aspects of sustainability that they experience as challenging. This can also be a win-win situation for all teachers, as certain teachers, for example, many humanistic and language teachers, will not grasp the value of teaching sustainability because they have difficulties identifying relevant content within the field of sustainability due to their disciplinary tradition (Borg et al. 2012; Sund and Gericke 2020; Uitto and Saloranta 2017). The understanding of scientific facts is the very expertise that science teachers could contribute in cross-curricular settings. Hence, a cross-curricular teaching approach of collaboration between teachers of different school subjects can offer a complementary perspective of sustainability that leverages the individual strengths of different school subject teachers; together, this can yield sustainability education that is holistic, pluralistic, and action-oriented (Sinakou et al. 2019).

This type of education should be considered in light of the four principles of sustainability teaching proposed in the review by Eilam and Trop (2010): the academic; the multidisciplinary; the multidimensional; and the emotional aspects. More insights of how to accomplish this in practice can be found in the following publications (Gericke 2022, 2025a, b; Gericke et al. 2020a; Sund and Gericke 2020, 2022; Sund et al. 2020). To conclude, science teachers need to embrace the sustainability perspective when planning their teaching even if they experience serious challenges. Luckily, there are various ways and strategies that science teachers can use to overcome these hurdles and provide students with fruitful, cross-curricular teaching.

References

Bencze L, Pouliot C, Pedretti E, Simonneaux L, Simonneaux J, Zeidler D (2020) SAQ, SSI and STSE education: defending and extending "science-in-context". Cult Stud Sci Educ 15(3):825–851. https://doi.org/10.1007/s11422-019-09962-7

Berglund T, Gericke N (2018) Exploring the role of the economy in young adults' understanding of sustainable development. Sustainability 10(8):2738. https://doi.org/10.3390/su10082738

Berglund T, Gericke N (2022) Diversity in views as a resource for learning? Student perspectives on the interconnectedness of sustainable development dimensions. Environ Educ Res 28(3):354–381. https://doi.org/10.1080/13504622.2021.1980501

Bernstein B (1999) Vertical and horizontal discourse: an essay. Br J Sociol Educ 20(2):157–173. https://doi.org/10.1080/01425699995380

Biglan A (1973) The characteristics of subject matter in different academic areas. J Appl Psychol 57(3):195–203. https://doi.org/10.1037/h0034699

Boeve-de Pauw J, Gericke N, Olsson D, Berglund T (2015) The effectiveness of education for sustainable development. Sustainability 7(11):15693–15717. https://doi.org/10.3390/su71115693
Bonnett M (2002) Education for sustainability as a frame of mind. Environ Educ Res 8(1):9–20. https://doi.org/10.1080/13504620600942683
Borg C, Gericke N, Höglund H-O, Bergman E (2012) The barriers encountered by teachers implementing education for sustainable development: discipline bound differences and teaching traditions. Res Sci Technol Educ 30(2):185–207. https://doi.org/10.1080/02635143.2012.699891
Borg C, Gericke N, Höglund H-O, Bergman E (2014) Subject- and experience-bound differences in teachers' conceptual understanding of sustainable development. Environ Educ Res 20(4):526–551. https://doi.org/10.1080/13504622.2013.833584
Cooke SJ, Lynch AJ, Piccolo JJ, Olden JD, Reid AJ, Ormerod SJ (2021) Stewardship and management of freshwater ecosystems: from Leopold's land ethic to a freshwater ethic. Aquat Conserv Mar Freshwat Ecosyst 31(6):1499–1511. https://doi.org/10.1002/aqc.3537
Costanza R, Alperovitz G, Daly H, Farley J, Franco C, Jackson T, Kubiszewski I, Schor J, Victor P (2013) Building a sustainable and desirable economy-in-society-in-nature. ANU Press. http://www.jstor.org/stable/j.ctt5hgz53
Dawe G, Jucker R, Martin M (2005) Sustainable development in higher education: current practice and future developments. Higher Education Academy
Eilam E, Trop T (2010) ESD pedagogy: a guide for the perplexed. J Environ Educ 42(1):43–64. https://doi.org/10.1080/00958961003674665
Engeström Y (1999) Activity theory and individual and social transformation. In: Perspectives on activity theory, vol 19, pp 19–30. https://doi.org/10.1017/CBO9780511812774.003
Environment (2024) Oxford English dictionary, 3rd edn. Oxford University Press. https://www.oed.com/environment
Erduran S (2023) Social and institutional dimensions of science: the forgotten components of the science curriculum? Science 381(6659):eadk1509. https://doi.org/10.1126/science.adk1509
Gericke N (2022) Teaching for the Anthropocene: Bildung-oriented education for sustainable development in a subject-specific curriculum. In: Krogh E, Qvortrup A, Ting Graf S (eds) Bildung, knowledge, and global challenges in education. Taylor & Francis, pp 53–69. https://doi.org/10.4324/9781003279365-5
Gericke N (2025a) Education for sustainable development in subject-specific education and the contribution of language teaching. In: Maijala M, Kuusalu SR, Ullakonoja R (eds) Transformative language teaching for sustainability. Palgrave Macmillan, pp 27–51. https://doi.org/10.1007/978-3-031-85493-4_2
Gericke N (2025b) Education for sustainable development from a comparative subject didactic perspective. Sammenlignende Fagdidaktik 2025(8):31–52. https://doi.org/10.7146/sammenlignendefagdidaktik.v2025i8.157361
Gericke N, Hudson B, Olin-Scheller C, Stolare M (2018) Powerful knowledge, transformations and the need for empirical studies across school subjects. Lond Rev Educ 16(3):428–444. https://doi.org/10.18546/LRE.16.3.06
Gericke N, Huang L, Knippels MC, Christodoulou A, Van Dam F, Gasparovic S (2020a) Environmental citizenship in secondary formal education: the importance of curriculum and subject teachers. In: Hadjichambis A, Reis P, Paraskeva-Hadjichambi D, Činčera J, Pauw J B-d, Gericke N, Knippels M-C (eds) Conceptualizing environmental citizenship for 21st century education: environmental discourses in science education, vol 4. Springer, pp 193–212. https://doi.org/10.1007/978-3-030-20249-1_13
Gericke N, Manni A, Stagell U (2020b) The green school movement in Sweden – past, present and future. In: Gough A, Lee JC, Tsang EPK (eds) Green schools movements around the world: stories of impact on education for sustainable development. Springer, pp 309–332. https://doi.org/10.1007/978-3-030-46820-0_17
Giddings B, Hopwood B, O'Brien G (2002) Environment, economy and society: fitting them together into sustainable development. Sustain Dev 10(4):187–196. https://doi.org/10.1002/sd.199

Glackin M, King H, Cook R, Greer K (2018) Understanding environmental education in secondary schools in England: the practitioners' perspective *(Report 2)*. King's College London

Hodson D (2014) Learning science, learning about science, doing science: different goals demand different learning methods. Int J Sci Educ 36(15):2534–2553. https://doi.org/10.1080/09500693.2014.899722

Holmes B, McLean M (2019) The curriculum: a comparative perspective. Routledge

Hopkins C, McKeown R (2002) Education for sustainable development: an international perspective. In: UNESCO chair in reorienting education towards sustainability. York University

Jickling B, Wals AE (2012) Debating education for sustainable development 20 years after Rio: a conversation between Bob Jickling and Arjen Wals. J Educ Sustain Dev 6(1):49–57. https://doi.org/10.1177/097340821100600111

Kollmuss A, Agyeman J (2002) Mind the gap: why do people act environmentally and what are the barriers to pro-environmental behavior? Environ Educ Res 8(3):239–260. https://doi.org/10.1080/13504620220145401

Kopnina H (2014) Revisiting education for sustainable development (ESD): examining anthropocentric bias through the transition of environmental education to ESD. Sustain Dev 22(2):73–83. https://doi.org/10.1002/sd.529

Kuhn TS (1996) The structure of scientific revolutions, 3rd edn. The University of Chicago Press

Levinson R, Turner S (2002) Valuable lessons: engaging with the social context of science in schools. The Wellcome Trust. www.wellcome.ac.uk/stellent/groups/corporatesite/@msh_peda/documents/web_document/wtd003446.pdf

Luke SH, Roy HE, Thomas CD, Tilley LA, Ward S, Watt A, Carnaghi M, Jaworski CC, Tercel MP, Woodrow C, Aown S, Banfield-Zanin JA, Barnsley SL, Berger I, Brown MJ, Bull JC, Campbell H, Carter RA, Charalambous M, Dicks LV (2023) Grand challenges in entomology: priorities for action in the coming decades. Insect Conserv Divers 16(2):173–189. https://doi.org/10.1111/icad.12637

Mackenzie FT, Ver LM, Lerman A (2002) Century-scale nitrogen and phosphorus controls of the carbon cycle. Chem Geol 190(1–4):13–32. https://doi.org/10.1016/S0009-2541(02)00108-0

Maijala M, Gericke N, Kuusalu SR, Heikkola LM, Mutta M, Mäntylä K, Rose J (2023) Conceptualising transformative language teaching for sustainability and why it is needed. Environ Educ Res 30(3):377–396. https://doi.org/10.1080/13504622.2023.2167941

Marcinkowski T, Reid A (2019) Reviews of research on the attitude-behavior relationship and their implications for future environmental education research. Environ Educ Res 25(4):459–471. https://doi.org/10.1080/13504622.2019.1634237

Maton K (2006) On knowledge structures and knower structures. In: Moore R, Arnot M, Beck J, Daniels H (eds) Knowledge, power and educational reform: applying the sociology of Basil Bernstein. Routledge, pp 44–59

Mogensen F, Schnack K (2010) The action competence approach and the 'new' discourses of education for sustainable development, competence and quality criteria. Environ Educ Res 16(1):59–74. https://doi.org/10.1080/13504620903504032

Muller J (2012) Forms of knowledge and curriculum coherence. In: Lauder H, Young M, Daniels H, Balarin M, Lowe J (eds) Educating for the knowledge economy? Critical perspectives. Routledge, pp 114–138

Munkebye E, Scheie E, Gabrielsen A, Jordet A, Misund S, Nergård T, Øyehaug AB (2020) Interdisciplinary primary school curriculum units for sustainable development. Environ Educ Res 26(6):795–811. https://doi.org/10.1080/13504622.2020.1750568

Nilsen HR (2010) The joint discourse 'reflexive sustainable development': from weak towards strong sustainable development. Ecol Econ 69(3):495–501. https://doi.org/10.1016/j.ecolecon.2009.11.011

Öhman J (2004) Moral perspective in selective traditions of environmental education. In: Wickenberg P, Axelsson H, Fritzén L, Helldén G, Öhman J (eds) Learning to change our world. Studentlitteratur, pp 33–57

Öhman J (ed) (2008) Values and democracy in education for sustainable development: contributions from Swedish research. Liber
Oulton C, Day V, Dillon J, Grace M (2004) Controversial issues: teachers' attitudes and practices in the context of citizenship education. Oxf Rev Educ 30(4):489–507. https://doi.org/10.1080/0305498042000303973
Pezzey J (1997) Sustainability constraints versus intertemporal concern, and axioms versus data. Land Econ 73(4):448–466
Purvis B, Mao Y, Robinson D (2019) Three pillars of sustainability: in search of conceptual origins. Sustain Sci 14(3):681–695. https://doi.org/10.1007/s11625-018-0627-5
Rittel HW, Webber MM (1973) Dilemmas in a general theory of planning. Policy Sci 4(2):155–169. https://doi.org/10.1007/BF01405730
Rockström J, Steffen W, Noone K, Persson Å, Chaplin FS, Lambin EF, Foley JA (2009) A safe operating space for humanity. Nature 46:472–475. https://doi.org/10.1038/461472a
Sachs J, Schmidt-Traub G, Kroll C, Durand-Delacre D, Teksoz K (2017) The sustainable development goals report. United Nations. https://unstats.un.org/sdgs
Sadler TD (2009) Situated learning in science education: socio-scientific issues as contexts for practice. Stud Sci Educ 45(1):1–42. https://doi.org/10.1080/03057260802681839
Sass W, Boeve-de Pauw J, Olsson D, Gericke N, De Maeyer S, Van Petegem P (2020) Redefining action competence: the case of sustainable development. J Environ Educ 51(4):292–305. https://doi.org/10.1080/00958964.2020.1765132
Sinakou E, Donche V, Boeve-de Pauw J, Van Petegem P (2019) Designing powerful learning environments in education for sustainable development: a conceptual framework. Sustainability 11(21):5994. https://doi.org/10.3390/su11215994
Sjöström J, Eilks I (2018) Reconsidering different visions of scientific literacy and science education based on the concept of Bildung. In: Dori YJ, Mevarech ZR, Baker DR (eds) Cognition, metacognition, and culture in STEM education. Innovations in science education and technology, vol 24. Springer, pp 65–88. https://doi.org/10.1007/978-3-319-66659-4_4
Smyth JC (2008) Environment and education: a view of a changing scene. In: Reid A, Scott W (eds) Researching education and the environment. Routledge, pp 1–30
Summers M, Childs A, Corney G (2005) Education for sustainable development in initial teacher training: issues for interdisciplinary collaboration. Environ Educ Res 11(5):623–647. https://doi.org/10.1080/13504620500169841
Sund P (2016) Discerning selective traditions in science education: a qualitative study of teachers' responses to what is important in science teaching. Cult Stud Sci Educ 11(2):387–409. https://doi.org/10.1007/s11422-015-9666-8
Sund P, Gericke N (2020) Teaching contributions from secondary school subject areas to education for sustainable development – a comparative study of science, social science and language teachers. Environ Educ Res 26(6):772–794. https://doi.org/10.1080/13504622.2020.1754341
Sund P, Gericke N (2022) Investigating the nature of powerful knowledge and epistemic quality in education for sustainable development. In: Hudson B, Gericke N, Olin-Scheller NC, Stolare M (eds) International perspectives on knowledge and curriculum: epistemic quality across school subjects. Bloomsbury Publishing, pp 177–196
Sund P, Gericke N, Bladh G (2020) Educational content in cross-curricular ESE teaching and a model to discern teacher's teaching traditions. J Educ Sustain Dev 14(1):78–97. https://doi.org/10.1177/0973408220930706
Sustainability (2024) Oxford English dictionary, 3rd edn. Oxford University Press. https://www.oed.com/sustainability
Sweller J (2010) Cognitive load theory: recent theoretical advances. In: Plass JL, Moreno R, Brünken R (eds) Cognitive load theory. Cambridge University Press, pp 29–47
Tryggvason Á, Öhman J, Van Poeck K (2023) Pluralistic environmental and sustainability education: a scholarly review. Environ Educ Res 29(10):1460–1485. https://doi.org/10.1080/13504622.2023.2229076

Uitto A, Saloranta S (2017) Subject teachers as educators for sustainability: a survey study. Educ Sci 7(1):8. https://doi.org/10.3390/educsci7010008

UNESCO (2017) UNESCO global action programme on education for sustainable development. UNESCO. https://unesdoc.unesco.org/ark:/48223/pf0000246270

Zalasiewicz J, Williams M, Haywood A, Ellis M (2011) The Anthropocene: a new epoch of geological time? Philos Trans R Soc A Math Phys Eng Sci 369(1938):835–841. https://doi.org/10.1098/rsta.2010.0339

Niklas Gericke, PhD, is Professor of Science Education and Dean of the Teacher Education Faculty Board at Karlstad University in Sweden, and visiting professor at University of Inland, Norway. He has a formal training as a science teacher, and have long teaching experience from secondary school and teacher education. His main research interests are biology education and sustainability education with more than 100 international publications. His area of research includes education for sustainable development from both conceptual and implementation perspectives. Much of his work relates to how disciplinary knowledge is transformed into school knowledge and what impact these transformations have on teachers' work and students' understanding.

Chapter 2
Higher Education's and Science Education's Contributions to a Critical Citizenry

Kerry Shephard

Abstract Higher education institutions around the world claim to educate students to think critically and independently. New Zealand's Education and Training Act (2020), for example, confirms that the principal aim of universities is to develop intellectual independence. But higher education has long-standing concerns about its ability to teach outside disciplinary boundaries and increasing pressure to address matters of environmental, social, and economic justice. This chapter will focus on science, science education, and science education's responsibility to graduate citizens who are able and willing to think critically about the problems of the world, even in circumstances that do not encourage them to do so.

Keywords Intellectual independence · Critical thinking · Dispositions to think critically · Roles and responsibilities of universities · Science education's contribution to teaching critical thinking

2.1 On the Concept of a Critical Citizenry

While I was imagining the contents of this chapter, a news item[1] challenged me to address the complexity of the sustainability issues considered not only in my chapter but in this book and on our planet. Apparently, Aotearoa New Zealand's[2] Ministry of Education has decided to put nylon carpet tiles in 600 rural schools as part of an ongoing refurbishment program: nylon rather than wool. And this in a nation that prides itself on having many more sheep than people. The news item included

[1] https://www.odt.co.nz/business/school-floored-govt's-carpet-choice

[2] New Zealand is a small multi-island nation in the South Pacific. Its name is gradually changing to Aotearoa New Zealand in recognition of its pre-colonial history.

K. Shephard (✉)
Higher Education Development Centre, University of Otago, Dunedin, New Zealand
e-mail: kerry.shephard@otago.ac.nz

F. Skarstein, L.-A. Wolff (eds.), *Sustainability and Science Education*, Environmental Challenges and Solutions,
https://doi.org/10.1007/978-3-032-24408-6_2

outrage from several educational, environmental, and farming stakeholder groups. What are Aotearoa New Zealand's citizens to think of this? What are they to do? What science, issues, facts, and unknowns should they incorporate in their thinking process? My first thought considered the fact that Aotearoa New Zealand produces some great wool, and that the sheep industry makes an important contribution to our economic and social well-being. I am an enthusiast for sheep wool. My wife and I have a small farm. We keep sheep (just a few). My mother-in-law taught me how to spin, and I taught myself how to weave. Our house is insulated with many centimeters of wool. I put it there. Science confirms what a great insulator it is. I wear woolly jumpers in winter. I am clearly an enthusiast for wool, so proclaim my bias. Even so, our coarse wool industry is in crisis. It costs farmers more to shear their sheep than the wool is worth, financially, and much of our pastoral economy is supported by the international price of meat and milk, rather than wool. Nylon carpet tiles, however, come from the oil industry and these particular nylon carpet tiles come from another country, whose economy is doing fine without us supporting it more than we do at present. On the other hand, science tells us that producing wool, even in Aotearoa New Zealand's low-density pastoral settings, is far from carbon neutral. And some object to animal farming altogether. Those with an interest in agricultural and ecological sciences might also wish to consider the beneficial impacts of low-density pastoral farming on Aotearoa New Zealand's landscape and ecosystem, accepting of course that our broad expanses of grass paddocks and upland grazing are far from natural. (It seems likely that prior to the arrival of humans, most of Aotearoa New Zealand was covered in trees. Many of our iconic species may have thrived after historical deforestation, but some other species no doubt did not.) But what will our children think? Many are educated in our Enviroschools[3] and will have grown up thinking about such things. Perhaps they will think of this as yet another reason to mistrust adults who say one thing but do another. And how did the Ministry of Education people come to this decision? What factors were involved and what were their objectives? How will the other citizens of Aotearoa New Zealand react to this decision, or will they make decisions of their own? What role does education, and science, have in how people come to decisions?

Does everyone struggle as I do to think about so many different things and with so many around us trying to influence what we think, how we think, and what we do? Or do most people prefer not to think so much, or so deeply? I am forced to remember that Harari, summarizing much academic progress in psychology, tells us that "most human decisions are based on emotional reactions and heuristic shortcuts rather than on rational analysis" (Harari 2018, p. 222).

I am not aware that the term "Critical Citizenry"[4] has been formally defined, but perhaps for the purposes of this chapter the essence of such a definition would insist that critical citizens make decisions based on something more than emotional reac-

[3] Schools with a pronounced sustainability focus.

[4] I do not know who first used this term, but I first encountered it in Williams (2014). The concept of critical citizens is, of course, long-standing in political discourse, as is criticality and critical thinking in education and philosophy.

tions and heuristic shortcuts and that this quality extends beyond individual citizens to characterize the population as a whole. This chapter goes further to suggest that the principal aim of a nation's education system should be to prepare its citizenry to make such decisions. And given that higher education educates so many people on the planet nowadays, and that decisions made based on something more than emotional reactions and heuristic shortcuts are likely to be troublesome for those who are obliged to make them and so demanding a higher education, rather than a lower one, we might imagine that the institution of higher education has a particular role here. I have no doubt that some will regard this role as quixotic, but also every confidence that it remains a core function for higher education in functional democracies. Accordingly, this chapter explores progress and limitations in higher education's contribution to a critical citizenry, where science and science education fits in this enterprise, and where relevant draws from current developments and discourses in Aotearoa New Zealand.

2.2 On the Nature of Science and of Being Scientific

Nowadays, it behoves the authors of any book with "science" in its title to define the word. If science ever was a simple concept, it is far from that in the twenty-first century. Everyday use, nowadays, might identify science as "systematic study of the physical and living world through observation or experimentation, and development and testing of theories against evidence." Being systematic is an important attribute of science, and in everyday usage being systematic is probably synonymous with being scientific. Beyond everyday use, universities and their disciplines have limited the physical and living world to afford particular status to the human parts, relegating, or advancing, these human and social bits to the social sciences; still anticipating development to be scientific, but with altogether different rules and processes involved. Noting that everyday humans do not necessarily appreciate boundaries between them and the physical and living world, much interdisciplinary confusion exists both within higher education's disciplines and in more everyday circumstances. The UK-based Science Council (struggling to straddle the divide between academic and everyday) suggests that "*science is the pursuit and application of knowledge and understanding of the natural and social world following a systematic methodology based on evidence.*" (Science Council 2023, n.p). This is likely in an attempt to bring the sciences into the twenty-first century by accepting and owning the inevitable overlaps between people and everything else (although in the process labeling people as something other than natural). Inconvenient as it may be, historically (dictionaries tend to use the term archaic), but arguably more sensibly, the word "science" referred to any knowledge or knowledge system, however developed. In Aotearoa New Zealand, we celebrate Mātauranga Māori, helpfully defined by ScienceLearn NZ as " ... a modern term for the combined knowledge of Polynesian ancestors and the experiences of Māori living in the environment of Aotearoa" (ScienceLearn NZ 2023, n.p). Aotearoa New Zealand is currently struggling, in the context of its Treaty of Waitangi[5] and notions of partnership, to determine the place of Mātauranga Māori in formal education. Seven professors from the University of Auckland shook Aotearoa New Zealand's university world recently, by questioning parity between Mātauranga Māori with other bodies of knowledge.[6] The issues are complex and clearly relevant to this book, this chapter, and this chapter's section. If we cannot agree what the words "science" and "scientific" mean in the context of sustainability, this book may serve no purpose other than continued obfuscation. For the purposes of this chapter, I prefer to emphasize the "systematic" aspect of science and identify with notions of being "scientific" as being systematic according to rules defined within particular and defined (to a degree) "ways of knowing." In this respect, what appears to be important in any discourse on science and sustainability is to be specific about the ways of knowing involved.

2.3 On the Role of Science in Achieving Sustainability: Focus on the SDGs

For the purposes of this chapter, I am happy to recruit the 17 Sustainable Development Goals,[7] agreed by the majority of nations in 2015, as representative of the aim of our shared sustainability mission, and suggest that the agreed goals, targets, and indicators paint a fairly good picture of what sustainable development looks like, if it was achievable, and a base from which to argue about the practicalities of higher-education-teaching contributing to this broad aim. Although individual ethnic groups, nations, and academics may prefer to prioritize particular knowledge systems, the SDGs' goals do not overtly depend on particular knowledge systems, or academic disciplines, but rather seek to establish common ground, as if all shared a common knowledge system. At a grand level, the SDGs expect nations to "End poverty in all its forms everywhere" (SDG#1), "Ensure inclusive and equitable quality education and promote lifelong learning opportunities for all" (SDG#4), "Ensure sustainable consumption and production patterns" (SDG#12), and "Take urgent action to combat climate change and its impacts" (SDG#13). Its targets are similarly prescriptive, but no more discriminatory than the goals with respect to the roles of higher education or the knowledge systems involved. Indeed, most academic analysis must identify the naivety of their conceptual underpinning by any systematic analysis. Target 4.7 (By 2030, ensure that all learners acquire the knowledge and skills needed to promote sustainable development, including, among

---

[5] A treaty between many of the chiefs of the first human inhabitants of Aotearoa, Māori, and the constitutional monarch of Great Britain, signed in 1840.

[6] https://www.rnz.co.nz/news/te-manu-korihi/447898/university-academics-claim-matauranga-maori-not-science-sparks-controversy

[7] https://sdgs.un.org/goals

others, through education for sustainable development and sustainable lifestyles, human rights, gender equality, promotion of a culture of peace and nonviolence, global citizenship, and appreciation of cultural diversity and of culture's contribution to sustainable development), for example, suggests that learned knowledge and skills will be enough for individuals to set forth in promoting sustainable development themselves, and ignores what other forms of learning might influence their personal agency or be necessary for its functioning. Indicator 4.7.1 [and also 12.8.1 and 13.3.1; perhaps emphasizing the extent to which the United Nations (UN) appreciates education as a contributor to the achievements of the SDGs] highlights the "Extent to which (i) global citizenship education and (ii) education for sustainable development are mainstreamed in (a) national education policies; (b) curricula; (c) teacher education; and (d) student assessment." The UN clearly has confidence that combinations of "global citizenship education" and "education for sustainable development" will lead to these desirable goals but leaves decisions about the extent to which science education, or other forms of education, and the diverse knowledge systems pertaining to them, should be involved, to individual nations to decide for themselves, and for academics to discuss and argue about in books like this one. The UN is not prescriptive about how important basic science (likely involving Dalton and Darwin, or perhaps emphasizing fundamental elements of indigenous knowledge systems) might be to education *for* sustainable development (ESD) and the achievement of, say, SDG#13. Nor does the UN highlight the roles of the social sciences in helping people, say, understand why they feel the need to consume so much "stuff." Even so, I am not particularly convinced that any of the models of science depicted above currently work in the context of sustainable development and at the level of higher education, or are likely to do so in the future. We need to consider the limitations that Science Education in higher education currently operates within and consider ways to address them.

2.4 Limits to the Role of Science Education in HE, as It Is Now, in Achieving Sustainable Development

Higher education internationally has been built on liberal traditions of not only academic freedom (allowing university teachers some autonomy to decide for themselves what to teach and how to teach it) but also the freedoms of students in HE to make choices about what to learn. Sweden may be one notable exception, legislating since 2006 that its universities are to promote sustainable development (Finnveden et al. 2020). Aotearoa New Zealand may be an exception in another way, legislating since 1989 that the principal aim of its universities should be to develop intellectual independence, with possible consequences to universities promoting anything else that might stand in the way of achieving their principal aim (Shephard 2022). Such aims should also be seen in the context of the expansion of higher education, and the accepted roles of higher education in democratic nations.

Trow (1973), for example, argued that once higher education reaches 50% of a nation's young people, its focus becomes enabling whole populations to adapt to social and technological change. But, at present and in general, students have choices, and they do not all choose to be educated *for* sustainable development or choose to study science at the university level. Indeed, it appears that some of our students have not studied much science at all in preparation for their higher education, making its further study at university particularly problematic. In promoting ESD, some claim that students ask for sustainability education at university, but the data available is problematic to interpret (reviewed by Shephard 2023a). In the United Kingdom, for example, the National Union of Students (NUS) has been surveying students every year since 2005. Since 2011, a particular survey has asked students' views about their experiences of and expectations for education for sustainable development at their universities, and since 2015, this has been organized by Students Organising for Sustainability (SOS). In 2020/2021, more than 80% of the more than 8000 students who responded thought that their place of study should actively incorporate and promote sustainable development through all courses (Students Organising for Sustainability 2022). Some students clearly appreciate sustainability-related experiences integrated into their studies, but most students did not respond to the survey. More than two million students typically study at any one time in the United Kingdom's universities. ESD applied systematically to all students, willing or not, provides a different image of the roles and responsibilities of higher education from the one that we are used to. Similar concerns about the extent to which university teachers "educate for sustainable development" have been recorded by Brown et al. (2019) and, in the context of academic development, by Persson et al. (2023). What students choose to study at university may, of course, reflect what they have been taught, and what opportunities they have received during their education prior to university. Concerns about the limited extent to which university-based teacher-educators teach trainee teachers to be *for* sustainable development have been noted by Goller and Rieckmann (2022). It is important to note that particular forms of ESD, applied in particular contexts, can have a positive impact on learners "sustainability attributes." Research in Sweden, in the particular context of compulsory education in a nation that legislates that its schools promote sustainable development (Olsson et al. 2016, 2022; Pauw et al. 2015), is highly positive in this regard, but finding similarly positive conclusions in post-compulsory education is more difficult.

It appears that another challenge that limits the capacity of higher education to educate *for* sustainability is that the learning involved in being *for* sustainability, or indeed *for* anything in particular, is essentially affective in nature and so dependent on students being taught specific values, attitudes, or dispositions (Shephard 2008). This much is explicit within modern-day descriptions of so-called sustainability competencies (Wiek et al. 2011). The use of words such as competence, competencies, and competency to describe educational outcomes is highly contested in the international education literature (Schaffar 2021) and calls into question the fundamental purposes of education, as opposed to training. Essentially, for education to be *for* sustainability or *for* sustainable development, students need to emerge from it being willing to undertake a range of sustainability-related behaviors. As defined by Rieckmann (2012):

> Competencies may be characterised as individual dispositions to self-organisation which include cognitive, affective, volitional (with deliberate intention) and motivational elements; they are an interplay of knowledge, capacities and skills, motives and affective dispositions. Consequently, these components are part of each competency, not having to be regarded independently, but in their interaction. Competencies facilitate self-organised action in various complex situations, dependent on the given specific situation and context. (Rieckmann 2012, p 4)

ESD practitioners do not agree on the definitions of disposition (Shephard 2022) but few would argue with its essential affective nature or that those who do not successfully learn to be disposed to particular actions are unlikely to perform these actions in challenging circumstances. Unfortunately, higher education teachers are not necessarily either competent or willing to teach their students to be *for* a wide range of outcomes (Shephard and Furnari 2012; Persson et al. 2023), noting that being *for* a wide range of *professional* values is a particular situation (Shephard and Egan 2018). One of the first research-based indications that higher education was finding the affective elements of the mission of ESD problematic (those relating to concern, commitment, and willingness to act) came from institutional research in the United States. Using both quantitative and qualitative research approaches directed at student learning, this research found; "... no evidence that, as students move through [the University], they became more concerned about various aspects of sustainability or more committed to acting in environmentally responsible ways, either in the present moment or in their adult lives" (Schoolman et al. 2016, p. 498). Similar expressions of concern internationally were reviewed by Brown et al. (2019).

As a sign of the times, we should also be concerned about our children's declining interest in science, university students' absenteeism in some nations, declining recruitment into higher education science programs and, notably in Aotearoa New Zealand, reports of imminent redundancy of science academics. At the time of writing, six of our eight universities are facing profound financial difficulties, and, in some, traditional science programs are threatened. At the same time, university interest in indigenous worldviews and decolonized curricula is increasing around the world. Concerns about the relative roles of science and beliefs in ESD are heightened by research that finds a negative relationship between religious belief and sustainability concern (Hope and Jones 2014) and challenges posed by religious belief on the measurement of sustainability concern (Shephard et al. 2021). Also in Aotearoa New Zealand, those who design our national school curriculum appear set on replacing traditional biology, physics, and chemistry with issues-based curriculum elements.

Given the rather challenged scheme of teaching science to university students as a means to educate them to be *for* sustainable development, we had better focus our attention on an alternative plan of action. This investigates where broader imaginations of how science and science education may fit within being *for* sustainable development and explores the concept of "being scientific" in a little more detail.

2.5 The Principal Aim of Universities Is to Develop Intellectual Independence

Aotearoa New Zealand's legislation insists that the principal aim of universities should be to develop intellectual independence (New Zealand Legislation 2020); not to educate for sustainable development, or to teach science, or to train professionals, or to address the ills of colonization, or to undertake a myriad of other activities that individuals or groups of stakeholders may think most important. The legislation does not define what intellectual independence is or how universities are to develop it, but I shall attempt to do both in the paragraphs that follow [in summary of more extensive considerations made elsewhere by Shephard (2020, 2022)].

Common usage would suggest that intellectually independent people tend to think for themselves. A great deal of philosophical thought on the topic emphasizes that there are bound to be limits to the extent to which individuals could think for themselves, so accepting some degree of epistemic authority is generally helpful even to independent thinking (Gaon and Norris 2001). Nevertheless, intellectually independent people feel the need to ensure that their beliefs and actions are primarily based on their own intellectual endeavors, and not on those of their parents, siblings, neighbors, professors, politicians, or religious leaders, no matter how convenient, helpful, or persuasive these may be.

The question of what guides the beliefs and actions of individuals from within was addressed by Scriven and Paul in 1987:

> Critical thinking is the intellectually disciplined process of actively and skillfully conceptualizing, applying, analyzing, synthesizing, and/or evaluating information gathered from, or generated by, observation, experience, reflection, reasoning, or communication, as a guide to belief and action (Scriven and Paul 1987, n.p.)

The question of what critical thinking comprises, beyond Scriven and Pauls' helpful definition, was addressed in detail by an expert panel, in the late 1980s (Facione 1990, 2000). In essence, critical thinking requires a particular range of skills, or abilities. Facione (1990) lists these as Explanation, Interpretation, Analysis, Evaluation, Inference, and Self-regulation.

Underpinning the use of these abilities are a particular range of dispositions, or expressions of willingness to use thinking skills and how best to use them. Facione (1990) included Clarity in stating the question or concern; Precision to the degree permitted by the subject and the circumstance; Orderliness in working with complexity; Persistence though difficulties are encountered; Prudence in suspending, making, or altering judgments; Willingness to reconsider and revise views where honest reflection suggests that change is warranted; Reasonableness in selecting and applying criteria; Honesty in facing one's own biases, prejudices, stereotypes, egocentric, or sociocentric tendencies; Open-mindedness regarding divergent world views; and Fair-mindedness in appraising reasoning, among a longer list of dispositions.

Both skills and dispositions can be learned, and taught, but importantly, while critical thinking skills relate strongly to learning in Bloom's cognitive hierarchy

(Bloom 1956), the dispositions do not. Dispositions are affective in nature and relate strongly to Bloom and Krathwohl's affective domain of learning (Krathwohl et al. 1964). They require specific learning and teaching situations to achieve (Shephard 2020) and are notoriously difficult to assess (Bloom et al. 1971). A significant body of research and analysis nowadays suggests that affective attributes provide the link between what people know, what skills people learn to put their knowledge to effect, and what people choose to do with the knowledge and skills that they have learned (Shephard et al. 2015). Many university teachers think that they can and do assess critical thinking skills, even though the challenges involved are significant (Liu et al. 2014; Loyalka et al. 2021), but much less is known about teaching dispositions to critical thinking.

Taking critical thinking skills and dispositions together suggests one possible plan for higher education functioning. This chapter suggests that higher education is to address its principal aim of developing intellectual independence by ensuring that its graduates have learned the skills and dispositions to think critically about everything, including the challenging decisions that they will face as members of a critical citizenry. Critical citizens are critical thinkers, and they do not base their decisions *on emotional reactions and heuristic shortcuts*. In this imaginary world, those charged with the difficult decision that led to imported nylon carpet tiles being laid in 600 rural schools, in Aotearoa New Zealand (arguably the wool capital of the world) would not only have had the skills to think critically about this decision, they would have been fair-minded in appraising their reasoning, flexible in considering alternatives and opinions, reasonable in selecting and applying criteria, diligent in seeking relevant information, and willing to reconsider (where honest reflection suggest that change was warranted). Of course, this may have been the case (how would we know?) and importantly we cannot say what they should have decided, as the point about being intellectually independent is that decisions are based on what goes on inside a person's head, not on what those around the person proclaim should. If we want to be sure of a particular outcome, we in higher education had better not educate our young people to be intellectually independent!

2.6 Diverse Conceptualizations of Critical Thinking in Higher Education

To be clear, Facione's (1990) skills and dispositions are not the only conceptualizations of critical thinking. Indeed, these concepts and their components are highly contested in academic discourse, as are the abilities and willingness of universities and of university teachers to teach them (Davies and Barnett 2015) and the purposes of doing so (Jiménez-Aleixandre and Puig 2022). Lists of skills and dispositions may be best interpreted as theoretical frameworks, ready to be tested and improved by processes dependent on the academic discipline, field of enquiry or knowledge system involved. "Traditional" scientists may prefer to develop with hypotheses, experimentation, and control groups. Social scientists, and the discipline of education, may prefer more qualitative approaches and perhaps accept limitations to the extent to which we can ever know something as complex as human thinking, or its links to action. In recent years, fields of enquiry known as ESD and Sustainability Science have tried to combine cognition and affect as new and particular conceptualizations of competence (as in the definition recorded in the pages above (see also Brundiers et al. 2020). In another conceptualization of critical thinking, proponents of critical theory identify critical thinking as one step toward critical action (Brookfield 2012). Taking the academic literature as a whole, it might be possible to rationalize critical thinking as almost anything at all, and its achievement made possible by diverse pedagogical approaches. Our collective inability to agree what it is, how to teach it, how to assess, evaluate, or measure it, or rationalize the consequences of its operation, makes it the universal get-out clause in the social contract between education and society. Nearly everyone claims to not teach their students what to think, but how to think. Nearly everyone endorses critical thinking as an attribute of graduateness. But even among those who celebrate critical thinking in higher education, few consider its affective, dispositional dimensions.

I prefer to consider that both skills and dispositions are necessary for people to think critically and so be intellectually independent, but that dispositions rather than skills require particular attention in this chapter. I suggest that different pedagogical processes may be necessary to enable our students to learn critical thinking skills from those that will lead them to be disposed to use them in particular ways; perhaps in particular to be bothered to think critically in circumstances that make *emotional reactions and heuristic shortcuts rather than rational analysis* attractive. I wonder if being "open-minded regarding divergent world views," "willing to reconsider and revise views where honest reflection suggests that change is warranted," "honest in facing one's own biases, prejudices, stereotypes, egocentric or socio-centric tendencies," and "understanding of the opinions of other people" is, to a degree, missing in our societies at present, missing in lists of graduate outcomes to be assessed or evaluated, but somehow lurking unseen in the targets and indicators of the SDGs. I wonder who would have noticed had Indicator 4.7.1 appeared as; "Extent to which dispositions (i) to be understanding of the opinions of other people and (ii) 'honest in facing one's own biases, prejudices, stereotypes, egocentric or socio-centric tendencies' are mainstreamed in (a) national education policies; (b) curricula; (c) teacher education; and (d) student assessment."

My preference opens up three lines of questioning: how do we teach and assess intellectual independence (and in the process assess, evaluate, monitor, or measure its development); what might get in the way of us teaching and assessing it; and what has it got to do with sustainable development? For those following the argument building in this chapter, it may be clear that we are attempting to develop clarity about the meaning of the adjectives "scientific" and "systematic" in these contexts and the academic circumstances in which being scientific or systematic make sense.

2.7 How Do We Teach the Skills and Dispositions to Be Intellectually Independent?

I remember a particular physics lesson when I was 13. Mr. Jones had us looking through a microscope at smoke particles dancing. He proposed two of perhaps many explanations. He encouraged us to consider the observation of "Brownian motion" as a form of evidential support for atomic theory, suggesting that matter, such as air, is made from molecules whose random collisions with smoke particles caused them to dance randomly. Importantly for me he proposed another theory; that the particles were being moved by energetic goblins, too small to be seen by the microscope, as indeed were the air molecules. Thank you, Mr. Jones. I have no doubt that although you taught me much that rightly belongs in the physics textbook, the lasting lesson for me was to always look for the goblins in all of my observations, and in the assertions of others. I often encounter academics who appear to me to be putting too much trust in ideas and explanations and to be oblivious to other explanations or alternative ways of understanding. I put my personal dispositions (in as much as I have them) to be: prudent in suspending, making, or altering judgments; inquisitive with regard to a wide range of issues; concerned to become and remain generally well-informed; and willing to reconsider and revise views where honest reflection suggests that change is warranted, down to Mr. Jones and to other science teachers like him. And I acknowledge that such dispositions did not suddenly appear in a physics lesson, but as a gradual process punctuated by discussions around Miss Gardner's nature table in my primary school, overcoming challenges of experimental repeatability during my PhD, and in encouraging other academics to address affect in science and more widely, since then.

It is possible that my schoolteachers in other subjects helped me to learn some other dispositions to think critically, but I am less clear about this. Soon after Mr. Jones taught me physics, my history teacher was a military man who talked about nothing other than his own experiences in the Second World War. My English teacher appeared to me to be obsessed by phallic symbolism in everything that we read. I failed examinations in history and in English literature at that time but always did well in the sciences. Perhaps I was simply not receptive to the undertones of what I was being taught at the time, but I would need to leave it to others to argue that the study of subjects other than science help learners to become disposed to think critically. For my part, I am not convinced that my science background helped me to develop "open-mindedness regarding divergent world views," but then I am not confident that I am as disposed to be "open-minded to divergent world views," or to be "understanding of the opinions of other people," as I am disposed to other things. For me these are works in progress. Perhaps science is good at teaching some dispositions, such as "Clarity in stating the question or concern," "Precision to the degree permitted by the subject and the circumstance," and "Orderliness in working with complexity" but other disciplines excel in teaching other dispositions, such as "Open-mindedness regarding divergent world views." A much-cited exploration of science teaching does suggest the possibility that our science graduates care less about the social implications of their science than many would hope (Cech 2014).

With respect to higher education's principal aim of developing intellectual independence, if our students graduate capable of thinking critically, and disposed to do so with persistence, with willingness to change their minds, fair-mindedness in appraising their reasoning, and open-mindedness, does it really matter if we achieve this primarily in our chemistry department, sociology department, or art department? It may be more important that all disciplines and departments agree to pursue intellectual independence, that all agree (as far as is possible in the institution of higher education) what it is, and that they do not simultaneously attempt to teach other things that get in the way of its achievement.

In addition, however, perhaps all will agree that we, academic people, need to be systematic, or scientific, in our exploration of how best to teach young people to be disposed to think critically; and similarly scientific or systematic in our efforts to assure our societies that we achieve these things. While the above paragraphs may have been descriptive, even reflective, they were far from systematic. Being systematic may require us to take note, for example, of the Improvement Science assertion that we cannot improve something if we cannot measure it (Carnegie Foundation 2023). Higher education prides itself on the extent to which we do understand the skills involved in critical thinking, and how to teach them, and to a degree, how to measure their attainment. The same cannot be said for being disposed to think critically. Approaches to teach and assess dispositions to being willing to reconsider, open minded, and being fair-minded in appraising reasoning have been considered by Shephard (2020) and Shephard et al. (2021). What research there is on this topic suggests that we have much to learn about teaching affective outcomes such as values, attitudes, and dispositions. I suggest that our unwillingness to address these matters systematically, or scientifically, is preventing us from rationalizing which affective outcomes we should teach in higher education, and which we should not.

2.8 What Might Get in the Way of Teaching Intellectual Independence?

If developing intellectual independence is a principal aim, rather than just one of many aims, it seems rational to ensure that other educational objectives do not in some way limit the attainment of intellectually independence. Intellectually independent people will find their own beliefs and values from within, and do not need to be taught them by well-meaning academics and academic institutions. Intellectually independent people will decide for themselves what actions they will take, and do not need to be told how to behave. Education that is *for* anything in particular, other than *for* intellectual independence, may simply not be compatible with education's principal aim of developing intellectual independence. Institutional efforts to instill a competitive spirit, or a cooperative one, traditional values, or professional values, indigenous ways of thinking or working, or modern-day constructively aligned-thinking or working, all appear potentially contradictory to supporting higher education learners to think for themselves. My own experience of higher education's learning objectives, aspirational graduate attributes, and expressions of values-based education suggest to me that higher education nowadays does not expect or trust its teachers to teach intellectual independence, or expect or trust its learners to use their intellectual independence wisely.

An expanding social concept explores how the power and resilience of persistent hierarchies of privilege in universities are embodied and maintained as academic identity. The concept suggests that although rationality and affect work together to generate individual agency, in higher education rationality dominates emotion to the extent that affect is virtually erased (Clegg 2013). Knowledge and skills, of course, occupy important positions in university research and teaching but become disembodied from their affective, emotional, self-aware underpinnings. Shephard (2023b) linked academia's aversion to teaching in the affective domain to its unwillingness to research or to monitor university impacts on social justice agendas. The review process for this chapter expressed concern about this chapter's first-person narrative (apparently the pronoun "I" is used more than 30 times, and this is, ostensibly, too much for an academic text). "I" argue that intellectual independence requires first-person affective engagement in our planet's and societies' problems, and in our learning.

2.9 What Has Intellectual Independence to Do with Sustainable Development?

If we cannot be sure what decisions intellectually independent people will make in the context of social, environmental, and economic justice, should we be teaching intellectual independence? What if our graduates systematically use their intellectual independence to protect their own interests and those of their descendants, rather than attend to the needs of others, as some economic and evolutionary theories predict? Surely it would be better to teach our students the knowledge, skills, and values that they need to be sustainable, and to live sustainably, rather than to empower them to use their education for something else. Perhaps, but such ideas belong elsewhere. This chapter argues that science education may be a great way to teach young people to be disposed to think critically about the problems of the world, that we need to address this possibility systematically, or scientifically, and that currently we fall far short of this. This chapter must end with a significant item for a research agenda. Are those graduates and citizens capable of thinking critically and disposed to do so with persistence, with willingness to change their minds, fair-mindedness in appraising their reasoning, and open-mindedness, more likely to be concerned about issues of social, environmental, and economic injustice, than those who are not? And if they are, are they more likely to do something about them? If the answer to either is, or may be, yes, then surely, we need to be scientific in our attempts to teach these things. At present, we appear to be too uncertain about the purposes of science education to ask these questions or to reasonably distinguish between what we should be teaching, and what we should not.

References

Bloom BS (1956) Taxonomy of educational objectives, handbook: the cognitive domain. David McKay

Bloom BS, Hastings JT, Madaus GF (1971) Handbook on the formative and summative evaluation of student learning. McGraw-Hill

Brookfield SD (2012) Teaching for critical thinking: tools and techniques to help students question their assumptions. Jossey-Bass

Brown K, Connelly S, Lovelock B, Mainvil L, Mather D, Roberts H, Skeaff S, Shephard K (2019) Do we teach our students to share and to care? Res Post-Compuls Educ 24(4):462–481. https://doi.org/10.1080/13596748.2019.1654693

Brundiers K, Barth M, Cebrián G, Cohen M, Diaz L, Doucette-Remington S, Dripps W, Habron G, Harré N, Jarchow M, Losch K, Michel J, Mochizuki Y, Rieckmann M, Parnell R, Walker P, Zint M (2020) Key competencies in sustainability in higher education: toward an agreed-upon reference framework. Sustain Sci. https://doi.org/10.1007/s11625-020-00838-2

Carnegie Foundation (2023) The six core principles of improvement. https://www.carnegiefoundation.org/our-ideas/six-core-principles-improvement/

Cech EA (2014) Education: embed social awareness in science curricula. Nature 505(7484):477–478. https://doi.org/10.1038/505477a

Clegg S (2013) The space of academia: privilege, agency and the Erasure of affect. In Maxwell C, Aggleton P, Privilege, agency and affect (pp 71–87). Palgrave Macmillan UK. https://doi.org/10.1057/9781137292636_5

Davies M, Barnett R (eds) (2015) The Palgrave handbook of critical thinking in higher education. Palgrave Macmillan US. https://doi.org/10.1057/9781137378057

Facione PA (1990) Critical thinking: a statement of expert consensus for purposes of educational assessment and instruction. The California Academic Press

Facione PA (2000) The disposition toward critical thinking: its character, measurement, and relation to critical thinking skill. Informal Log 20(1):61–84

Finnveden G, Friman E, Mogren A, Palmer H, Sund P, Carstedt G, Lundberg S, Robertsson B, Rodhe H, Svärd L (2020) Evaluation of integration of sustainable development in higher education in Sweden. Int J Sustain High Educ 21(4):685–698. https://doi.org/10.1108/IJSHE-09-2019-0287

Gaon S, Norris SP (2001) The undecidable grounds of scientific expertise: science education and the limits of intellectual independence. J Philos Educ 35(2):187–201. https://doi.org/10.1111/1467-9752.00220

Goller A, Rieckmann M (2022) What do we know about teacher educators' perceptions of education for sustainable development? A systematic literature review. J Teach Educ Sustain 24(1):19–34. https://doi.org/10.2478/jtes-2022-0003

Harari YN (2018) 21 lessons for the 21st century. Spiegel and Grau

Hope ALB, Jones CR (2014) The impact of religious faith on attitudes to environmental issues and carbon capture and storage (CCS) technologies: a mixed methods study. Technol Soc 38:48–59. https://doi.org/10.1016/j.techsoc.2014.02.003

Jiménez-Aleixandre MP, Puig B (2022) Educating critical citizens to face post-truth: the time is now. In: Puig B, Jiménez-Aleixandre MP (eds) Critical thinking in biology and

environmental education: contributions from biology education research. Springer. https://doi.org/10.1007/978-3-030-92006-7_1
Krathwohl D, Bloom B, Masia B (1964) Taxonomy of educational objectives, Handbook II: the affective domain. David McKay Co.
Liu OL, Frankel L, Roohr KC (2014) Assessing critical thinking in higher education: current state and directions for next-generation assessment. ETS Res Rep Ser 2014(1):1–23. https://doi.org/10.1002/ets2.12009
Loyalka P, Liu OL, Li G, Kardanova E, Chirikov I, Hu S, Yu N, Ma L, Guo F, Beteille T, Tognatta N, Gu L, Ling G, Federiakin D, Wang H, Khanna S, Bhuradia A, Shi Z, Li Y (2021) Skill levels and gains in university STEM education in China, India, Russia and the United States. Nat Hum Behav. https://doi.org/10.1038/s41562-021-01062-3
New Zealand Legislation (2020) Education and Training Act 2020, Establishment of institutions. https://www.legislation.govt.nz/act/public/2020/0038/latest/LMS202213.html
Olsson D, Gericke N, Chang Rundgren S-N (2016) The effect of implementation of education for sustainable development in Swedish compulsory schools: assessing pupils' sustainability consciousness. Environ Educ Res 22(2):176–202. https://doi.org/10.1080/13504622.2015.1005057
Olsson D, Gericke N, Boeve-de Pauw J (2022) The effectiveness of education for sustainable development revisited – a longitudinal study on secondary students' action competence for sustainability. Environ Educ Res 28(3):405–429. https://doi.org/10.1080/13504622.2022.2033170
Pauw J, Gericke N, Olsson D, Berglund T (2015) The effectiveness of education for sustainable development. Sustainability 7(12):15693–15717. https://doi.org/10.3390/su71115693
Persson C, Einarson D, Melén M (2023) Educating the educators to be a driving force in higher education towards sustainable development. Int J Sustain High Educ 24(9):197–212. https://doi.org/10.1108/IJSHE-10-2022-0332
Rieckmann M (2012) Future-oriented higher education: which key competencies should be fostered through university teaching and learning? Futures 44(2):127–135
Schaffar B (2021) Competent uses of competence: on the difference between a value-judgment and empirical assessability. Nord J Stud Educ Policy 7:55–64. https://doi.org/10.1080/20020317.2021.1958993
Schoolman ED, Shriberg M, Schwimmer S, Tysman M (2016) Green cities and ivory towers: how do higher education sustainability initiatives shape millennials' consumption practices? J Environ Stud Sci 6(3):490–502. https://doi.org/10.1007/s13412-014-0190-z
Science Council (2023) Our definition of science. https://sciencecouncil.org/about-science/our-definition-of-science/
ScienceLearn NZ (2023) Mātauranga Māori and science. https://www.sciencelearn.org.nz/resources/2545-matauranga-maori-and-science
Scriven M, Paul R (1987) Cited by the foundation for critical thinking. Defining critical thinking. https://www.criticalthinking.org/pages/defining-critical-thinking/766
Shephard K (2008) Higher education for sustainability: seeking affective learning outcomes. Int J Sustain High Educ 9(1):87–98. https://doi.org/10.1108/14676370810842201
Shephard K (2020) Higher education for sustainability: seeking intellectual independence in Aotearoa New Zealand. Springer Singapore. https://doi.org/10.1007/978-981-15-1940-6
Shephard K (2022) On intellectual independence: the principal aim of universities in New Zealand. N Z J Educ Stud 57(1):269–284. https://doi.org/10.1007/s40841-022-00250-7
Shephard K (2023a) Student experiences of higher education for sustainability and for intellectual independence. In: Baik C, Kahu ER (eds) Research handbook on the student experience in higher education. Edward Elgar Publishing, pp 524–537. https://doi.org/10.4337/9781802204193.00047
Shephard K (2023b) Academic identity and "education for sustainable development": a grounded theory. Front Educ 8. https://doi.org/10.3389/feduc.2023.1257119
Shephard K, Egan T (2018) Higher education for professional and civic values: a critical review and analysis. Sustainability 10(12):4442. https://doi.org/10.3390/su10124442

Shephard K, Furnari M (2012) Exploring what university teachers think about education for sustainability. Stud High Educ 38(10):1577–1590. https://doi.org/10.1080/03075079.2011.644784

Shephard K, Harraway J, Lovelock B, Mirosa M, Skeaff S, Slooten L, Strack M, Furnari M, Jowett T, Deaker L (2015) Seeking learning outcomes appropriate for 'education for sustainable development' and for higher education. Assess Eval High Educ 40(6):855–866. https://doi.org/10.1080/02602938.2015.1009871

Shephard K, Kalsoom Q, Gupta R, Probst L, Gannon P, Santhakumar V, Ndukwe IG, Jowett T (2021) Exploring the relationship between dispositions to think critically and sustainability concern in HESD. Int J Sustain High Educ 22(5):1166–1185. https://doi.org/10.1108/IJSHE-07-2020-0251

Students Organising for Sustainability (2022) Sustainability skills survey 2020–2021. https://www.sos-uk.org/research/sustainability-skills-survey

Trow M (1973) Problems in the transition from elite to mass higher education. In: Carnegie Commission on higher education. McGraw-Hill

Wiek A, Withycombe L, Redman CL (2011) Key competencies in sustainability: a reference framework for academic program development. Sustain Sci 6(2):203–218. https://doi.org/10.1007/s11625-011-0132-6

Williams R (2014) There's no fooling about impact. Times Higher Education April 17, 2014. https://www.timeshighereducation.com/comment/opinion/rowan-williams-theres-no-fooling-about-impact/2012699.article

Kerry Shephard, PhD, is Emeritus Professor at the University of Otago, New Zealand. He has an honors degree in zoology and botany, a PhD in environmental physiology, and went on to teach and research in the life sciences for many years. He progressively developed an interest in university learning and teaching and, after a period as learning technologies coordinator at the University of Southampton's Centre for Learning and Teaching, went on to become professor of higher education development at the University of Otago. His research interests include education for sustainable development, education for intellectual independence, and the future development of higher education.

Chapter 3
Uncertainty, Science Education, and Sustainability: A Brief Introduction

Fiona van Schaik

Abstract In working toward sustainability, people face several interconnected problems in relation to which an inexhaustible number of questions might be posed. These questions may reflect uncertainties. This chapter addresses the pervasiveness of uncertainty in science within the context of sustainability issues and suggests some implications for science education on such topics. First, the chapter explores uncertainty and its various types. Second, the chapter describes which terms reflect some uncertainty in the context of both science and science education. Third, after a brief consideration of sustainability education and building on the rest of the chapter, the chapter presents three suggestions for how science educators can take uncertainty into account.

Keywords Uncertainty in science · Sustainability science education · Nature of science · Socioscientific issues · Post-normal science

3.1 Introduction

> To learn is to question. (Koefoed and Burø 2022, p. 193)
>
> To learn is to question how to question. (Koefoed and Burø 2022, p. 201)

This chapter concerns uncertainty in science education about sustainability. The uncertainties of relevance in this context are versatile, but one kind is perhaps the most central: the questions that are asked. Questions, with all their uneasiness and uncertainty, form the basis of all learning according to Koefoed and Burø (2022), who describe learning as a quest. By asking questions, people venture out into unknown, unexplored terrain. Similarly, Rosen indicates the value of uncertainty in "coming to know" (2019, p. 2), by pointing to the complexity of pedagogical

F. van Schaik (✉)
Department of Education and Sports Science, Faculty of Arts and Education, University of Stavanger, Stavanger, Norway

F. Skarstein, L.-A. Wolff (eds.), *Sustainability and Science Education*, Environmental Challenges and Solutions,
https://doi.org/10.1007/978-3-032-24408-6_3

processes. To learn, one engages with questions, and as such, with the uncertainties surrounding these questions.

That people ask questions is, quite probably, true of all times. Nevertheless, some developments in society may be argued to have enabled a wide variety of questions to be posed. In the book *Risk Society*, sociologist Ulrich Beck (1986/1992)[1] suggests that modern industrial developments have had unintended and undesired side effects, as risks that are often imperceptible. Beck (1986/1992, p. 22) refers to the risks associated with radioactivity and pollution of air, food, and water. For instance, pharmaceuticals in rivers can impact environmental and human health (Wilkinson et al. 2022). Through scientific activities, a range of otherwise imperceptible issues can be identified, or indeed be seen as problems as such (Beck 1986/1992, pp. 26–27). Beck argues how risks of this kind not only require an appreciation of probability, but also a stance on what is deemed as ethically acceptable, and with it, consideration of "*how we wish to live*" (Beck 1986/1992, p. 28; original emphasis).

In essence, the sustainable development project can be understood as the United Nations' idea of how "we" wish to act. This is defined as "development that meets the needs of the present without compromising the ability of future generations to meet their own needs," a definition that the World Commission on Environment and Development (1987, p. 41) formulated in the Brundtland Report, *Our Common Future*, published in 1987. The report was a response to the General Assembly of the United Nations' call for "[a] global agenda for change" (World Commission on Environment and Development 1987, p. 5). In the following decades, the work on sustainable development continued through several international gatherings such as the 1992 Earth Summit, where Agenda 21 was adopted; the Millenium Summit in 2000 that formulated the Millenium Development Goals to end poverty; the World Summit on Sustainable Development in 2002; and the United Nations Conference on Sustainable Development in 2012 (see UN n.d.). This work formed the basis for the United Nations' 2030 Agenda for Sustainable Development, which announced the 17 Sustainable Development Goals (SDGs) (UN n.d.). The SDGs concern environmental challenges, as well as themes such as poverty, well-being, sanitation, education, inequality, justice, and biodiversity (UN n.d., p. 17), and serve to *safeguard* the needs of present and future generations. The goals present the normative ambitions of contemporary societies and reflect the challenges that these societies face, including the risks, the "*projected dangers of the future*" (Beck 1986/1992, p. 34; original emphasis) that they wish to prevent. One of these goals is SDG 13 Climate Action, which strives for mitigation and adaptation to contemporary climate change, an issue induced by the emission of greenhouse gases by, predominantly, wealthy, industrialized countries. Likewise, some elements of SDG 6 Clean Water and Sanitation may be suggested to address the risks induced by modern societies, for instance, the potential effects of pharmaceutical pollution.

[1] Also see the work by Giddens on risk society and reflexive modernity (Ekberg 2007; Heiskala 2011).

Many of the challenges addressed by the SDGs can be seen in relation to each other. For example, the availability and temperature of water are vital for hydropower and thermoelectric power production, but will be impacted by climate change (van Vliet et al. 2017). At the same time, energy demands are expected to increase with economic growth, and water resources are also required for meeting other human demands and for supporting ecosystems. It is thus vital to consider interactions between economic development, energy production, water availability, biodiversity, and climate change to ensure adequate supplies of both water and energy (van Vliet et al. 2017). These and other interactions between the SDGs are often context-dependent: hydropower stations impact Swedish freshwater resources, and wind turbines affect Denmark's land and ocean systems (Nilsson et al. 2018; Schwanitz et al. 2017). The various geographical and time scales involved add to the complexity (Nilsson et al. 2018). Some consequences of an action or lack of action might be experienced locally and immediately (e.g., improved energy supply improves heating of houses), whereas others might not be noticed until many years later, but might have a global impact (e.g., detrimental health effects of climate change) (Nilsson et al. 2018). The work on a particular SDG may barely to greatly impact other SDGs, and this impact may be both negative and positive (Pham-Truffert et al. 2020).

Sustainability challenges are thus highly complex and can involve many uncertainties. This is signified by the inexhaustible number of questions that might be posed in relation to sustainability issues. Such questions may indicate a "knowledge gap" and can be explored or investigated in scientific research projects, for instance, the research regarding the interactions between energy and water supply previously mentioned (van Vliet et al. 2017). Similarly to the interactions of energy and water sectors, the various uncertainties associated with these sectors may relate to each other, too. For example, both the energy and water sectors are affected by weather conditions. More importantly, striving for sustainability is marked by an anticipative dimension, as attempts to make improvements toward a different state. In other words, as the future is inherently uncertain, decision-makers and other actors striving for sustainability are bound to face uncertainty in pursuing a sustainable future. As Beck (1986) points out, the notion of risk reflects a preoccupation with both the future and with safety. Sustainable development may be seen as an attempt to "control" risks, as potential future threats to safety.

Arguably, however, "the challenge is not that we must find ways to 'know' the future, rather we need to find ways to live and act with not-knowing the future" (Miller 2011, p. 24). People may respond to uncertainty in various ways. For some, it seems quite common to attempt to minimize uncertainty as much as possible, while others may focus more on the hope and opportunities that uncertainty may entail. Some politicians rather appear to wait-and-see when dealing with uncertainty, preferring to wait for research to resolve the uncertainty (e.g., Grunwald 2007). A problem with this approach is that this may limit future opportunities. An example from the Netherlands illustrates this. In an interview with a Dutch national newspaper (Schreuder 2022), environmental scientist Marjolijn Haasnoot, known for developing the Dynamic Adaptative Policy Pathways method that accommodates uncertainty in management plans (Haasnoot et al. 2013), argues that the Netherlands should take long-term climate adaptation scenarios into account when investing in new neighborhoods. The building of houses in low-lying areas near the coast could limit future opportunities for water managers to develop flooding areas or force people to leave their homes in the future (Haasnoot and Diermanse 2022; Schreuder 2022). How countries deal with the uncertainties they face could impact their future options.

This chapter seeks to contribute to the question of how science education could deal with uncertainty in the context of sustainability. So far, I have sketched how science, uncertainty, and the pursuit of sustainability relate to each other. This forms a backdrop against which I suggest some implications for science education.

3.2 The Concept of Uncertainty

In the introduction to this chapter, I used various terms that may be associated with uncertainty: questions, unknown, unexplored, complexity, probability, potential, risks, knowledge gap, anticipation, future, threats, not-knowing, opportunities, indeterminate. Although these terms each give an idea of my perception of uncertainty, they do not necessarily define it. What is uncertainty? How can it be recognized?

In this chapter, I define uncertainty as any lack of knowledge or confidence about the state of something (e.g., future technological developments) or actions (e.g., ways to address pharmaceutical pollution), which may be a characteristic of both the subject and the object of (not) knowing. However, there are many ways to define or characterize uncertainty, and I doubt that this definition is robust.

A search in the *Oxford English Dictionary* yields a variety of definitions, with the common definitions being associated with a state or a quality:

> The quality of being uncertain in respect of duration, continuance, occurrence, etc.; liability to chance or accident. Also, the quality of being indeterminate as to magnitude or value; the amount of variation in a numerical result that is consistent with observation. […] The state of not being definitely known or perfectly clear; doubtfulness or vagueness. […] The state or character of being uncertain in mind; a state of doubt; want of assurance or confidence; hesitation; irresolution.
> (Oxford English Dictionary n.d.)

The dictionary also provides a more specific use of the term in economics, as a "business risk which cannot be measured and whose outcome cannot be predicted or insured against" (Oxford English Dictionary n.d.). This definition is derived from Knight (1921), who contrasts risk as a measurable, quantifiable uncertainty, with "true" uncertainty, which would not be quantifiable.

This versatility in the dictionary definitions is also reflected in research literature. Multiple studies report on the inconsistent use and definition of the term *uncertainty* (Bevan 2022; Hanke et al. 2025; Wakeham 2015). For example, uncertainty could be interpreted as "a characteristic of the knowledge itself or a characteristic of the

Fig. 3.1 Uncertainty as a subjective experience (i.e., psychological) versus uncertainty as a (socially constructed/objective) part of reality (e.g., sociological) (Bevan 2022; Wakeham 2015). (Own figure)

knower" (Wakeham 2015, p. 720). This refers to the site where the uncertainty would be "located" (Bevan 2022) and is illustrated in Fig. 3.1. Underlying ideas about the absence of uncertainty, as in an "idealised yet unachieved state" (Bevan 2022, p. 4), is another factor that influence how uncertainty is understood. For instance, an absence of uncertainty could refer to complete confidence, knowledge, or determinism (Bevan 2022).

Perhaps unsurprisingly, scholars have attempted to categorize uncertainties in many ways. Such distinctions can be based on the extent of supposed knowability, measurability, ignorance, indeterminacy, or the object (Bevan 2022). A literature review by Bevan (2022) identifies 156 conceptual frameworks for uncertainty in a range of disciplines focused on environmental change, including economics, risk studies, and environmental modeling. Most of these frameworks have been developed since the 1980s. According to Bevan's (2022) review, there are similarities in the ways the frameworks classify various uncertainties. For example, uncertainties may be categorized depending on the extent of ignorance or the extent of indeterminacy. Another common distinction is made between aleatory and epistemic uncertainty. These uncertainties describe statistical uncertainty and a lack of knowledge, respectively, and may be differentiated on the basis of their different extent of measurability, "meta-uncertainty (described uncertainty versus *uncertainty about that uncertainty*)," nature, and reducibility (Bevan 2022, pp. 4–5; original emphasis). However, terms are often used variably in different frameworks and in many of these frameworks, there are inconsistencies regarding the underlying philosophies (Bevan 2022). In educational research, many studies only define uncertainty implicitly (Hanke et al. 2025).

3.3 Word(s) About Uncertainty in Science and Society

According to Ravetz (1997, p. 13), "[t]he closer we approach to policy issues, the more that the physical science recedes into the framing of the problem, and the more that issues of policy, equity and lifestyle obtrude." Beyond the uncertainties that stimulate scientific research, the uncertainties and complexities in dominant policy discourses have led scholars to wonder about the role of science in addressing policy issues. Some particularly challenging issues are "wicked problems" (Rittel and Webber 1973). According to Rittel and Webber (1973), a conventional search for a science-based approach to solving these issues can be expected to fail due to characteristics such as an indeterminate problem description, an absence of ultimate, right solutions, and an interrelatedness with other issues. The quotation of Ravetz (1997) above and Beck's (1986, p. 28) treatise on risk society reflect on something similar. The problem of framing problems can be recognized in relation to an issue such as climate change (see, e.g., Hulme 2009, 2022). Should climate change be interpreted as a matter of, for example, technological innovation, overpopulation, overconsumption, and/or justice?

Like Rittel and Webber (1973), Funtowicz and Ravetz (1993) identify the changing demands for scientific activities, which now have to engage with ethical and epistemological uncertainties and pluralism. They describe that "[t]he activity of science now encompasses the management of irreducible uncertainties in knowledge and in ethics, and the recognition of different legitimate perspectives and ways of knowing" (Funtowicz and Ravetz 1993, p. 754). Funtowicz and Ravetz (1993) call this reconceptualized form of science for "post-normal," a term that refers to both postmodernism and to Kuhn's "normal science." In normal science, "uncertainties are managed automatically, values are unspoken, and foundational problems are unheard of" (Funtowicz and Ravetz 1993, p. 740). Post-normal science is described as a problem-solving strategy taking place in a context of conflicting goals, and ethical or epistemological uncertainties. As the "normal" approach to science collapses, post-normal science would offer an alternative that clarifies assumptions, actively deals with uncertainties, and can be understood to take the form of a dialogue, with multiple knowledge perspectives. While place, time, and process could have counted as irrelevant in modern normal science, post-normal science would start to embrace these dimensions (Funtowicz and Ravetz 1993). A new kind of leading question would become more important in science: "what-if?" (Ravetz 1997, p. 11). Besides placing more weight on scientific uncertainties, these kinds of questions may also be associated with the previously mentioned focus of both society and science on safety, and are perhaps especially suited to confront unknown threats (Ravetz 1997). Ravetz (1997) appeals to the field of science education to take note of these new kinds of questions in scientific practice.

3.4 Word(s) About Uncertainty in Science and Sustainability Education

Followed by their use to describe the complexities, disagreements, and uncertainties in the interplay between science and society, terms such as "risk" and "risk society," "wicked problems," and "post-normal" have been applied to various issues in educational research (Agustian 2025; Block et al. 2018). Scholars such as Kolstø (2006), Christensen (2009), Levinson et al. (2011), Hansen and Hammann (2017),

Schenk et al. (2019), and Pietrocola et al. (2021) write about risk in science education, typically with reference to the term "wicked problems" or "socioscientific issues." The term socioscientific issue (SSI) points to issues with both scientific and societal elements that are applied in science education, not only to demonstrate the societal relevance of science, but also to stimulate the development of scientific literacy and reasoning skills. To this end, the issues are, to some degree, controversial and demand reasoning (Zeidler and Nichols 2009); they are "complex, open-ended, often contentious dilemmas, with no definitive answers" (Sadler 2004, p. 514). Using SSIs entails "the deliberate use of scientific topics that require students to engage in dialogue, discussion, and debate" (Zeidler and Nichols 2009, p. 49). Although both "wicked problems" and "socioscientific issues" may be used to point to the same kind of problems and may be used in similar contexts, the terms have different roots and connotations. In the context of education, a variety of terms[2] may be understood to point to the relevance of asking the question "what kind of pedagogy does justice to the experience and challenge of living in a complex world?" (Wessels et al. 2024, p. 759).

3.5 Sustainable Development and Education

In the context of sustainable development, education has been granted several roles. Perhaps the most evident of these is the role granted by Sustainable Development Goal (SDG) 4 Quality Education, which aims for free primary and secondary schools, affordable vocational training, and access to higher education for all, irrespective of gender or wealth (UN 2023). Through education, the UN strives to achieve reduced inequalities, citizens that can live healthy and sustainably, as well as societies that are more peaceful and citizens that are more tolerant toward each other (UN 2023). Although quality education for all is thus a goal in itself, it also serves to support other SDGs, such as SDG 3 Good Health and Well-being and SDG 16 Peace, Justice and Strong Institutions. When referring to education that is meant to contribute to sustainable development by educating citizens about themes of relevance to sustainable development, one may refer to education *about* and *for* sustainable development (Sinnes 2021). Education could take place *in* contexts that strive for sustainable development, and may directly contribute to sustainable development by implementing changes: education *as* sustainable development (Sinnes 2021). It may be argued that students could, and perhaps should, encounter uncertainty in all four modes (*about*, *for*, *in*, and *as*) of sustainability education.

Similar to the four approaches to sustainability in education, scholars identify three visions on scientific literacy and science education (see, e.g., Liu 2013;

[2] Other terms include (ref: examples from educational research): VUCA (volatile, uncertain, complex, and ambiguous) (Stein 2021), hyperobject (Bengtsson and Van Poeck 2021; Saari and Mullen 2020), Anthropocene (Gilbert 2016), and risk-related issues (RRIs) (Hansen and Hammann 2017; Van Schai 2024).

Roberts 2007; Sjöblom et al. 2023; Sjöström and Eilks 2018). According to Sjöström and Eilks, vision I of scientific literacy emphasizes knowledge of science and scientific processes, which they call "science learning for later application" (2018, p. 80). Vision II, "meaningful science education for all" (2018, p. 80), focuses on the everyday contexts in which science can be encountered and is of use. Vision III, "science education for transformation" (Sjöström and Eilks 2018, p. 80), aims for transformative action and is also referred to as "critical scientific literacy" (Hodson 2003, 2009). Although uncertainty seems of relevance to each of these visions, vision III of scientific literacy appears to encourage consideration of uncertainty in deciding what action to take on the basis of scientific knowledge and societal/ethical reasons (see, e.g., Sjöström and Stenborg 2014).

Drawing on the idea of "post-normal science" by Funtowicz and Ravetz (1993), Block et al. (2018) formulate three principles for wicked sustainability problems in education. The first of these principles is "be modest" (Block et al. 2018, p. 1434) and functions as a recommendation to not suggest that issues are simple when they are not, to entertain multiple perspectives, and to take into account that facts and knowledge are products of specific approaches and contexts. The second principle, "be immodestly modest" (Block et al. 2018, p. 1434), warns against relativism and advocates for educators to draw attention to sustainability issues that they find of importance. Block et al. write:

> By taking responsibility for sustainability issues, thereby revealing their own, singular perspectives, concerns, value judgements, attachments, etc. educators can stimulate learners to be attentive to it [them], to explore it [them] further, to confront it [them] with other perspectives and, finally, to relate to it [them] by formulating their own, singular perspective, concern and judgement. (Block et al. 2018, p. 1434)

The second principle may thus be seen as an attempt to acknowledge the necessity of addressing sustainability issues and therefore of taking a stance, but also to gain an apprehension of how a particular account is a specific, context-dependent way of viewing these problems, and of how other perspectives may provide other ideas for solutions. In this sense, the second principle builds upon the first principle of being modest. Lastly, the third principle formulated by Block et al. is to "[d]are to experiment" (Block et al. 2018, p. 1435). It advocates an openness to trying out, and an openness to "venturing into the unknown" (Block et al. 2018, p. 1436).

3.6 Suggestions for Science Educators

For science education, the principles in the article by Block et al. (2018), as well as other literature of relevance, may, in consideration of uncertainty, be translated into a few suggestions:

1. Clarify the Nature of Science: Science Cannot Give All *the* Answers

"[A] critical goal of science education should be to help students understand how science may be employed as an uncertain and limited, yet still useful tool for informing decisions about socioscientific problems" (Covitt and Anderson 2022, p. 1155). With regard to issues such as climate change, some perspectives and claims for action appear to be presented as entirely science-based (Hulme 2022). However, as, among others, Hulme (2022), Beck (1986/1992), and Covitt and Anderson (2022) have argued, science does not dictate the implications of its findings. While scientific inquiry may reveal and describe what is going on, the question of what to do with this information remains to be answered. Elements such as people's worldview and risk perception also matter (Hulme 2009). In other words, the mere communication of scientific findings, uncertainties, and risks does not alone determine what actions should be undertaken, although they can indeed provide useful contributions to such decisions. The expectation of science as a provider of "true" or "right" answers or strategies is thus challenged in relation to socioscientific issues or wicked problems (e.g., Covitt and Anderson 2022; Rittel and Webber 1973). The first recommendation therefore echoes Block et al.'s (2018) principle of being humble and relates clearly to vision I of scientific literacy: clarify that uncertainty is a part of scientific processes and that in the case of socioscientific issues or wicked problems, a single "correct" answer or way to respond cannot be expected of scientific research.

In this context, one may also think of the potential consequences of a specific perception about the nature of science for how uncertainty is perceived in the communication of (school) science. A study by Rabinovich and Morton (2012) assessed people's perception of uncertainty in scientific messages in relation to their perception of science. Some participants held the traditional view that scientific questions have a single correct answer, whereas others saw science more as a debate, with multiple answers that can be discussed. Such perceptions of science appeared to guide participants' responses to the ways in which uncertainty was communicated in messages about climate change risks. In particular, those participants who saw science as debate seemed "more accepting of scientific uncertainty and more willing to act in response to risk messages" (Rabinovich and Morton 2012, p. 1001). Although it is important to acknowledge uncertainty as a part of scientific processes and findings, such uncertainties should not be misrepresented or over-emphasized, as has been the case with the topic of human-induced climate change in American school books (Román and Busch 2016).

2. Use Activities in Which Students Actively Engage with Uncertainty

In education *about* sustainability and associated issues, it seems inevitable to encounter some uncertainty, if only because of the future dimension that comes with caring for future needs. To avoid "aboutism" (see, e.g., Gilbert 2016), however, it may be necessary to go one step further. In education *for* and *as* sustainable development, students should, perhaps, also become familiar with the idea of uncertainty more directly and develop skills to understand, assess, and deal with such uncertainties. Such skills can be called "uncertainty competences" (Tauritz 2012, 2016, 2019), and these may serve to embrace, accept, or reduce uncertainty, depending on what is appropriate in a particular context.

It may not be a big challenge to stimulate such skills to deal with uncertainty, although more research is needed to examine this. An acknowledgment of uncertainty in the classroom does not need to entail a radical break with previous practices. As some form of uncertainty may be argued to be present in many learning processes (Rosen 2019), various educational approaches may in some way use or engage with uncertainty. Both already existing and new practices might provide students with familiarity and skills in relation to uncertainty. The more open tasks included in textbooks, for example, might stimulate exploration of uncertainties (e.g., Andersson-Bakken et al. 2020). Another practice is that of inquiry-based learning activities (e.g., Butcher et al. 2023; Paseka et al. 2025) and citizen science (e.g., Aridor et al. 2023). Approaches using fiction, such as games, could also be helpful in raising questions on science-related uncertainties (Caracciolo 2022; van Schaik 2023a, b).

3. Explore the Opportunities That Uncertainties Can Entail

"[U]ncertainty comprises not only risks for directing activities but simultaneously opens up spaces for reflection, learning and open deliberation" (Grunwald 2007, p. 258). Besides uncertainty serving as a starting point for scientific research and indicating uncertainties in research dissemination typically being regarded as a mark of quality, uncertainty can also be understood in a positive way in relation to the possibilities that exist. The future is open and there are options to influence it (Grunwald 2007). The inherent uncertainty of the future can also invite contributing toward building a desired future. Besides risks, there are also opportunities.

One of these opportunities is the opportunity to ask questions. What are the students' "wondering questions" (Sezen-Barrie et al. 2020)? Such engagement with the uncertainties that students express may contribute to their sensemaking of the scientific content and simultaneously to their understanding of the nature of science (Chen 2022; Chen et al. 2019; Chen and Qiao 2020; Watkins et al. 2018; Watkins and Manz 2022). In accordance with visions II and III of scientific literacy, science education may foster change through active engagement with the questions that contemporary societies face.

3.7 Conclusion

In the introduction of this chapter, learning was described as a quest, an exploration of questions (Koefoed and Burø 2022). Such questions can be faced in the context of sustainable development, also within science education. Although science may not provide *the* answer to all the questions relating to sustainable development, it can contribute to knowledge about the issues at hand and clarify what remains uncertain. The idea of safeguarding the needs of present and future generations is entangled with uncertainties, and science education could help students find their way in such challenging circumstances. I wrote this chapter hoping to inspire both educators and scholars to venture out to explore the way in which the terrains of uncertainty are, could, and should be shaped in science education in the context of sustainability. Indeed: *what if* we (scholars, educators, and students) directed our attention to uncertainty? What are the opportunities? What are the questions to raise?

References

Agustian HY (2025) The critical role of understanding epistemic practices in science teaching using wicked problems. Sci Educ 34:485–510. https://doi.org/10.1007/s11191-023-00471-2

Andersson-Bakken E, Jegstad KM, Bakken J (2020) Textbook tasks in the Norwegian school subject natural sciences: what views of science do they mediate? Int J Sci Educ 42(8):1320–1338. https://doi.org/10.1080/09500693.2020.1756516

Aridor K, Dvir M, Tsybulsky D, Ben-Zvi D (2023) Living the DReaM: the interrelations between statistical, scientific and nature of science uncertainty articulations through citizen science. Instr Sci 51(5):729–762. https://doi.org/10.1007/s11251-023-09626-8

Beck U (1986/1992) Risk society: towards a new modernity. Sage

Bengtsson S, Van Poeck K (2021) What can we learn from COVID-19 as a form of public pedagogy? Eur J Res Educ Learn Adults 12(3):281–293. https://doi.org/10.3384/RELA.2000-7426.3386

Bevan LD (2022) The ambiguities of uncertainty: a review of uncertainty frameworks relevant to the assessment of environmental change. Futures 137:102919. https://doi.org/10.1016/j.futures.2022.102919

Block T, Goeminne G, Poeck KV (2018) Balancing the urgency and wickedness of sustainability challenges: three maxims for post-normal education. Environ Educ Res 24(9):1424–1439. https://doi.org/10.1080/13504622.2018.1509302

Butcher KR, Hudson M, Lane M, Larson M (2023) Critical thinking during science investigations: what do practicing teachers value and observe? Teach Teach 29(6):594–614. https://doi.org/10.1080/13540602.2023.2191186

Caracciolo M (2022) Contemporary fiction and climate uncertainty: narrating unstable futures. Bloomsbury Academic. https://doi.org/10.5040/9781350233928

Chen Y-C (2022) Is uncertainty a barrier or resource to advance science? The role of uncertainty in science and its implications for science teaching and learning. Sci Educ 31:543–549. https://doi.org/10.1007/s11191-021-00244-9

Chen Y-C, Qiao X (2020) Using students' epistemic uncertainty as a pedagogical resource to develop knowledge in argumentation. Int J Sci Educ 42(13):2145–2180. https://doi.org/10.1080/09500693.2020.1813349

Chen Y-C, Benus MJ, Hernandez J (2019) Managing uncertainty in scientific argumentation. Sci Educ 103(5):1235–1276. https://doi.org/10.1002/sce.21527

Christensen C (2009) Risk and school science education. Stud Sci Educ 45(2):205–223. https://doi.org/10.1080/03057260903142293

Covitt BA, Anderson CW (2022) Untangling trustworthiness and uncertainty in science. Sci Educ 31:1155–1180. https://doi.org/10.1007/s11191-022-00322-6

Ekberg M (2007) The parameters of the risk society: a review and exploration. Curr Sociol 55(3):343–366. https://doi.org/10.1177/0011392107076080

Funtowicz SO, Ravetz JR (1993) Science for the post-normal age. Futures 25(7):739–755. https://doi.org/10.1016/0016-3287(93)90022-L

Gilbert J (2016) Transforming science education for the Anthropocene: is it possible? Res Sci Educ 46(2):187–201. https://doi.org/10.1007/s11165-015-9498-2

Grunwald A (2007) Working towards sustainable development in the face of uncertainty and incomplete knowledge. J Environ Policy Plan 9(3–4):245–262. https://doi.org/10.1080/15239080701622774
Haasnoot M, Diermanse F (eds) (2022) Analyse van bouwstenen en adaptatiepaden voor aanpassen aan zeespiegelstijging in Nederland (Report no. 11208062-005-BGS-0001). Deltares
Haasnoot M, Kwakkel JH, Walker WE, ter Maat J (2013) Dynamic adaptive policy pathways: a method for crafting robust decisions for a deeply uncertain world. Glob Environ Chang 23(2):485–498. https://doi.org/10.1016/j.gloenvcha.2012.12.006
Hanke M, Paseka A, Sprenger S (2025) Understanding the meaning of uncertainty in geography education: a systematic review. Teach Teach 31(2):174–200. https://doi.org/10.1080/13540602.2023.2263375
Hansen J, Hammann M (2017) Risk in science instruction. Sci Educ 26(7):749–775. https://doi.org/10.1007/s11191-017-9923-1
Heiskala R (2011) From modernity through postmodernity to reflexive modernization. Did we learn anything? Int Rev Sociol 21(1):3–19. https://doi.org/10.1080/03906701.2011.544177
Hodson D (2003) Time for action: science education for an alternative future. Int J Sci Educ 25(6):645–670. https://doi.org/10.1080/09500690305021
Hodson D (2009) Teaching and learning about science: language, theories, methods, history, traditions and values. BRILL. https://doi.org/10.1163/9789460910531
Hulme M (2009) Why we disagree about climate change: understanding controversy, inaction and opportunity. Cambridge University Press. https://doi.org/10.1017/CBO9780511841200
Hulme M (2022) Climate change. Routledge. https://doi.org/10.4324/9780367822675
Knight FH (1921) Risk, uncertainty, and profit. Houghton Mifflin
Koefoed O, Burø T (2022) The name of the world is chaos: learning in the Anthropocene. In: Bjerg Petersen K, von Brömssen K, Hellesdatter Jacobsen G, Garsdal J, Paulsen M, Koefoed O (eds) Rethinking education in light of global challenges: Scandinavian perspectives on culture, society, and the Anthropocene. Routledge, pp 193–203. https://doi.org/10.4324/9781003217213-19
Kolstø SD (2006) Patterns in students' argumentation confronted with a risk-focused socio-scientific issue. Int J Sci Educ 28(14):1689–1716. https://doi.org/10.1080/09500690600560878
Levinson R, Kent P, Pratt D, Kapadia R, Yogui C (2011) Developing a pedagogy of risk in socio-scientific issues. J Biol Educ 45(3):136–142. https://doi.org/10.1080/00219266.2011.576260
Liu X (2013) Expanding notions of scientific literacy: a reconceptualization of aims of science education in the knowledge society. In: Mansour N, Wegerif R (eds) Science education for diversity: theory and practice. Springer Netherlands, pp 23–39. https://doi.org/10.1007/978-94-007-4563-6_2
Miller R (2011) Being without existing: the futures community at a turning point? A comment on Jay Ogilvy's "facing the fold". Foresight 13(4):24–34. https://doi.org/10.1108/14636681111153940
Nilsson M, Chisholm E, Griggs D, Howden-Chapman P, McCollum D, Messerli P, Neumann B, Stevance A-S, Visbeck M, Stafford-Smith M (2018) Mapping interactions between the sustainable development goals: lessons learned and ways forward. Sustain Sci 13(6):1489–1503. https://doi.org/10.1007/s11625-018-0604-z
Oxford English Dictionary (n.d.) Uncertainty, n. In Oxford English dictionary. Retrieved 6 Nov 2023, from https://www.oed.com/dictionary/uncertainty_n?tab=meaning_and_use#17063467
Paseka A, Hinzke J-H, Boldt V-P (2025) Learning through perplexities in inquiry-based learning settings in teacher education. Teach Teach 31(2):281–296. https://doi.org/10.1080/13540602.2023.2266379
Pham-Truffert M, Metz F, Fischer M, Rueff H, Messerli P (2020) Interactions among sustainable development goals: knowledge for identifying multipliers and virtuous cycles. Sustain Dev 28(5):1236–1250. https://doi.org/10.1002/sd.2073
Pietrocola M, Rodrigues E, Bercot F, Schnorr S (2021) Risk society and science education. Sci Educ 30:209–233. https://doi.org/10.1007/s11191-020-00176-w

Rabinovich A, Morton TA (2012) Unquestioned answers or unanswered questions: beliefs about science guide responses to uncertainty in climate change risk communication. Risk Anal 32(6):992–1002. https://doi.org/10.1111/j.1539-6924.2012.01771.x

Ravetz JR (1997) Simple scientific truths and uncertain policy realities: implications for science education. Stud Sci Educ 30(1):5–18. https://doi.org/10.1080/03057269708560101

Rittel HWJ, Webber MM (1973) Dilemmas in a general theory of planning. Policy Sci 4(2):155–169. https://doi.org/10.1007/BF01405730

Roberts DA (2007) Scientific literacy/science literacy. In: Abell SK, Lederman NG (eds) Handbook of research on science education. Lawrence Erlbaum Associates, pp 729–780

Román D, Busch KC (2016) Textbooks of doubt: using systemic functional analysis to explore the framing of climate change in middle-school science textbooks. Environ Educ Res 22(8):1158–1180. https://doi.org/10.1080/13504622.2015.1091878

Rosen D (2019) Being in uncertainties: an inquiry-based model leveraging complexity in teaching-learning. Northeast J Complex Syst 1(1):5. https://doi.org/10.22191/nejcs/vol1/iss1/5

Saari A, Mullen J (2020) Dark places: environmental education research in a world of hyperobjects. Environ Educ Res 26(9–10):1466–1478. https://doi.org/10.1080/13504622.2018.1522618

Sadler TD (2004) Informal reasoning regarding socioscientific issues: a critical review of research. J Res Sci Teach 41(5):513–536. https://doi.org/10.1002/tea.20009

Schenk L, Hamza KM, Enghag M, Lundegård I, Arvanitis L, Haglund K, Wojcik A (2019) Teaching and discussing about risk: seven elements of potential significance for science education. Int J Sci Educ 41(9):1271–1286. https://doi.org/10.1080/09500693.2019.1606961

Schreuder A (2022) Wat kan en moet Nederland doen bij stijgende zeespiegel? 'We kunnen zo niet doorgaan. Bouw liever in hogere gebieden'. NRC. https://www.nrc.nl/nieuws/2022/09/29/we-kunnen-zo-niet-doorgaan-bouw-in-hogere-gebieden-2-a4143645

Schwanitz VJ, Wierling A, Shah P (2017) Assessing the impact of renewable energy on regional sustainability: a comparative study of Sogn og Fjordane (Norway) and Okinawa (Japan). Sustainability 9(11):11. https://doi.org/10.3390/su9111969

Sezen-Barrie A, Miller-Rushing A, Hufnagel E (2020) 'It's a gassy world': starting with students' wondering questions to inform climate change education. Environ Educ Res 26(4):555–576. https://doi.org/10.1080/13504622.2019.1610158

Sinnes AT (2021) Utdanning for bærekraftig utvikling: Hva, hvorfor og hvordan? 2nd edn. Universitetsforlaget

Sjöblom P, Wolff L-A, Sundman J (2023) Climate change as a socio-scientific issue in upper secondary education: addressing wicked problems through crosscurricular approaches. In: Klausen SH, Mård N (eds) Developing a didactic framework across and beyond school subjects. Routledge, pp 182–196. https://doi.org/10.4324/9781003367260-17

Sjöström J, Eilks I (2018) Reconsidering different visions of scientific literacy and science education based on the concept of Bildung. In: Dori YJ, Mevarech ZR, Baker DR (eds) Cognition, metacognition, and culture in STEM education: innovations in science education and technology. Springer, pp 65–88. https://doi.org/10.1007/978-3-319-66659-4_4

Sjöström J, Stenborg E (2014) Teaching and learning for critical scientific literacy: communicating knowledge uncertainties, actors interplay and various discourses about chemicals. In: Eilks I, Markic S, Ralle B (eds) Science education research and education for sustainable development. Shaker Verlag, pp 37–48. https://urn.kb.se/resolve?urn=urn:nbn:se:mau:diva-9727

Stein S (2021) Reimagining global citizenship education for a volatile, uncertain, complex, and ambiguous (VUCA) world. Glob Soc Educ 19(4):482–495. https://doi.org/10.1080/14767724.2021.1904212

Tauritz RL (2012) How to handle knowledge uncertainty: learning and teaching in times of accelerating change. In: Wals AEJ, Corcoran PB (eds) Learning for sustainability in times of accelerating change. Wageningen Academic Publishers, pp 299–316

Tauritz RL (2016) A pedagogy for uncertain times. In: Lambrechts W, Hindson J (eds) Research and innovation in education for sustainable development. Environment and School Initiatives, pp 90–105

Tauritz RL (2019) Certain you're not sure? An inquiry into pedagogical strategies for teaching children how to manage uncertain knowledge about sustainability challenges. Doctoral dissertation, The University of Edinburgh. ERA. https://era.ed.ac.uk/handle/1842/35629

UN (2023) What is goal 4: quality education. Department of Global Communications. https://www.un.org/sustainabledevelopment/wp-content/uploads/2023/09/Goal-4_Fast-Facts.pdf

UN (n.d.) The 17 goals: sustainable development. Retrieved 10 Sept 2021, from https://sdgs.un.org/goals

van Schaik F (2023a) In-game uncertainties and climate change challenges as identified by pre-service teachers. Simul Gaming 54(6):680–707. https://doi.org/10.1177/10468781231199262

van Schaik F (2023b) What happens if …? Uncertainty in games and climate change education. Environ Educ Res 29(12):1891–1910. https://doi.org/10.1080/13504622.2023.2225811

Van Schaik F (2024) Playing wicked games: climate change-related uncertainty in games and education (publication no. 796). Doctoral dissertation, University of Stavanger. PhD thesis UiS. https://hdl.handle.net/11250/3153696

van Vliet MTH, Flörke M, Wada Y (2017) Quality matters for water scarcity. Nat Geosci 10(11):11. https://doi.org/10.1038/ngeo3047

Wakeham J (2015) Uncertainty: history of the concept. In: Wright JD (ed) International encyclopedia of the social & behavioral sciences, 2nd edn. Elsevier, pp 716–721. https://doi.org/10.1016/B978-0-08-097086-8.03175-5

Watkins J, Manz E (2022) Characterizing pedagogical decision points in sense-making conversations motivated by scientific uncertainty. Sci Educ 106(6):1408–1441. https://doi.org/10.1002/sce.21747

Watkins J, Hammer D, Radoff J, Jaber LZ, Phillips AM (2018) Positioning as not-understanding: the value of showing uncertainty for engaging in science. J Res Sci Teach 55(4):573–599. https://doi.org/10.1002/tea.21431

Wessels KR, Bakker C, Wals AEJ, Lengkeek G (2024) Rethinking pedagogy in the face of complex societal challenges: helpful perspectives for teaching the entangled student. Pedagog, Cult Soc 32(3):759–776. https://doi.org/10.1080/14681366.2022.2108125

Wilkinson JL, Boxall ABA, Kolpin DW, Leung KMY, Lai RWS, Galbán-Malagón C, Adell AD, Mondon J, Metian M, Marchant RA, Bouzas-Monroy A, Cuni-Sanchez A, Coors A, Carriquiriborde P, Rojo M, Gordon C, Cara M, Moermond M, Luarte T et al (2022) Pharmaceutical pollution of the world's rivers. Proc Natl Acad Sci 119(8):e2113947119. https://doi.org/10.1073/pnas.2113947119

World Commission on Environment and Development (1987) Our common future. Oxford University Press. https://sustainabledevelopment.un.org/content/documents/5987our-common-future.pdf

Zeidler DL, Nichols BH (2009) Socioscientific issues: theory and practice. J Elem Sci Educ 21(2):49–58. https://doi.org/10.1007/BF03173684

Fiona van Schaik, PhD, works with stormwater management at the municipality of Sandnes, Norway. Her educational background includes degrees in educational sciences, environmental sciences, and liberal arts and sciences. As a PhD student, she focused on the fields of sustainability education and climate change education. The question of how uncertainty could be treated in educational contexts was of particular interest in the work for her PhD thesis. In her thesis, she explored the use of games in this context. Still focusing on climate change uncertainty, she is currently employed in the municipal sector to help facilitate climate change adaptation.

Chapter 4
Covid-19 Vaccine Hesitancy as an Imbricated Social, Cognitive, and Emotional Controversy

Claire Polo and Kristine Lund

Abstract The Covid-19 pandemic raised the challenge of building sustainable societies able to overcome health crises "leaving no one behind" (LNOB). In France, vaccine hesitancy disrupted the society and led to strong caricatured pro/anti-vax opposition in the mass media. It is only the mandatory health pass that resulted in having 79% of the population fully vaccinated. In this chapter, we argue that unpacking the imbricated social, cognitive, and emotional nature of these types of controversies, and developing education to socio-scientific issues, accordingly, may help future citizens reach critical and constructive argumentation. In particular, making the affective background of diverging claims explicit could help mutual understanding and avoid stigmatization of others. To illustrate our argument, we draw on the deep analysis of three widespread utterances in the French media about vaccine hesitancy. We show that, even if they might sound only cognitive, only social or only emotional at first glance, they in fact all rely on these three dimensions. We conclude by discussing how such a perspective can inform science education targeting an inclusive, sustainable society.

Keywords Socio-scientific education · Social, cognitive, and emotional reasoning · Argumentation analysis · Multimodality · Vaccine hesitancy

---

C. Polo (✉)
Lyon 2 University, ECP Laboratory, Lyon, France
e-mail: claire1.polo@univ-lyon2.fr

K. Lund
CNRS, École Normale Supérieure de Lyon, Lyon 2 University, UMR 5191 ICAR, Lyon, France

F. Skarstein, L.-A. Wolff (eds.), *Sustainability and Science Education*, Environmental Challenges and Solutions,
https://doi.org/10.1007/978-3-032-24408-6_4

4.1 Introduction

The Covid-19 pandemic raised the challenge of overcoming crises "leaving no one behind" (LNOB[1]). In this context, the French government invested in health education and vaccination policy. Nevertheless, such emergency public action developed in authoritative top-down ways without engaging in critical socio-scientific debate. We argue that such debate has social aspects if identity, human interaction, past experience, community, or public action is involved, and it is scientific if knowledge produced by science is at stake. When available, far fewer people than expected volunteered to receive vaccine injections. Actually, vaccine hesitancy disrupted the French media and led to caricatured pro/anti-vax opposition as being "against obscurantism or against a repressive policy" (Olivier 2023, p. 143). Finally, it was rather the health pass than being convinced by genuine argumentation that ended up having 79% of the French population vaccinated[2] (Gans 2023). It can be interpreted as a failure of the dialogic strategy aiming at convincing people to receive the vaccine through education instead of coercion. A more provocative, alternative view would be that part of the French was skeptical about such policy thanks to critical thinking associated with scientific literacy (see Holbrook and Rannikmae 2007). Is not doubting essential to inquiry and reasoning (Dewey 1938)? The matter is that, as science education turned to authoritative teaching of what to do, the media only left a place for doubt in conspiracy theories. The controversy about Covid-19 vaccine then polarized into two groups mutually depicted as emotions-driven and irrational. The pro-vaccination side was caricatured as "driven by a fear of the pandemic leading to blind obedience," whereas the vaccination hesitancy side was "driven by an egocentric desire of freedom and irrational mistrust in science" (Andolfatto 2023; Guerrini 2024). Disregarding others' position on the basis of emotional involvement corresponds to the typical accusation of *ad passiones* fallacy (Plantin 2018, pp. 185–190).

The public issue described above is a *socio-scientific controversy* (SSC), defined in the literature by four key features summarized by Polo et al. (2023). First, it is an interdisciplinary problem, involving different knowledge fields, here both science, health, and civic education. Second, it combines different sources of information with distinct epistemic status (daily experience, school knowledge, media discourse, etc.). Third, judgments about it are necessarily subjective, based on people's

[1] The United Nations declare: "Leave no one behind is the central, transformative promise of the 2030 Agenda for Sustainable Development (…). It represents the unequivocal commitment of all UN Member States to eradicate poverty in all its forms, end discrimination and exclusion, and reduce the inequalities and vulnerabilities that leave people behind and undermine the potential of individuals and of humanity as a whole" (https://unsdg.un.org/2030-agenda/universal-values/leave-no-one-behind, March 3rd, 2023).

[2] Figure delivered by the French authorities: https://sante.gouv.fr/soins-et-maladies/maladies/maladies-et-infections-respiratoires/covid-19/vaccin-covid-19/article/le-tableau-de-bord-de-la-vaccination, https://sante.gouv.fr/grands-dossiers/vaccin-covid-19/article/le-tableau-de-bord-de-la-vaccination, March 2nd, 2023.

socio-ethical beliefs guiding their hierarchization among competing priorities, such as individual freedom and common good. Fourth, this issue is by nature controversial, opposing several groups in a given society, like those with distinct material interests, values, and worldviews, such as health authorities, pharmaceutical firms, consumers, and people working in various sectors. A constructive dialogue on such *socio-scientific controversies* (SSC) requires learning to overcome the Western reason-emotion dichotomy and to apprehend them with a socio-cognitivo-emotional reasoning. Following the above definition of socio-scientific debate, an SSC occurs when, for social reasons, individuals, groups, or communities hold different opinions, beliefs, or promote conflicting arguments about scientific knowledge. A current consensus considers learning as a process involving all these three dimensions (e.g., Baker et al. 2013), but little research has analyzed them jointly in education to SSC.

4.2 Theoretical Background

In what follows, we defend theoretical foundations regarding reasoning, their application to our topic, and their necessity for teaching. First, we argue that reasoning during discussions of socio-scientific controversies is a process that imbricates social, cognitive, and emotional aspects. Second, we illustrate how the Covid-19 vaccine controversy is specifically imbricated in this way. Third, we argue that understanding the imbricated nature of such controversies is crucial for inclusive, sustainable health education, and teaching regarding socio-scientific controversies.

4.2.1 Reasoning About SSC as Social, Cognitive, and Emotional Interplay

We argue that accomplishing a unified way to reason about socio-scientific controversies requires meeting two goals. First, there is a need to overcome the dichotomy between reason and emotion. Second, there is a need to define a conceptual structure to apprehend imbricated reasoning.

4.2.1.1 Overcoming the Reason/Emotion Dichotomy with 4^E Learning

Learning sciences still tends to oppose cognitive activity to affective experience. Little research (but see Avry et al. 2020) addresses the intertwining of the cognitive, affective, and social dimensions of learning. Many studies focus only on socio-cognitive, or only on the socio-emotional aspects of educational processes. Even when the three dimensions are considered, analyses are focused on their alternation and not on their interplay (e.g., Baker et al. 2009) or separate socio-emotional aspects from cognitive ones (Isohätälä et al. 2017). Cognition somehow appears as a more noble, possibly standalone activity independent from the socio-emotional dynamics that may disturb it. This position manifests in defining emotional regulation only as reacting to affective arousals that might threaten learning instead of also as favoring learning. The ghost of this Cartesian dichotomy also prefers performing dialogue analysis on speech with little account of bodily communicative acts, as if reason were associated with the brain and emotions with the rest of the body.

However, the latest views on cognition as 4^E as embedded, extended, embodied, and enactive (Newen et al. 2017) offer a strong theory for integrating social and emotional dimensions as the core components of learning. Learning can then be defined as both a socially extended process (fully socio-cognitive), and an embodied and enactive process (fully cognitivo-emotional).

4.2.1.2 Concepts to Apprehend Imbricated Social, Cognitive, and Emotional Reasoning

Dialogic education targets learning through socio-cognitive conflict (Roschelle and Teasley 1995). Students then use three types of argumentative resources: beliefs and knowledge; affects and emotions; norms and values (Polo 2023). Polo distinguishes between three *logics* in students' dialogues about SSC. *Descriptive logic* addresses what *is* (knowledge and beliefs considered as *true*), *prescriptive logic* addresses what *should be* (norms and values considered axiologically *good), and affective logic* addresses what *feels good or right* (positive or negative emotional valency attributed to things). Therefore, socio-scientific reasoning results from socially shared views and emotional apprehension of the world. Then apprehending education to SSC requires to account for the *interplay* of the cognitive with the social and the emotional. Following Lund (2016, 2022), we argue for a model of the co-construction of knowledge in human interactions articulating the socio-cognitive, the socio-emotional, and the cognitivo-emotional dimensions of learning.

Regarding argumentation, Guerrini uses a tetrahedron visualization to present the values associated with a given public controversy, organized along three main poles: the object, the self, and others. He notably summarizes this way the key axiological principles referred to in the French controversy about Covid-19 vaccination (Guerrini 2024). Building upon Guerrini's work, we designed a simplified, triangular visualization of this controversy, adding three "filters" to emphasize the socio-cognitive-emotional nature of reasoning (cf. Fig. 4.1). Such a picture makes it easier to specify the definition of the categories of the social, the cognitive, and the emotional. All of them are actually cognitive in an extensive sense, corresponding to 4^E cognition (cf. Sect. 4.2.1.1). But "cognitive" here stands in a restrictive sense as the definition of the objects at stake and of their relations with other things and beings. In turn, the emotional refers to the affective relation between the self and the objects and between the self and the community. The social dimension corresponds to two

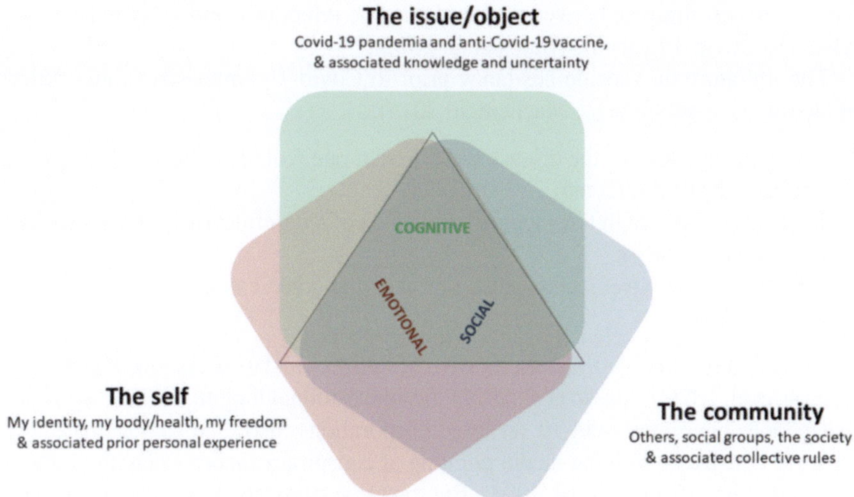

Fig. 4.1 The social, cognitive, and emotional nature of the Covid-19 vaccine controversy

phenomena: (1) belonging to one or several communities and (2) the appropriation of the issue by the relevant communities (here standing for human groups of various scales).

4.2.2 The Covid-19 Vaccine as an Imbricated Social, Cognitive, and Emotional Controversy

In this section, we first define vaccine hesitancy in general and present its cognitive determinants. Next, we describe the specificities of the hesitations of the French population regarding the Covid-19 vaccine. This allows us to present both the social side and the emotional side of the controversy.

4.2.2.1 Vaccine Hesitancy: Definition and Cognitive Determinants

Vaccine hesitancy did not appear with Covid-19, and neither did pure vaccine refusal. It is "a kind of decision-making process that depends on people's level of commitment to healthism/risk culture and on their level of confidence in the health authorities and mainstream medicine" (Peretti-Watel et al. 2015, p. 7). As it comes to Covid-19, studies emphasize the need for distinguishing between a "hard" anti-vax position, a "soft" anti-vax position, and a hesitancy position. Galasso et al. (2022) specify the three groups by referring to their intention to take vaccinated as assessed by a ten-point scale. A correct understanding of vaccination attitude also requires to discriminate between general vaccine reluctance and specific hesitancy related to Covid-19 vaccine (Bajos et al. 2022).

The literature on vaccine hesitancy prior to Covid-19 emphasizes three factors hindering vaccination (e.g., MacDonald 2015):

1. Complacency toward the disease (feeling that one would not become infected or not get very ill if infected)
2. Lack of confidence in the vaccine (avoidance of side-effect risks and assessment of vaccine efficacy)
3. Vaccination inconvenience—cost of receiving vaccination, either in money, time, or pain

It seems that the first two factors mostly explain Covid-19 vaccine hesitancy (e.g., Lazarus et al. 2021; Galasso et al. 2022). An international longitudinal survey shows that people blaming Covid-19 on large corporations also tend to vaccinate less, which is consistent with the media pointing at conspiracy theories playing a role in the controversy (Galasso et al. 2022). But the same study also notes safety concerns regarding possible negative side effects of the vaccine. This idea appears as a central part of specific Covid-19 vaccine hesitancy. Such a result is consistent with Malova's (2021) work on Twitter conversations revealing that the main discussed topic was vaccine safety. It also converges with a qualitative online study in Arkansas (Moore et al. 2021).

4.2.2.2 France as a Case of High Covid-19 Vaccine Hesitancy

France has long been a specific case with high vaccine hesitancy and vaccine opposition, notably since the 1990s (e.g., Larson et al. 2018). In 2015, only 52% of French people thought that the flu vaccine was safe while the corresponding rate reached 85% in the United Kingdom (Larson et al. 2018). Unsurprisingly, French opinion was hostile regarding Covid-19 vaccine, with only 44% of the population intending to be vaccinated, if possible, at the end of 2020, as opposed to 81% in the United Kingdom (Wouters et al. 2021). In December 2020, only 38% of the French population was pro-vax, the lowest rate among nine similar countries (Galasso et al. 2022). France had the higher hesitancy rate in another international study implemented between March 2020 and January 2021 (47.3% as compared to 27% in average). In a longitudinal survey conducted among a stratified random sample of 135,000 French people in 2020 (Bajos et al. 2022), 75% of strong, generally pro-vax people, had the intention to take vaccination against Covid-19. The same study revealed that less people were hesitating about Covid-19 vaccine than about vaccination in general (52% instead of 65%), and that mistrust in the government was more significant for determining vaccination attitudes when it came to Covid-19.

4.2.2.3 The Social Side of the Controversy

As the World Health Organization (WHO) declared in 2019[3] that vaccine hesitancy was among the main threat to global health, it is worth reminding that it is also generally determined by typical sociodemographic attributes. Studies dedicated to Covid-19 confirmed most of these trends, showing that refusal or hesitancy was higher among women, minorities, less-educated, rural, poorer, inactive, or young people, blue collars, service workers, and people mistrusting the government (e.g., Lazarus et al. 2021; Galasso et al. 2022). Still, the age effect is mostly due to older people feeling less complacency toward Covid-19, while adults seemed less likely to be vaccinated than 18–34-year-olds (Galasso et al. 2022). Interestingly, vaccination intentions were also lower for people informed through social media rather than traditional media, those who comply less with public health rules, who do not declare a political ideology, and who are more risk-averse (Galasso et al. 2022).

In France, social differentiation of attitudes regarding vaccination is also higher than usual for Covid-19 in terms of age (after 65 years) and gender (although women declare they follow more preventive behaviors). French women tend to be more afraid than men of being infected by Covid-19 (higher risk perception). Other typical factors (social hierarchy, ethno-racial, or minority status) play a similar role as usual (Bajos et al. 2022). A specific trend is noted: while first-generation Africa/Asia immigrants are generally the most reluctant (27% compared to 10% for the whole population), it is the French Overseas Departments (FOD) and their children who are the most anti-vax regarding Covid-19 (23.7% compared to 13.3% for the whole population). Living with a child is also associated with higher vaccine reluctance. Besides, Schwarzinger et al. (2021) pointed out that vaccine efficacy was a key determinant of attitudes of the working-age French population for Covid-19, but consisting of safety concerns rather than vaccine performance. More specifically, a questionnaire-based study reports that only 53% of hospital workers were in favor of vaccination in December 2020, and more than 66% of them worried about adverse effects due to unusually rapid research (Navarre et al. 2021).

The public debate was largely guilt-inducing, specifically through authorities' messages emphasizing one's responsibility, if unvaccinated, regarding endangering vulnerable people (Matuszewicz 2023). Such an altruistic view seems more effective than self-protection messages, especially since living with one's family is a key positive determinant of vaccination (Galasso et al. 2022). But some experts stress that alternative policies could have enabled to reach mass vaccination by strengthening French people's confidence in the authorities (Cambon et al. 2022, p. 181).

Medical mistrust is not purely irrational but based on historical events that concern marginalized communities (Griffith et al. 2021), or based on a documented vulnerability of specific groups. Women, for instance, report higher antibody responses (e.g., Flanagan et al. 2017), and more adverse effects for influenza,

---

[3] World Health Organization (2019) Ten threats to global health in 2019. World Health Organization. Retrieved on May 16, 2025, from https://www.who.int/news-room/spotlight/ten-threats-to-global-health-in-2019

hepatitis B, and yellow fever vaccines. Moreover, two widely used Covid-19 vaccines had more side effects among women, which may explain that women declare higher safety concerns than men toward vaccination (e.g., Flanagan et al. 2017; Galasso et al. 2022). Similarly, the history of the FOD consists of joint medical and state violence, which may explain why this population tend to be skeptical toward the injection of a Covid-19 vaccine. Back in the 1960s and 1970s, local authorities imposed demographic control illegally. They notably injected Depo-Provera, a contraceptive with severe side effects to black, poor, or isolated women more than 20 years before its commercialization (Sanseigne 2020). Despite such history, the French government managed the Covid-19 crisis in an authoritative way instead of trying to restore FOD confidence in health policies. This strategy increased the divide between such territories and mainland France and partially explains the high score of extreme right-wing candidates in the FOD in the following presidential elections in 2022.

4.2.2.4 The Emotional Side of the Controversy

Two major factors influencing vaccination attitudes are highly affective: complacency toward the disease and lack of confidence in the vaccine. The latter factor is mostly associated with the unusually short experimentation phase in the case of Covid-19. Among the basic emotions (Ekman 1999), the one at stake in this dilemma is fear: "Am I more afraid of being infected by Covid-19 or am I more afraid of the vaccine's potential side effects?". It is worth noting that such a dilemma is not irrational. We can hypothesize that the answer relies on one's perception of these two forms of risks, and it can be influenced by the experience of friends and family with Covid-19 and the vaccine, which would shorten the emotional distance to the issue and increase related affective tonality. Nevertheless, the only study gathering such data found no striking significant results about these variables, maybe because the survey was implemented only after two epidemic waves and before vaccination became widespread (Bajos et al. 2022). Surprisingly, though, the authors chose not to include people who tested positive for Covid-19, precisely because they thought that it could influence their attitude. But data from Greece tend to confirm such hypothesis: 64.7% of unvaccinated, hospitalized Covid-19 patients would be vaccinated if they could turn back the clock (Ioannou et al. 2022).

In the literature, positive emotions are little discussed in relation to Covid-19 vaccine hesitancy. Mayer et al. (2021) showed that hope was "a predictor of Covid-19 vaccine uptake" and linked significantly with vaccination intention in their online, exploratory survey in Israel. They defined hope as "a state of positive motivation based on determination to achieve one's goals rather than fear of the disease" (Mayer et al. 2021, p. 4941). Such an approach focuses on the "positive" side of the coin regarding the second factor determining vaccine hesitancy: confidence in the vaccine. An Italian study also emphasized such determinant of the intention to be vaccinated among health care workers as well as the internal locus of control of behavior as positively influencing their tendency to recommend the vaccine. Authors conclude that less negative messaging would improve health care workers' willingness and engagement to apply the vaccination campaign policies (Rapisarda et al. 2021).

4.2.3 Inclusive, Sustainable Health Education, and Teaching SSC

We argue for inclusive and sustainable health education that necessitates an understanding of science. In this section, we explain the challenges involved. First, they relate to fostering inclusive education for sustainable societies. Second, they arise from understanding how emotional, personal experiences play a role in SSC in health education.

4.2.3.1 The Challenge of Inclusive Education for Sustainable Societies

As the United Nations plead in favor of an LNOB policy, science education should develop an inclusive approach toward vaccine hesitancy to restore the social dialogue about health policy. Governance is one of the pillars of sustainability that can be defined as transitioning to ways of living and socioeconomic systems that respect ecological constraints so that each generation find in its environment the resources necessary for its long-lasting development (e.g., Barth and Michelsen 2013). Therefore, effective education needs to integrate all social groups in a renewed, critical, but constructive dialogue to apprehend health issues democratically. Inclusive health education has an interest in following two main directions.

First, it is essential that teaching gives an appropriate image of the nature of science to reach scientific literacy (Holbrook and Rannikmae 2007), endorsing fully the doubt associated with any research perspective. Indeed, during the present controversy, the French authorities used a particular image of science to justify public actions as *the* unquestionable source of truth. The government created and referred to a specific, small, scientific council as embodying the voice of science. But the politicians and main media rarely reported neither the work of traditional scientific institutions nor the inevitably contrasted results of scientific research (Andolfatto 2023). The public missed this opportunity to understand the uncertain nature of frontier science. Moreover, the safety and health council changed its recommendations many times, but always presenting them as unquestionable, which increased the mistrust in research and paved the way for putting scientific results on an equal footing with conspiracy theories. In such a context, people can count on science education to foster inquiry skills to help future citizens discriminate between more or less robust views.

Second, inclusiveness implies to fight against educational inequalities in order to achieve scientific literacy for all. The Covid-19 pandemic has strengthened existing social inequalities, impacting people depending on their identities (age, gender, ethnicity, etc.), place of residence or experiences (e.g., forced migration) (Samman et al. 2021). However, sustainability is impossible without fighting against inequalities:

> Tackling inequality matters not only because of issues involving social justice, but also for more instrumental reasons. (…) Inequality has been shown to increase ill health, damage social cohesion, foster instability and violence, fuel poverty and undermine environmental sustainability. (Mackie and Allwood 2022, p. 2)

4.2.3.2 Dealing with Emotional Experience of SSC in Health Education

Moreover, effective health education also faces the challenge of addressing the affective dimension of diverging claims in a critical but not contemptuous way. Inclusiveness supposes understanding diverse life experiences without preconceptions denying how emotions may rationally shape attitudes. Indeed, personal experience has more influence than formal knowledge on opinion-building and behaviors regarding SSC (Albe 2006). Therefore, the pedagogical approach to SSC must take into account cognitive appraisal of prior direct or indirect experience. Of course, such an approach can be applied to fear-oriented facts converging with the concerns on the pandemic risks and vaccine safety, but it may also include other emotions. The emergency Covid-19 crisis led to authoritative fear-based health education telling students what they should do to avoid terrible consequences. In a sustainable perspective, health education must shift into dialogic teaching exploring and comparing alternative views in a critical way. This is consistent with Sadler et al.'s (2007) first two operations of socio-scientific reasoning: (1) analyzing the complexity of the issue—here becoming aware of its affective side; and (2) considering it from diverse viewpoints—each one embedded in a lived, authentic cognitive-emotional experience. In turn, their last two operations echo the directions that we pointed out in Sect. 4.2.3.1: further, scientific but also social inquiry (3), expressing skepticism toward information.

4.3 Methodological Implications

The tridimensional nature of reasoning or learning to reason about SSC requires following two key analytical principles that we apply to the present case. The first is accounting for both cognitive and metacognitive discourse levels. The second is respecting the multifunctional nature of communication. We have developed our own method, based on these principles.

4.3.1 Embracing Cognitive and Metacognitive Levels of Discourse

Humans use language to communicate about what they discuss (cognitive-discursive level) and about their activity (metacognitive-metadiscursive level). They (self-) regulate on the meta-level concerning their ongoing activity (Lund 2019). Nevertheless, most cognitive-discursive-level action is never discussed at the meta-level, as inferences achieve implicit regulation. This is a strong motivation for the method we describe in this chapter. Isohätälä et al. (2019) considered emotional regulation at the meta-level in educational argumentative dialogues, when an affective matter was brought to the forefront. We argue in favor of also integrating the study of implicit mechanisms of the discursive-cognitive level for a more comprehensive picture of how emotion regulation works in controversial discourse. Part of such implicit emotion regulation takes place in an anticipatory way in a group discussion, before conflict arises, while reactive emotion regulation tends to be more explicit, often embodied in meta-level discourse (Polo et al. 2023).

4.3.2 Respecting the Nature of Human Communication

Discourse analysis techniques are often based on mutually exclusive codes where a word or a sentence cannot both be coded as "affective" *and* "conceptual," for instance. We rather align ourselves with pragmatics and conversation analysis researchers who have theorized multifunctionality of utterances (Bunt 2011). In order to understand the complexity of learning how to reason about a controversial issue, we need nonmutually exclusive analytical categories. Consequently, we do not use a predetermined coding scheme, often used in discourse analysis. We prefer to keep an emic outlook (Headland et al. 1990) with open interpretations that can emerge during analysis.

Language sciences also remind educational researchers of the multimodal nature of human communication (Kress 2010). Nevertheless, much of today's analyses of educational activities, on the one hand, and controversial discourse, on the other hand, focus only on speech. Yet, nonverbal communicative acts and modes of expression are essential for making sense of controversial messages. A multimodal approach to discourse is necessary to follow the interplay of the social, the cognitive, and the emotional as people reason about SSC and/or learn to apprehend them.

4.3.3 Methods for the Present Case Study

The case of the Covid-19 vaccine controversy illustrates our argument about the tripod nature of socio-scientific reasoning applied to an authentic health policy dilemma. Our purpose is not to report an empirical study of the associated public debate, but to point out the need for science education to embrace the socio-affective dimensions of SSC together with a critical-cognitive stance to train students to deal with them. We chose three widespread messages in this debate space, and they form our corpus. Our method is original. We applied it to philosophical dialogues with children (Polo and Lund 2021). It was also based on Lund (2016, 2022) where the analysis of the imbricated dimensions occurs in relation to the individual itself, to an interaction in which the individual participates, and to the wider cultural context of which the individual is part. We target varied semiotic productions (i.e., text from media sources, graffiti, and reported speech) and analyze them at micro, macro, and mesolevels. Our micro level concerns a linguistic and semiotic analysis of the message (e.g., content and grammar), but if applicable, also paraverbal (intonation, pitch, cadence, pauses), and nonverbal elements (e.g., gesture, posture, facial mimicry, and use of touch). Here, we rigorously identify linguistic and semiotic markers that justify analytical inferences. Our macro level focuses on long-term societal tendencies that have a bearing on the message (e.g., political, historical, medical contexts, and trajectories). Here, we bring in relevant societal narratives that give a wider context to the controversy. Our meso level reflects the space–time meeting of the macro and micro levels where relevant societal tendencies and linguistic and semiotic traces manifest within a specific controversy. Here, we account for the local context of production and diffusion of the messages, embracing the ethnomethodological principle of the indexicality of sense-making (Gumperz 1999). As a meeting place, the meso level does not have strict boundaries with micro and macro levels.

In applying our method, we first note if the message seems only cognitive, social, or emotional at a first glance. Such an interpretation can be gleaned by any of the levels (micro, meso, or macro), and we unpack the most explicit one. Interestingly enough, the most explicit message is often not at the heart of the controversy. We must explore the other two dimensions of reasoning, embodied in more subtle implicit mechanisms to uncover what deeply opposes the diverging groups on the issue.

The use of these three levels (micro, macro, and meso) in exploring the three dimensions of reasoning (cognitive, social, emotional) employs both etic (researcher external focus) and emic (participant-centered focus) perspectives (Headland et al. 1990).

4.4 Reasoning About Covid-19 Vaccine: Three Illustrations

We provide three illustrations of how our argument applies to authentic messages that emerged in the French public debate about the Covid-19 vaccine, during the most heated period of the controversy, in the fall of 2021. The first illustration is a headline in the media. The second is a graffiti on a primary school building. The third is a comment made by the French president, Emmanuel Macron.

4.4.1 Pfizer Vaccines Protect Up to 70% from Severe Forms of Omicron (South African Study)

Our first illustration seems to mainly be a cognitive message. In what follows, we unpack why this is the case. We then unveil the social and emotional dimensions that are either implicit or need more investigation to uncover.

4.4.1.1 Context of Production and Diffusion of an A Priori Cognitive Message

Ryan Noach, president of *Discovery*, the first private health insurance in South Africa, declared to the press: "A double dose of Pfizer vaccine proved to be efficient up to 70% in reducing hospitalizations." The information was published on December 14, 2021, by the French media *France 24*, and then in the well-known daily newspaper *Le Figaro* with the same title: "Le vaccin Pfizer protège à 70% des cas graves d'Omicron (étude sud-africaine)" [Pfizer vaccines protect up to 70% from severe forms of Omicron (South African study)]. It then went quite viral in the French media: in the newspaper *Capital*, on the website of one of the major national radio news stations, *France Info*, and on its rival radio station *France Inter*.

But what actual information does such study release? It reported a result based on 78,000 polymerase chain reaction (PCR) tests made from November 15 to December 7, 2021. What was at stake was to ensure the efficacy of the vaccine regarding the new form of Covid-19, *Omicron*, compared to the former one, *Delta*. The great majority of the media and the authorities emphasized the vaccine efficacy against the severe forms of the disease, even though efficacy was lower than it was against *Delta* (93%). The average age of the sample population participating in the study was 34, while older people are more likely to develop severe forms of Covid-19. Indeed, efficacy against such severe forms was only 67% for 60–69 year-olds and 60% for 70–79 year-olds.

Regarding Figaro's utterance itself, several indicators frame it as belonging to the cognitive dimension. First, it defines precisely what is at stake, mentioning the form of the virus tested, and its relationship with specific types of symptoms (severe ones), a relation even quantified by a statistical result. Moreover, a scientific source is provided, with a clear geographical framework, and the claim is written with indicative present tense, the usual verbal form to assert general truth. In sum, the cognitive content released is that someone who had two doses of Pfizer vaccine is 70% less likely to develop a severe form of *Omicron* than someone who is not vaccinated. But such a message also uses social and emotional dimensions to argue in favor of vaccination.

4.4.1.2 Unveiling Its Social and Emotional Dimensions

On the social plane, the identity of the locutor is not neutral. In the short term, *Discovery* has an interest in lowering the financial costs of the health crisis. Given that the vaccine costs less than a hospitalization, there is no surprise that Noach is arguing in favor of vaccination. Besides, the 70% figure, as any scientific result, was determined by the social process of making strategic choices in the study design. Actually, the studies on vaccine efficacy end when a given number of participants are infected by the virus. Pfizer initially chose 170 cases, for instance, while Moderna stopped when 196 participants got infected. But the number may be changed during the experiment in order to reach better results. This is precisely why AstraZeneca was singled out by health authorities. They restricted the firm's communication to exploratory studies with higher efficacy than their last results.

The fact that Noach's assertion was then widely spread in the French media can be explained by the ongoing public debate at this time. It actually answered to those questioning vaccine hesitancy regarding the new forms of the virus. Implicitly, all the public messages emphasizing the figure of 70% do not only release the cognitive information, but try to convince that it is a good, rational choice to be vaccinated because the benefits of the vaccine are higher than its risks. Nevertheless, Noach does not discuss any potential side effects. Surprisingly, no journalist addresses the question of the risks either. Thus, a divide appears between distinct communities: those who have the power to produce and release the scientific facts, and those who receive the information without the possibility to set the agenda on problems to be investigated (in this case their concern about the risks of the vaccine).

The message released by *Le Figaro* has also an emotional dimension, mainly relying on the affective relations between the self and the Covid-19 pandemic on the one side, and the self and the Pfizer vaccine on the other side. The use of the verb "to protect," moreover regarding "severe forms" has a very positive emotional tonality, depicting the virus as associated with death (that could result from severe illness), and in turn the vaccine associated to life (protecting against such risk). The life–death continuum is a key parameter of the emotional framing of discourse objects along the valence axis (positive/negative emotions) (Polo et al. 2017). We argue that most of the media displaying this information are delivering a message both argumentatively and emotionally orientated toward building confidence in the vaccine. Indeed, *France Inter*, a media that was more critical about the government policy during this period, precisely turns such argument around by emphasizing the limitation of vaccine efficacy in this study, stating that 33% people are still becoming infected.

Table 4.1 summarizes the elements above, showing that the cognitive dimension is more readily recognizable, due to its numerous explicit elements. The emotional dimension is also explicit through vocabulary, but it needs more context through analytical framing to fully understand its importance. Finally, seeing the social dimension requires having contextual knowledge of insurance company ownership and their research, and how government influences communication on medical health. These deeper analyses show where the controversy really lies.

Table 4.1 Summary table of the three analytical levels crossed with the dimensions of reasoning for Noach's message

	Micro	Macro	Meso
Cognitive	Specific percentage, virus variant, severe symptoms, indicative, present tense	Scientific source	Geographical frame
Social		Viral news in French mainstream media— who makes what result public?	Speaker's private interest in lowering costs, strategic choices in study design
Emotional	Use of "protects" and "severe cases" evokes the life–death continuum		

4.4.2 "NOT OUR CHILDREN"

Our second illustration seems to mainly be an emotional message. In what follows, we unpack why this is the case. We then unveil the social and cognitive dimensions that are either implicit or need more investigation to uncover.

4.4.2.1 Context of Production and Diffusion of an A Priori Emotional Message

On January 4, 2022, the local journal *Le Dauphiné Libéré* published the picture reproduced above (Fig. 4.2) in a paper entitled: "Villette-d'Anthon : des tags contre la vaccination des enfants sur la façade d'une école" [Villette-d'Anthon (a municipality of the *Isère* department in the South East of France): messages against children vaccination on a school building]. The information is then released by the news television channel *BFM-TV*, a media known—and sometimes criticized—for selecting content with high emotional load in order to attract the audience. During such controversy, many similar messages warning against vaccinating children were written in other schools.

The emotional dimension of the message in Fig. 4.2 is quite obvious. First, it relies on its context: it impacts the familiar, daily environment, and is directed at the parents as they are dedicated to their children. In terms of affective distance, a key parameter in building emotional intensity (Polo et al. 2017), such a setting frames the message as very strong. The affective proximity is also embodied in the chosen words: NOT OUR CHILDREN, with the use of the first person and the possessive "our." Moreover, the symbol of prohibition filled with a syringe and a drop of liquid conveys a very negative depiction of vaccination by two mechanisms: (1) insisting on the act of injection, which can be stressful or painful, and (2) associating the red color and the drop, which can evoke bleeding. On the valence axis, opposition to child vaccination is in turn very positively called, as an appeal to "RESISTANCE," the word used to refer to the fight against Nazism, a reference that is well known for

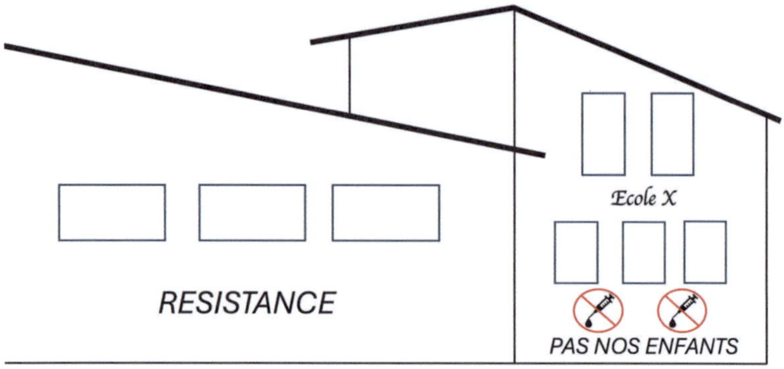

Fig. 4.2 Drawing adapted from the pictures released by *Le Dauphiné Libéré* and *BFM-TV* showing a message against children's Covid-19 vaccination on a school building

the French. Such an analogy is clearly on the "life" side of the continuum distinguishing positive from negative emotions.

4.4.2.2 Unveiling Its Social and Cognitive Dimensions

On the cognitive plane, such messaging relies on a logical process of categorization pointing at "children" as a particularly vulnerable human group. Somehow, it depicts a situation that would be the opposite of the usual common-sense priorities in cases of emergency as embodied by the principle: "women and children first." The authors of these inscriptions are using a widely shared definition of children as weak but precious beings, representing the future of societies that adults have the duty to protect against dangers. But they are also providing a specific understanding of what the current danger is: they only present the vaccine itself as threatening the children, not the pandemic. Indeed, considering that the virus is not very dangerous for children is consistent with the available scientific data at this time: they usually do not develop severe forms. Such a message builds on shared facts and beliefs to convince people of a more controversial idea, that the Covid-19 vaccine is dangerous.

Moreover, the syntax "not our children" implicitly refers to a sort of enumeration that must come to an end: "the enemy did this and that, but we will not accept to give them our children." Such an argumentation is cognitive: it draws a situation of offense that is going too far, and against which the victims have the right to enforce a limit. The fact that no clear enemy is actually specified, nor what they actually have done, allows the readers to subjectively infer such a list depending on their own experience. The recent times of trouble could actually activate a wide range of negative experiences in the reader's mind, which may make them likely to say "it's enough now." Here again, the authors take advantage of a shared experiential knowledge that the last months were not desirable forms of living to argue their claim about avoiding what they depict as a new danger. In the spatiotemporal context when the vaccine just became publicly available, along with authorities' agreement for it to be commercialized for children, their concern is clearly not on forbidding children's access to it, but rather on avoiding it to become compulsory, either directly or through an extension of the ongoing health pass. The main issue for parents opposed to their children's vaccination against Covid-19 is then not to lose their right to access school—a right already limited in France to children proving to have received 11 compulsory vaccines. Consequently, displaying such a message on a school wall appears as a very logical place for such a struggle.

On the social plane, the possessive "our" also allows the reader to distinguish between communities, opposing "our children" to "your children" or "their children." Expressing their view on the street frames the authors as members of the lower-class groups, who have no easy access to mainstream media to display their view. The "other" group (and their children) would therefore be the powerful leaders, those who decide, control higher levels of government and administrations, and display their vision of the world in mass media. But such a distinction between groups of parents also refers, more simply, to the fact that all families may not have the same opinion and strategy in this context of health crisis and that each one should keep the right to decide for their children. Here, the underlying axiological principle used is the defense of individual freedom against a potential authoritarian public policy.

Table 4.2 summarizes the elements above, showing that the emotional dimension is the most readily recognizable due both to the explicit nature of the message content and of its location. The social dimension part of the message requires understanding that communities are pitted against each other in different ways. Whose children are in danger, who has to make the decisions concerning them, and what communication tools do parents have to voice their fears? The cognitive dimension also needs more analysis to grasp it. The message relies on knowing that children are vulnerable. As a result, the danger is displaced from stopping the pandemic per se, to stopping children from being vaccinated, based on a misunderstanding of science. In addition, given that so many have suffered during the pandemic, there is an idea to stop any further suffering, especially for children (even if science tells us this is misguided). These cognitive and social dimensions illustrate the deeper, value-based nature of the controversy.

Table 4.2 Summary table of the three analytical levels crossed with the dimensions of reasoning for the message *NOT OUR CHILDREN*

	Micro	Macro	Meso
Cognitive	Implication of a list with children being the last straw	Children are precious and vulnerable, must be protected	Danger is vaccine and not pandemic, people have suffered enough
Social	Idea of us versus them	Families choose or families obey government?	Lower-class (graffiti) versus upper-class (media)
Emotional	Use of capital letters, drawing of needles, blood	RESISTANCE referring to WWII and fight against Nazis	Familiar, everyday context (school), media source is known for emotional content

4.4.3 *The Nonvaccinated People, I Very Much Want to Annoy Them*

Our third illustration seems to mainly be a social message, even if within a political context. In what follows, we unpack why this is the case. We then unveil the emotional and cognitive dimensions that are either implicit or need more investigation to uncover.

4.4.3.1 Production and Diffusion of an A Priori Socio-political Message

On January 4, 2022, the French president declared in an interview to *Le Parisien*:

> Moi, je ne suis pas pour emmerder les Français. Je peste toute la journée contre l'administration quand elle les bloque. Eh bien là, **les non-vaccinés, j'ai très envie de les emmerder**. Et donc on va continuer de le faire, jusqu'au bout. C'est ça, la stratégie. [I am not in favor of annoying the French people. I damn the administration all day long when it blocks them. Well here, **the non-vaccinated people, I very much want to annoy them**. And so we'll keep on doing it, until the end. This is the strategy.]

The sentence in bold was then repeated in many media and gave birth to much discussion on the French political scene. Some commentators considered it as a communication mistake driven by the man's emotion; others perceived it as a strategical way to attract the pro-vax electors by speaking up their frustration pointing at anti-vax. At this time, the health pass was actually restricting the daily life of the people who were not vaccinated, notably to access to public places and health care jobs. Three months later, on April the 11th, as he was campaigning for his reelection, a woman blamed him for having declared "*que les devoirs étaient au-dessus des droits. Vous avez dit 'emmerder les non-vaccinés', on est des sous-citoyens* [that duties were above rights. You said that you 'annoyed the non-vaccinated people', we are sub-citizens"]. Macron then replied:

> Je n'ai jamais dit "sous-citoyen" (…). Je l'ai dit, entre guillemets, de manière affectueuse *(…)* ce n'était pas une insulte. [I never said 'sub-citizens' (…). I said so, in air quotes, with a tender intention (…) it was not an insult].

Such dialogue was reported the same day by the famous daily newspaper *Le Monde* in an article entitled "*Emmerder les non-vaccinés*"*: Emmanuel Macron, en campagne dans le Nord, se defend* ["Annoying non-vaccinated people": Emmanuel Macron, campaigning in the North, is defending himself].

The social dimension of Macron's message *les non-vaccinés, j'ai très envie de les emmerder* [the nonvaccinated people, I very much want to annoy them] is clear. At first sight, his main intention is precisely to distinguish between the "good" French citizens who were vaccinated and the "bad" ones who refuse to do it, positioning himself as belonging to and supporting the first group, and condemning and fighting against the second one. Indeed, he switches at the following sentence from the singular to the plural first person, clarifying that he is speaking as part of the same group. The vocabulary that he uses is also very socially significant. Talking about "a strategy" to be followed "until the end" is a direct metaphor of war, the one he is declaring on the nonvaccinated people. He makes no distinction here between the anti-vax and those who hesitate. The term "emmerder" ("to annoy") is very rude, and using it breaks the rules of usual public communication of a president. Moreover, the French Constitution attributes to the president a supposed neutrality, and, once elected, the duty to represent all the French people. Here, voicing what many vaccinated people were thinking allowed them to identify with the exasperation that he was displaying in colloquial terms that they could have used.

Last but not least, by promising a punishment to the "bad" citizens, he is acting as a teacher or a parent who would know better than them what is good for France. Such paternalism is typical to the French Republic, the symbol of the providential man saving the country being recurrent in its history. Not only noting but emphasizing the divide between the vaccinated and nonvaccinated people, Macron stigmatizes the second group. The words themselves are a way to annoy the nonvaccinated people by adding social pressure on them: either you take the vaccine and are on our side; or you are our enemy, not really part of the French community anymore. This is actually what the woman has understood when she mentions "sub-citizens."

4.4.3.2 Investigating the Emotional and Cognitive Dimensions of Macron's Message

Macron uses emotions to a great extent in his argumentative message. First, he is publicly displaying exasperation and even acting as if he could not control himself anymore due to how much he is mad at the people who are not vaccinated. His own emotions would explain that he is using a vocabulary that is not conventional for a president. While some commentators blame him for such lack of self-control, his emotional behavior is also a way to frame the situation as exceptional and particularly grave. This is a very clear instance of how the affective and the cognitive are intertwined: if I'm being rude, it's not because I am out of control, it's because the danger is so big that I have to react firmly. Here, Macron is both displaying fear of Covid-19 and determination to achieve widespread vaccination, as the only way out of the crisis ("the strategy"). He also warmly identifies with the side of the vaccinated people as peers and distinguishes them from the community of the nonvaccinated, that he overtly despises.

But the affective content of such message also relies on the emotions that Macron is attempting to provoke in the audience. Quite generally, he uses a provocative tone, designed to raise emotions either positive or negative. At any rate, people and the media will talk about him, conferring him a central role in the public debate, a few months before the presidential campaign starts. This type of communication strategy was described as typical to the French extreme right-wing politicians such as Jean-Marie Lepen (Bernard 2007). Moreover, his message aims at creating sympathy from most vaccinated people and confirming them that they made the right choice. Finally, he also aims of course at frightening the nonvaccinated people by threatening them implicitly with new, stronger restrictions, within an unlimited time frame.

Still, Macron's message has a strong cognitive content. As previously explained, the way he delivered it with high emotions conveys a cognitive sense in itself: the one of an immediate, severe danger. By not talking about anything else than vaccination and the nonvaccinated people, he depicts the vaccine as the only way to fight the pandemic, and the nonvaccinated as the agents responsible for the current situation. This is precisely because they are implicitly depicted as selfish that the president can blame them so offensively. They are not just defined as opposed to the vaccine or hesitant to take it, but as dangerous for other people. Such a framing is consistent with the available scientific data on collective immunity, which show that it is necessary to reach a minimal vaccination rate.

Macron here explains that he will indirectly force reluctant and hesitant people to be vaccinated, and such a "strategy" seems likely to work based on two characteristics of the situation that his speech reflects upon. First, the fact that nonvaccinated people are described as not respecting the collective interest somehow justifies to "punish" them. Second, the health pass has already proved efficient to bring many hesitant people to vaccination. Therefore, this strategy appears as realpolitik. Still, one can wonder why he doesn't just make Covid-19 vaccine mandatory? Is Macron concerned by the reaction of the anti-vax? Does he stop short of endorsing such a strong restriction because he somehow values the freedom of choosing for one's own health? But those afraid of the vaccine's potential side effects could easily argue that he is not imposing the vaccine to avoid being responsible for them. Here, Macron's ambiguous position benefits conspiracy theorists.

Table 4.3 summarizes the elements above, showing why the social dimension is the most readily recognizable. This is mainly due to the vulgar vocabulary, shocking for a president. Macron also uses war-like language that has the effect of dividing the French population, marginalizing the unvaccinated. This is also shocking, as the president should represent all people. The emotional dimension comes through in video-clips, not always available when only his words are cited, but where his anger and exasperation is conveyed through voice tone. Emotions are also provoked in the listeners: confidence for the vaccinated, fear for the unvaccinated. The cognitive

Table 4.3 Summary table of the three analytical levels crossed with the dimensions of reasoning for Macron's message

	Micro	Macro	Meso
Cognitive	Vaccine refusal is a severe and immediate danger, people are selfish if they are not vaccinated	Macron relies on current science	No distinction between vaccine refusal and hesitancy, but will punish all with a stricter health pass
Social	Vocabulary is populist, language evokes war, distinguishes between good and bad citizens	French presidents always display paternalism, but should represent everyone	Marginalizes nonvaccinated, making media cover him during election time
Emotional	Tonality shows exasperation, cannot control his anger, problem is grave		Gives confidence to the vaccinated, provokes fear in others

dimension reveals a missing element, often overlooked: no distinction is made between those who refuse the vaccine and those who hesitate. Vaccine refusal is portrayed as dangerous and implicitly those who refuse as selfish. The next section (Discussion) will explain why the situation is more complicated, but we can see here that the cognitive and emotional dimensions require deeper analysis that involve distinguishing between populations, understanding their values, while weighing this against current scientific knowledge.

4.5 Discussion: Implications for Education to SSC

Macron and the anti-vax's messages illustrate a great divide opposing two groups mutually mistrusting and caricaturing each other. Yet, in the media very little space remains for the large group of hesitant people to debate. Some of them were finally "convinced" by the health pass (Gans 2023), while others too greatly doubted the reliability of the accelerated process of vaccine production. But both shared the reality of being unsure about the decision to take beforehand, and about whether they took the right one in the end. The vaccination policy then failed in reaching social sustainability. The more a group was marginalized, the more they were likely to lack confidence in the authorities, and not to take the vaccine. They were then further marginalized with health pass restrictions. A survey points out that in France confidence in science decreased by 20% between March and October 2020. Such a vicious circle should end, as the literature on vaccine hesitancy emphasizes the need to restore citizens' confidence in scientific and political authorities (Peretti-Watel et al. 2015; Bajos et al. 2022). Regarding the Covid-19 vaccine, it seems more fruitful to provide a framework in which people can develop voluntary acceptance and self-determined decisions rather than to be forced to follow authoritative policies (Rapisarda et al. 2021). Health education could play a role in meeting such challenges if it provided an inclusive space for the expression of all experiences. Having unpacked the imbricated social, cognitive, and emotional nature of messages about Covid-19 vaccine hesitancy, our analyses help us to better understand its controversial nature without simplistically disregarding divergent groups. Our results suggest that a didactic approach to SSC leaving no one behind may benefit from taking all opinions into account, including emotions, values, interests, social norms, political attitudes together with technoscientific knowledge in an open, though critical, dialogue.

More specifically, to better conform to the ideal of developing socio-scientific literacy for a sustainable society, we propose that science education focuses on the following five goals as implications of our study:

1. Epistemology. Understanding the nature of scientific results as provisionary though robust knowledge is essential to restore citizens' confidence in science, as mentioned by the French Parlementary Office Assessing Scientific and Technological Choices (OPECST) in six reports about the Covid pandemic. Malgorn (2023, p. 44) summarizes it as " the necessity to restore confidence and

to do so, that the authorities display a discourse of truth, based on a scientific truth that does not exclude mitigated results and that might evolve."

2. Inquiry skills. Citizens need to be able to undertake information-seeking and critical assessment of the gathered elements using scientific and social criteria, notably in order to avoid the "misinformation climate" to blur any thinking on SSC, as it did with the studied controversy (e.g., Patriat 2023).
3. Collaborative attitude. Openness to diverse perspectives, exploring challenging views and questioning one's own opinion is a prerequisite for deeply understanding SSC. This allows participants to reach democratic and fruitful decisions that incorporate a balance between common good and individual freedom. Unfortunately, our case study shows that such a dialogue did not emerge during the Covid-19 vaccine hesitancy controversy in France. Rather, each side caricatured the other in a detrimental way.
4. Integrative content. Making sense of complex SSC supposes reflection on how scientific facts relate to daily cognitive-affective experience. While pro-vax activists claimed "science" was supporting only their views, the scientific method can also be applied to the concerns of those who hesitate to be vaccinated, for example, due to vaccine side effects.
5. Individual and collective reasoning. Articulating arguments through both descriptive, prescriptive, and affective logics through exploratory talk is crucial to enable engagement, as a group, in constructive reasoning (Polo 2023).

Following the agenda above would not only allow educational action to be more inclusive, but it would also enrich the debate. Students would then collectively investigate their doubts, fears, and hopes to clarify what they rely on and build more robust arguments supporting their view, either clear or hesitant.

To meet these goals, we recommend to design and orchestrate pedagogical practices engaging students to

- Become aware of their own emotional position and opinion on the issue
- Understand SSC content by both learning scientific knowledge and analyzing media discourse
- Debate about SSC relating facts to their personal experiences, expressing doubts, and opinion shifts
- Reflect on the debate inclusiveness, on the quality of the arguments, and on the potentially missing information
- Investigate the missing content
- Report on the socio-scientific reasoning that they have developed

4.6 Conclusion

Through the case of the French controversy about Covid-19 vaccination, we showed that socio-scientific argumentation is always social, cognitive, and emotional in a way that is fully imbricated. Even messages that may seem at first glance as consisting of one or the other of these dimensions actually convey the three of them. Therefore, accusing the rival party of displaying an irrational emotion-driven reasoning is not fair play, and misses the opportunity for a deep understanding of the controversy. Reaching strong socio-scientific reasoning requires acknowledging the role of socio-affective processes in shaping attitudes. The idea is not to learn how to manipulate others on the basis of such processes but rather to explore one's own socio-affective relation to the controversy to unveil how (well) it supports underlying arguments, since claims are consubstantial to emotional positions (Polo et al. 2017). In this regard, only education building up from people's full experience can really include the typically marginalized groups instead of just blaming them for their vaccine hesitancy. Recognizing the legitimacy of doubt is central to engage toward sustainability, combining critical thinking and empathetic social cohesion. We suggest a 5-goal agenda for science education to train future citizens regarding this ideal. We accordingly point recommendations for pedagogical practice, which we are currently implementing in higher education teaching. Future work on this project will help refining such principles for instructional design on SSC.

Acknowledgments The authors are grateful to the ASLAN project (ANR-10-LABX-0081) of the Université de Lyon for its financial support within the French program "Investments for the Future" operated by the National Research Agency (ANR).

References

Albe V (2006) Procédés discursifs et rôles sociaux d'élèves en groupes de discussion sur une controverse socio-scientifique. Revue française de pédagogie. Recherches en éducation, (157):103–118. https://doi.org/10.4000/rfp.604

Andolfatto D (ed) (2023) Citoyens dans la crise sanitaire. Classiques Garnier

Avry S, Chanel G, Bétrancourt M, Molinari G (2020) Achievement appraisals, emotions and socio-cognitive processes. Comput Hum Behav 107:106267. https://doi.org/10.1016/j.chb.2020.106267

Bajos N, Spire A, Silberzan L (2022) The social specificities of hostility toward vaccination against Covid-19 in France. PLoS One 17(1):e0262192. https://doi.org/10.1371/journal.pone.0262192

Baker M, Andriessen J, Lund K (2009) Socio-relational, affective and cognitive dimensions of CSCL interactions. In: O'Malley C, Suthers D, Reimann P, Dimitracopoulou A (eds) Computer supported collaborative learning practices (2). International Society of the Learning Sciences, pp 31–33

Baker M, Andriessen J, Järvelä S (eds) (2013) Affective learning together. Routledge

Barth M, Michelsen G (2013) Learning for change: an educational contribution to sustainability science. Sustain Sci 8:103–119. https://doi.org/10.1007/s11625-012-0181-5

Bernard M (2007) Le Pen, un provocateur en politique (1984–2002). Vingtième Siècle Rev Hist 93:37–45. https://doi.org/10.3917/ving.093.0037

Bunt H (2011) Multifunctionality in dialogue. Comput Speech Lang 25(2):222–245. https://doi.org/10.1016/j.csl.2010.04.006

Cambon L, Schwarzinger M, Alla F (2022) Increasing acceptance of a vaccination program for coronavirus disease 2019 in France. Vaccine 40:178–182. https://doi.org/10.1016/j.vaccine.2021.11.023

Dewey J (1938) Logic: the theory of inquiry. Southern Illinois University Press

Ekman P (1999) Basic emotions. In: Dalgliesh T, Power MJ (eds) Handbook of cognition and emotion 16. Wiley, pp 45–60

Flanagan KL, Fink AL, Plebanski M, Klein SL (2017) Sex and gender differences in the outcomes of vaccination over the life course. Annu Rev Cell Dev Biol 33:577–599. https://doi.org/10.1146/annurev-cellbio-100616-060718

Galasso V, Pons V, Profeta P, Becher M, Brouard S, Foucault M (2022) From anti-vax intentions to vaccination: panel and experimental evidence from nine countries (No. w29741). National Bureau of Economic Research

Gans JS (2023) Vaccine hesitancy, passports, and the demand for vaccination. Int Econ Rev 64(2):641–652. https://doi.org/10.1111/iere.12609

Griffith J, Marani H, Monkman H (2021) COVID-19 vaccine hesitancy in Canada: content analysis of tweets using the theoretical domains framework. J Med Internet Res 23(4):e26874. https://doi.org/10.2196/26874

Guerrini J-C (2024) The use of diagrams in argumentation. Logical, argumentative and rhetorical approach. In: Boogaart R, Garssen B, Jansen H, Leeuwen M, van Pilgram R, Reuneker A (eds) Proceedings of the tenth conference of the International Society for the Study of Argumentation. Sic Sat, pp 297–312

Gumperz JJ (1999) On interactional sociolinguistic method. In: Sarangi S, Roberts C (eds) Talk, work and institutional order: discourse in medical, mediation and management settings. De Gruyter Mouton, pp 453–472

Headland TN, Pike KL, Marvin H (1990) Emics and Etics: the insider/outsider debate. Sage Publications

Holbrook J, Rannikmae M (2007) The nature of science education for enhancing scientific literacy. Int J Sci Educ 29(11):1347–1362. https://doi.org/10.1080/09500690601007549

Ioannou GN, Locke ER, O'Hare AM, Bohnert AS, Boyko EJ, Hynes DM, Berry K (2022) COVID-19 vaccination effectiveness against infection or death in a national US health care system. Ann Intern Med 175(3):352–361. https://doi.org/10.7326/M21-3256

Isohätälä J, Näykki P, Järvelä S, Baker M (2017) Striking a balance: socio-emotional processes during argumentation in collaborative learning interaction. Learn Cult Soc Int 16:1–19. https://doi.org/10.1016/j.lcsi.2017.09.003

Isohätälä J, Näykki P, Järvelä S (2019) Cognitive and socio-emotional interaction in collaborative learning. Scand J Educ Res:1–21. https://doi.org/10.1080/00313831.2019.1623310

Kress G (2010) Multimodality: a social semiotic approach to contemporary communication. Routledge

Larson H, de Figueiredo A, Karafillakis E, Rawal M (2018) State of vaccine confidence in the EU 2018. European Commission Publications

Lazarus JV, Ratzan SC, Palayew A, Gostin LO, Larson HJ, Rabin K, Kimball S, El-Mohandes A (2021) A global survey of potential acceptance of a COVID-19 vaccine. Nat Med 27(2):225–228. https://doi.org/10.1038/s41591-020-1124-9

Lund K (2016) Modeling the individual within the group: an interdisciplinary approach to collaborative knowledge construction. HDR Université Grenoble Alpes

Lund K (2019) Building and regulating cognitive, linguistic, and interactional aspects of knowledge between the individual and the group. Comput Hum Behav:370–383. https://doi.org/10.1016/j.chb.2019.04.013

Lund K (2022) Modelling the co-elaboration of knowledge: connecting cognitive, linguistic, social and interactional systems. In: Lund K, Basso Fossali P, Mazur A, Ollagnier-Beldame M (eds) Language is a complex adaptive system: explorations and evidence. Language Science Press, pp 35–48

MacDonald NE (2015) Vaccine hesitancy: definition, scope and determinants. Vaccine 33(34):4161–4164. https://doi.org/10.1016/j.vaccine.2015.04.036

Mackie J, Allwood G (2022) The implementation of the 2030 Agenda's principles of 'leaving-no-one-behind' and 'addressing the needs of those furthest behind first' in the EU's development policy. European Union

Malgorn B (2023) Le pouvoir et ses conseils en état d'urgence. In: Andolfatto D (ed) Citoyens dans la crise sanitaire. Classiques Garnier, pp 39–60

Malova E (2021) Understanding online conversations about covid-19 vaccine on twitter: vaccine hesitancy amid the public health crisis. Commun Res Rep 38(5):346–356. https://doi.org/10.1080/08824096.2021.1983424

Matuszewicz R (2023) Les débats parlementaires sur le passe sanitaire. In: Andolfatto D (ed) Citoyens dans la crise sanitaire. Classiques Garnier, pp 115–134

Mayer Y, Etgar S, Shiffman N, Bloch Y, Mendlovic S, Lurie I (2021) Hope as a predictor for COVID-19 vaccine uptake. Hum Vaccin Immunother 17(12):4941–4945. https://doi.org/10.1080/21645515.2021.1989911

Moore R, Purvis RS, Hallgren E, Willis D, Hall S, Reece S, CarlLee S, Judkins H, McElfish PA (2021) Motivations to vaccinate among hesitant adopters of the COVID-19 vaccine. J Community Health:1–9. https://doi.org/10.1007/s10900-021-01037-5

Navarre C, Roy P, Ledochowski S, Fabre M, Esparcieux A, Issartel B, Dutertre M, Blanc-Gruyelle AL, Suy F, Adelaide L, Pariset C, Kisterman JP, Champagne H, Saison J (2021) Determinants of COVID-19 vaccine hesitancy in French hospitals. Infect Dis Now 51(8):647–653. https://doi.org/10.1016/j.idnow.2021.08.004

Newen A, de Bruin L, Gallagher S (2017) Oxford handbook of 4E cognition. Oxford University Press

Olivier L (2023) Les partis et le mouvement "antivax". In: Andolfatto D (ed) Citoyens dans la crise sanitaire. Classiques Garnier, pp 135–150

Patriat C (2023) En guise de conclusion. De quoi le covid est l'acronyme ? In: Andolfatto D (ed) Citoyens dans la crise sanitaire. Classiques Garnier, pp 383–392

Peretti-Watel P, Larson HJ, Ward JK, Schulz WS, Verger P (2015) Vaccine hesitancy: clarifying a theoretical framework for an ambiguous notion. PLoS Curr 7. https://doi.org/10.1371/currents.outbreaks.6844c80ff9f5b273f34c91f71b7fc289

Plantin C (2018) Dictionary of argumentation. College Publications

Polo C (2023) The fertile debate. College Publications

Polo C, Lund K (2021) La saisie émotive du kairos avec des enfants : entre acte philosophique et geste didactique. Studia UBB Philos 66(LXVI, 3):103–134. https://doi.org/10.24193/subbphil.2021.3

Polo C, Plantin C, Lund K, Niccolai GP (2017) Emotional positioning as a cognitive resource for arguing. Pragmat Soc 8(3):323–354. https://doi.org/10.1075/ps.8.3.01pol

Polo C, Simonian S, Chaker R (2023) Distant collaboration at university: anticipatory and reactive emotion regulation of argumentation-based learning. In: Damşa C, Borge M, Koh E, Worsley M (eds) Proceedings of the 16th international conference on CSCL. International Society of the Learning Sciences, pp 11–18

Rapisarda V, Vella F, Ledda C, Barattucci M, Ramaci T (2021) What prompts doctors to recommend COVID-19 vaccines: is it a question of positive emotion? Vaccines 9(6):578. https://doi.org/10.3390/vaccines9060578

Roschelle J, Teasley SD (1995) The construction of shared knowledge in collaborative problem solving. In: O'Maley C (ed) Computer-supported collaborative learning. Springer, pp 69–97

Sadler TD, Barab SA, Scott B (2007) What do students gain by engaging in socio-scientific inquiry? Res Sci Educ 37:371–391. https://doi.org/10.1007/s11165-006-9030-9

Samman E, Roche JM, Sarwar MB, Evans M (2021) 'Leave no one behind' – five years into agenda 2030. Overseas Development Institute

Sanseigne F (2020) L'État contre lui-même: Contraception et illégalismes d'État de la métropole aux DOM (1961–1967). Genèses 121:52–75. https://doi.org/10.3917/gen.121.0052

Schwarzinger M, Watson V, Arwidson P, Alla F, Luchini S (2021) COVID-19 vaccine hesitancy in a representative working-age population in France: a survey experiment based on vaccine characteristics. Lancet Public Health 6(4):e210–e221. https://doi.org/10.1016/S2468-2667(21)00012-8

Wouters OJ, Shadlen KC, Salcher-Konrad M, Pollard AJ, Larson HJ, Teerawattananon Y (2021) Challenges in ensuring global access to COVID-19 vaccines. Lancet 397:1023–1034. https://doi.org/10.1016/S0140-6736(21)00306-8

Claire Polo, PhD, is a permanent member of the Education, Cultures and Policies laboratory of Lyon 2 University. She develops her research at the crossroads of the fields of collaborative learning, argumentation studies and interaction analysis. She has a background in political sciences and was a social sciences and economics teacher at high school before getting a full position at university. During her PhD, she studied socio-scientific debates about drinking water management with high school students in Mexico, the USA, and France. She combines deep linguistic analysis of educational dialogues with a multi-dimensional, social, cognitive, and affective approach to learning processes and their mediation. She published the book *The Fertile Debate: Affective Exploration of a Controversy.*

Kristine Lund is director of the LabEx ASLAN (Advanced Studies on Language Complexity) and a member of the Interactions, Corpus, Apprentissages, Representations (ICAR) laboratory affiliated to Lyon 2 University, the French National Scientific Research Center, and the Ecole Normale Supérieure de Lyon. She mobilizes interdisciplinary perspectives on collaborative learning, studying the relations between individuals, groups, and communities. She is an elected fellow of the International Society of the Learning Sciences.

Chapter 5
Fostering Evidence-Based Hope and Engagement for the Ocean Through Education

Tuula H. Sarvas Skarstein and Frode Skarstein

Abstract Environmental and sustainability discourse is frequently dominated by narratives of disaster and loss. While such negatively charged messaging aims to prompt action, it can also lead to emotional disengagement and a sense of hopelessness. In contrast, recent research highlights the value of hope-centered narratives that are grounded in evidence and can foster constructive engagement to environmental action. This study explores how a shift from predominantly negative stories of disaster to a greater focus on positive, solution-focused narratives may be advanced within education about the ocean. It also seeks to develop educational tools to support such a shift. Inspired by educational design research, the study integrates previous research findings with active stakeholder involvement to ensure practical relevance of the results. All stakeholders contributing to this study are based in coastal areas of Norway and include teacher education students, lectures, and educational researchers working within teacher education, high school teachers, as well as guides and marine scientists cooperating within an eco-tourism company.

The results suggest that focusing on ocean narratives that highlight emerging knowledge and solutions, collective engagement, local agency, and policy action, and that address personally meaningful ocean issues, evoke awe and curiosity about the oceans, and support envisioning positive futures, may foster evidence-based hope and encourage engagement in ocean conservation efforts.

Keywords Evidence-based hope · Ocean literacy · Hope-centered environmental narratives · Student engagement and agency · Educational design research

T. H. S. Skarstein (✉)
Department of Early Childhood Education, Faculty of Arts and Education, University of Stavanger, Stavanger, Norway
e-mail: tuula.skarstein@uis.no

F. Skarstein
Department of Education and Sports Science, Faculty of Arts and Education, University of Stavanger, Stavanger, Norway
e-mail: frode.skarstein@uis.no

F. Skarstein, L.-A. Wolff (eds.), *Sustainability and Science Education*, Environmental Challenges and Solutions,
https://doi.org/10.1007/978-3-032-24408-6_5

5.1 Introduction

In the public discourse surrounding environmental and sustainability challenges, narratives emphasizing disaster and loss often prevail. Constant reports of environmental degradation contribute to a collective narrative of suffering, culminating in an overshadowing doom and gloom vision of the future. Stories of disaster and loss are often aimed at scaring people to act. However, such predominantly negative narratives may have the opposite effect if they make people feel that it is too late to do anything about these problems and thus foster a sense of hopelessness that discourages engagement (Kelsey 2020; Stoknes 2014). Examples of such narratives commonly connect to ocean science and include rising ocean temperatures, ocean acidification, overfishing, invasive species, loss of marine biodiversity, eutrophication, plastic pollution, contamination from environmental toxins, and seabed destruction resulting from deep-sea mining.

Research has shown that children, young people, and adults experience high levels of alarm about environmental problems, often accompanied by emotions such as worry, anxiety, sadness, frustration, anger, and fear (Chawla 2020; Ojala et al. 2021). Children and adolescents are particularly vulnerable to mental distress related to serious environmental problems, such as climate change (Burke et al. 2018). At the same time, environmental challenges are core issues in contemporary science education, which underscores the importance of how science education must not only convey the complexity of these challenges but also support students in processing their emotional responses and developing constructive forms of coping and engagement. This shift in framing is particularly relevant, as science education is increasingly called upon to equip learners not only with scientific knowledge, but also with the capacity to engage emotionally and ethically with sustainability challenges (Chawla 2020; Wals 2010).

While feelings of hope and concern can motivate action, despair and hopelessness may lead to the opposite (Chawla 2020; Kelsey 2020; Ojala 2023; Ojala et al. 2021). However, sources of hope can be both constructive and less constructive from an engagement perspective (Ojala 2023). This distinction is crucial in science education, where the goal is not simply to inspire optimism, but to deepen students' understanding of environmental issues and to develop their ability to critically assess solutions and envision actionable futures. Ojala's (2023) review of recent research shows that cognitive measures of hope where both pathways to a more sustainable future and a feeling of agency are included seem to be consistently positively related to engagement, whereas results of studies regarding hope as an emotion are less consistent.

People find different ways to cope with disturbing environmental information and their ways of coping affect their willingness to acknowledge such information and to act. People may cope through distraction, for instance, deliberately thinking about something else or avoiding disturbing information. Others focus on feelings of hopelessness and may conclude that action is pointless. Some deny that the environmental problem, such as climate change, exists or believe that it does not affect their own life. Such ways of coping seek to escape painful feelings and get rid of negative emotions with distancing strategies and are all negatively associated with environmental action (Ojala 2012a, b, 2013; Stoknes 2014).

Coping strategies that support action taking are problem-focused coping—such as thinking about, planning, and trying to do something to reduce the cause of the problem—and meaning-focused coping, which involves being able to switch perspective, finding positive value in confronting the problem, and cultivating hope that can exist side by side with worry. Problem-focused coping is often associated with individual environmental action. However, sole emphasis on individual action in addressing problems that are more than a person can solve alone may lead to feelings of helplessness and negatively impact well-being. Meaning-focused coping, involving positive reappraisal and reframing a problem to find positive meaning in addressing it, on the other hand, is associated with positive feelings and higher well-being. This is particularly true concerning problems that cannot be solved immediately but demand active engagement over the longer term (Chawla 2020; Ojala 2012a, b, 2013).

It is also essential that young people are given the opportunity to express their emotions about environmental problems, particularly since coping occurs not only at an individual level but also as a social process. Therefore, science education that includes socio-emotional learning opportunities, for example, through classroom dialogue about environmental grief or hope, can cultivate resilience and long-term engagement in sustainability issues (Ojala 2015).

To foster hope and help people cope constructively with ocean environmental issues, a change in the way communication and education about the oceans are carried out is needed. Research shows that when media coverage focuses on success stories of activism and actions within local contexts that readers can relate to, people feel more enthusiastic and optimistic about the possibility of solving environmental problems (Kelsey 2020). Also, when high school students felt that their teachers respected their emotions concerning societal issues and had a future-oriented, positive, and solution-oriented communication style, they felt more constructive hope concerning environmental problems such as climate change (Ojala 2015).

While in recent years, the number of marine conservation studies showing recovery, success, and improvement is growing, they are not widely communicated, and educational resources do not seem to reflect this change. Such studies include, among others, recoveries of previously threatened species, growing protection of marine habitats, decline in certain types of pollution, and accelerating restoration of degraded habitats (Knowlton 2021). Successful scientific case studies are useful pedagogical tools that can be integrated into science curricula to strengthen ocean literacy, demonstrate the efficacy of conservation science, and foster hope based on evidence (Asikin et al. 2025; Knowlton 2021).

By incorporating a greater focus on success stories, proven solutions, and positive trends into the curriculum, educators can provide students with more hopeful and empowering perspectives on marine environmental issues. This reflects what Hicks (2014) calls "educating for hope," that is, equipping learners to imagine and act upon a more sustainable future. Integrating more hopeful narratives into science education also supports the development of learners' action competence, a key outcome in transformative sustainability education that emphasizes democratic participation and critical reflection (Wals 2010). Furthermore, it addresses the well-documented disconnect between environmental knowledge and behavior, which Stevenson (2007) attributes to educational structures that neglect emotional engagement and contextual relevance.

Thus, embedding evidence-based hope in science education through success stories not only enhances students' understanding of complex marine ecological issues and ocean literacy, but also cultivates their sense of agency and motivation to engage in pro-environmental behaviors. This is particularly important in marine contexts, where studies have shown that ocean literacy improves when science education explicitly emphasizes human–ocean relationships, solutions to environmental problems, and positive examples of change (Asikin et al. 2025). Altogether, such approaches align with the broader goals of sustainability education, which seek to empower learners as active and informed contributors to societal transformation.

In this study, the term *evidence-based hope* is used to describe hope grounded in evidence of things that are moving in a positive direction (Kelsey 2020; Kelsey and Broadhurst 2024). By building upon the evidence of what works, the impact of these solutions can be amplified. Evidence-based sources of hope enable people to envision a better future, work toward it, and hold those in power accountable for implementing proven solutions that move issues in the right direction (Kelsey 2020). To shift ocean narratives toward solutions-orientated focus, people need tools at hand. These tools may include highlighting and promoting solutions to environmental communication challenges, like sharing success stories, and seeking out and supporting news and information that are solutions-based (Kelsey and Broadhurst 2024). Ocean education literature also emphasizes that presenting tangible examples of ecological recovery and successful conservation initiatives can enhance learners' critical thinking, problem-solving, and future-oriented capacity (Shellock et al. 2024; Mokos et al. 2020). These narrative strategies support ocean literacy development and have increased students' engagement and behavioral intentions by offering pathways to meaningful action (Stoll-Kleemann 2019).

The goal of this study is to contribute to the overall effort of shifting environmental narratives from pessimism to a more constructively optimistic vision of the future based on evidence-based hope. More specifically, this study aims to explore how a shift from negative ocean narratives of disaster to positive, solution-focused perspectives can be advanced within education. The research question of the study is: *What elements in educational narratives may foster evidence-based hope regarding the environmental status of the oceans?* The study also seeks to develop educational tools to support such a shift and fuel evidence-based hope and engagement in marine conservation and protection.

5.2 Methods

The design of the study (Fig. 5.1) was inspired by educational design research (Plomp 2013). Accordingly, the study was informed by an array of sources, including recent literature, a targeted survey, and inquiries conducted through several workshops in close co-operation with stakeholders in real-world settings. This ensured a comprehensive approach that combines scientific investigation with the development of practical solutions to educational challenges. A review of recent research literature, particularly focusing on suggested approaches to enhancing awareness, hope, and engagement in environmental issues, provided a robust theoretical foundation for the study, whereas the engagement of diverse stakeholders was essential to ensure the practical relevance of the results.

The data used in this study was collected during a larger project (NHK 240013) financed by the Nordic Council of Ministers. We conducted new analyses of the data for the current study.

5.2.1 Stakeholder Involvement

The study aimed to involve various stakeholder groups to ensure that the results were both wide-reaching and grounded in real-world interactions. We identified potential key stakeholders living in coastal areas of Norway and recruited them as informants for the study. These included student teachers from various teacher education programs (i.e., programs qualifying to teach in early childhood, primary and

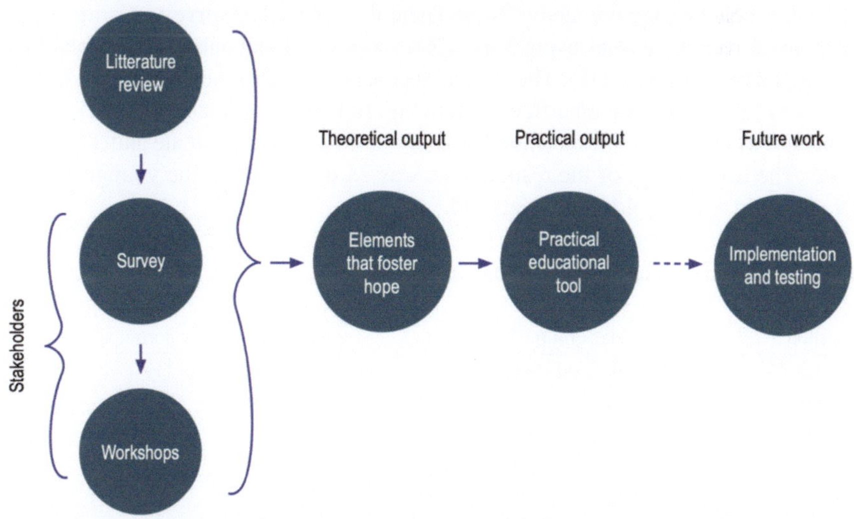

Fig. 5.1 A schematic representation of the design of the current study

lower secondary, or upper secondary education), lecturers and educational researchers working in teacher education, high school teachers, as well as guides and marine scientists collaborating within an ecotourism company. Student teachers were considered an important stakeholder group, as they represent young adults, future educators, and custodians of environmental engagement for future generations. Teacher education lecturers and educational researchers were included due to their important role in influencing educational practice and curricula. The participants in this group represented diverse educational fields, including science education, physical education, food and health education, visual and creative arts education, and language education. Most of them were both lecturers and educational researchers. Moreover, guides and marine scientists cooperating within an ecotourism company were included to encompass perspectives beyond the formal educational institutions and enabling a link to scientific and community-based approaches. Many of the stakeholders in this group had several roles, that is, they worked, or had previously worked, as both guides, scientists in research projects, and as teachers at school.

Table 5.1 presents information on the sampling methods, stakeholder groups, sampling locations, and number of participants. We collected information from the student teachers through an anonymous survey, while the other stakeholders were invited to express their views in physical workshops arranged at their workplaces. The workshops also enabled us to better understand the human–ocean contexts relevant for each stakeholder group.

5.2.2 *Survey*

We conducted an anonymous survey on three student teacher groups (see Table 5.1) attending a Norwegian university located near the ocean. The survey was run during ordinary university lessons using SurveyXact, a professional online survey platform developed by Ramboll (n.d.). The student teachers were informed about the study at the start of their lesson and participation in the study was voluntary. Responses were collected from a total of 153 respondents, of whom 78% were female and 22% were male. The average age of the respondents was 25,6 years, with the youngest being 19 years old and the oldest 55 years old.

Table 5.1 Information on participants and sampling methods

Sampling method	Stakeholder group	Sampling location	Number of participants
Survey	Early childhood student teachers	University	119
Survey	Primary and lower secondary student teachers	University	26
Survey	Upper secondary student teachers	University	8
Workshop 1	Lecturers and educational researchers	University	10
Workshop 2	Guides, marine scientists, and high school teachers	Ecotourism center	23

The survey included questions of the participants background (age, sex, study program) followed by 14 closed and 7 open-ended questions about their thoughts and feelings regarding the ocean and its environmental status. In the current study, we only analyzed the responses to the background questions and three of the other questions. The rest of the survey data will be used in another study.

The three questions analyzed in this study were

1. "To what extent do you feel hopeful that the environmental challenges facing the ocean can be solved?" This was a closed-ended question with a six-point Likert scale ranking, $n = 147$.
2. "In your own words, describe the main reasons why you feel hopeful about the possibility of solving or managing marine environmental problems in the future—in other words, why do you feel hopeful?" This was an open-ended follow-up question to Question 1. Only respondents who indicated feeling at least some degree of hopefulness received this question, $n = 121$.
3. "What do you think could make you feel more hopeful about the marine environment?" This was an open-ended question for all respondents, $n = 122$.

We included these three questions in order to quantify student teachers' baseline levels of hope regarding the ocean, identify the sources they themselves considered to underpin existing hope, and elicit factors that could increase hope.

5.2.3 Workshops

Two stakeholder workshops were held to gather insights from the stakeholder groups listed in Table 5.1. The invitations to the workshops were sent by email, and participation was voluntary. In total, 33 individuals participated in the workshops.

Each workshop started with a presentation of the project, including its aims, the theoretical perspectives obtained from the review of recent research literature and preliminary results of the student survey. After the presentation, we asked the participants to informally discuss their views and thoughts about the project and share examples of their own practice. One of us acted as a moderator of the discussion, introducing topics, facilitating interchange, and encouraging all the participants to actively contribute. The other made written notes of what was said, but to ensure anonymity, we excluded any information that could reveal the identity of the participants. The presentation lasted approximately 20 minutes, and the discussion afterward lasted 30–40 minutes. Participants were also invited to speak with us after the workshop if they wished to share further views or opinions.

5.2.4 Analyses

We analyzed the answers to the open-ended survey questions as well as the notes from the workshops by qualitative content analysis using an inductive approach (Elo and Kyngäs 2008). The analysis included open coding, creation of categories, and abstraction. To start with, each of us separately read all survey responses and workshop notes to obtain a sense of the whole, and then worked through them, independently coding all aspects of the content relevant for the study. Each of us independently grouped the initial coding into categories that could be useful when describing the participants' views and perspectives on what may foster hope in education about the oceans. We then compared, discussed, and negotiated the suggested categories to produce a set of categories that we agreed on.

One set of categories was produced for the survey results and separate sets for each workshop (Table 5.2). Finally, similar categories were combined to establish final higher order categories covering the entire data material. The process involved numerous stages of review of the material and discussions before the final set was established. We selected utterances from the workshops and quotations from the survey as descriptive examples for the categories. The presence of both of us at all workshops strengthened the reliability of the analysis of the workshop notes, as we could both agree that the final categories accurately reflected the participants' intended meanings and impressions. The frequency distribution of responses to the single closed-ended question was calculated and plotted as a bar chart (Fig. 5.2).

Table 5.2 Categories established in the analysis of the survey and the workshops. The different shades of grey illustrate which categories were combined to form the final categories encompassing the entire data material. Categories without color were kept as standalone categories-

Student teacher survey	Workshop with lecturers and researchers	Workshop with guides, scientists, and teachers
1. New knowledge, new solutions and increased awareness of ocean challenges	1. Knowledge about the ocean and existing solutions	1. Knowledge about the local ocean environment
2. Collective engagement for the ocean	2. Local agency for the ocean	2. Local engagement for local marine environments
3. More policy action for the ocean	3. Personally meaningful ocean issues	3. Direct experiences with ocean life
4. Experiencing positive change regarding the oceans	4. Focus on solutions	4. Awe of and curiosity about oceans and ocean life

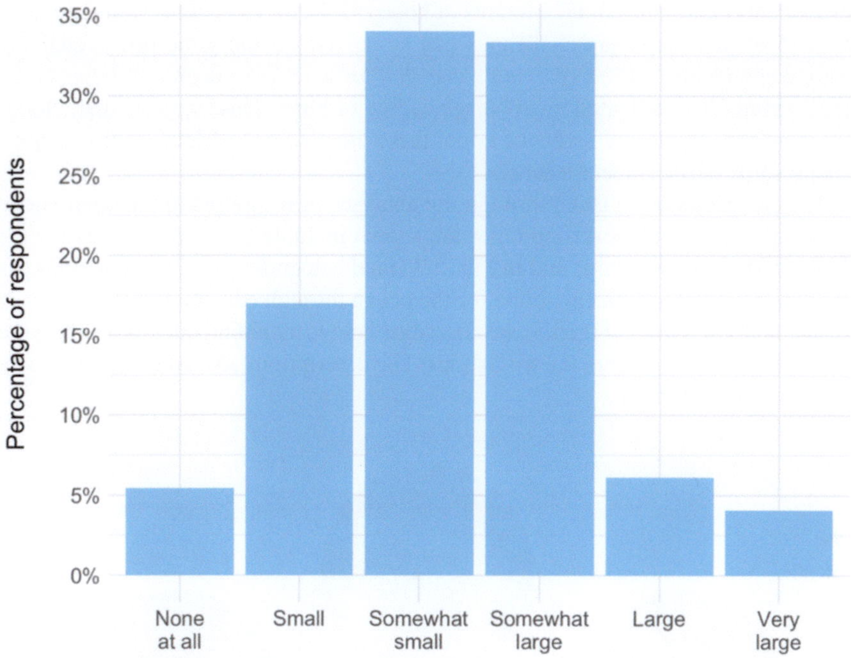

Fig. 5.2 A bar chart showing the frequency distribution of answers to the question "To what extent do you feel hopeful that the environmental challenges facing the ocean can be solved?", $n = 147$

5.3 Findings and Discussion

The results of this study are a synthesis of input from recent research literature, the results of the analyses of the student teacher survey, and the results of the analyses of the two stakeholder workshops. In addition, a practical tool for educators is presented, which was developed during the study and based on the interactions with the stakeholders, the results of the analyses, and our own expertise based on many years of experience as science teachers at schools and as lecturers and educational researchers at teacher education institutions.

5.3.1 Survey and Workshops

Figure 5.2 shows the distribution of responses to the question "To what extent do you feel hopeful that the environmental challenges facing the ocean can be solved?" Over half (56%) of the student teachers responded that they feel either no hope or only a *small* to *somewhat small* degree of hope, while less than half (44%) reported feeling a *somewhat large* to *very large* degree of hope. In other words, the distribution is skewed toward the lower end of the scale, with a greater proportion of respondents expressing small amounts of hope (51%), or, most concerning, no hope at all (5%) that the ocean challenges can be solved. At the same time, relatively many respondents (40%) expressed *somewhat large* or *large* degree of hope, and a small group (4%) indicated a *very large* degree of hope. This suggests that, alongside concerns about low levels of hope, there are also encouraging signs of hope among some of the student teachers.

The categories established through the analyses of responses to the open-ended survey questions and workshop notes are shown in Table 5.2 (separate sets for the survey and each workshop) and in Fig. 5.3 (final categories), and present elements that can be useful in fostering hope for the ocean through education.

The final categories (Fig. 5.3) are described below, including quotations from the survey and utterances from the workshops. The descriptions also address how recent literature supports them.

Fig. 5.3 An illustration of the final categories established through the analyses of the stakeholder survey and workshops

5.3.1.1 New Knowledge About the Ocean and Solutions to Its Challenges

All stakeholder groups emphasized the importance of both existing and emerging knowledge in fostering hope, as new knowledge expands understanding of the oceans as well as the causes and solutions to its challenges—in other words, highlighting the potential for progress through innovation and learning. This included experiencing increased awareness of ocean issues among the public. While the participants of both workshops pointed out that gaining more understanding about the ocean and existing solutions may foster hope, many of the student teachers particularly highlighted how increasing awareness, scientific progress, and new technology give them hope. The following quotes from the survey illustrate such views:

> Generations now and in the future are increasingly aware of how pollution in the ocean affects marine wildlife and the ocean environment and want to take care of it.
>
> We constantly gain new knowledge that can help us find alternatives or solutions to the problem.
>
> The little hope I have is basically that technology keeps getting better and better.

This category resonates well with similar findings in the research literature. Chawla (2020) emphasizes the need for young people to understand both the physical and social causes of environmental change, to be able to identify effective solutions, and to know what they can do to address these issues. In a review of ocean conservation successes, Knowlton (2021) highlights growing evidence of progress in marine conservation and stresses the importance of sharing knowledge about these solutions and successes, both to encourage broader application of their lessons and to cultivate hope through tangible, evidence-based examples. Similarly, Stoknes (2014) argues that hope and inspiration can be strengthened by an increased emphasis on stories of progress, such as scientific discoveries about ecological systems, opportunities emerging from smarter relationships between economic production and nature, and rapid shifts toward a greener economy in technology, business, and lifestyle.

5.3.1.2 Collective Engagement and Local Agency for the Ocean

All stakeholder groups also regarded collective efforts and shared responsibility as important in fostering hope, since experiencing, seeing, or hearing about many people working together may overcome challenges too great for any one individual to solve alone. The participants of the workshops specifically underscored the value of collective engagement for local marine environments, which brings large, global issues into more relatable and relevant contexts and makes them feel more manageable and their solutions more achievable. Several of the student teachers also highlighted that they feel some hope since they experience others being engaged in ocean issues, while others noted that they would feel more hopeful if they saw greater collective engagement, as illustrated by the following quotations:

> Collaboration between many people can give me some hope about how we will solve marine environmental problems.
>
> If more people contribute and show their engagement.
>
> I do have a hope that we will eventually wake up and come together to address this issue.

Collective engagement and local agency are suggested as fostering hope also in recent literature. Ojala et al. (2021) highlight that building social trust and a sense that the challenges are at least somewhat controllable is important for fostering constructive forms of hope. According to them, this may be promoted by creating opportunities to talk about emotions connected to environmental change with others, participating in intergenerational community efforts and engaging collectively through organizations. Likewise, Chawla (2020) stresses the importance of enabling young people experience that they are not alone but allied with others who are working to protect nature. Kelsey (2020) points out that hope emerges through interactions with those around us and emphasizes the importance of recognizing that, while participation in collective action fuels feelings of hope, hope also plays a key role in mobilizing individuals to take part in such action. Ojala (2012a) offers a similar perspective, on hope being evoked by the actions themselves, suggesting that hope then becomes embodied.

5.3.1.3 Personally Meaningful Ocean Issues

The stakeholders participating in the workshops highlighted that finding ocean issues personally meaningful plays a significant role in fostering hope for the oceans. When people can, either individually or collectively, engage with ocean environmental issues that matter to them, including simple, impactful choices in their daily lives, the challenges feel more personal and possible to tackle. The guides and marine scientists specifically highlighted that it was important that people have personal experiences with ocean life to help them understand the ocean ecosystems and feel that they can be a part in solving its challenges. Similarly, several of the teacher education lecturers emphasized the importance of firsthand experiences, such as visiting local ocean areas or recalling childhood memories of specific places, as powerful ways to find personal meaning in addressing ocean issues.

Ojala (2012a) argues that trust in one's own ability to make a difference highlights the importance of personal actions for fostering hope. She emphasizes the importance of encouraging a sense of agency by reinforcing the idea that everyone's effort is worth something and helping especially the young to find satisfaction also in small successes. Likewise, Chawla (2020) highlights the importance of providing young people with opportunities to experience agency by focusing on environmental problems that concern them and to engage them to determine actions and ideas that are personally meaningful for them and which they can accomplish individually or in partnership with others.

5.3.1.4 More Policy Action for the Ocean

Many of the student teachers expressed that seeing more engagement and action from policymakers would bring them more hope. This included the establishment of more regulations to protect the ocean, increased financial support to implement measures, greater international cooperation, and more requirements for large companies to take responsibility for the well-being of the ocean. It was evident that the student teachers wished to see greater action taken by politicians and others responsible for policy. This is illustrated by the following quotes:

> That concrete measures, goals, and regulations are established which everyone must follow.

> It has to be something big, like all the world's leaders introducing laws and sanctions that can prevent not only individuals but also companies from harming the marine environment.

> The only way this can be slowed down is through boycotts, but more importantly, through heavy fines and restrictions. This is an issue that politicians MUST take seriously.

The importance of policy action and engagement of various societal actors is also highlighted in research literature. Ojala (2023) regards trust in the idea that societal actors at least try to do their part regarding environmental challenges as an important source of hope. Examples of societal actors that have worked for the environment for a long time and shown persistence in the face of challenge, may be a way to build trust (Ojala et al. 2021). Ojala and co-workers (2021) also discuss how information regarding policy support and putting demands on politicians and other more powerful actors may promote constructive coping with environmental challenges.

5.3.1.5 Experiencing Positive Change Regarding the Oceans

The teacher education lecturers and educational researchers highlighted the importance of solutions orientation in fostering hope. Providing concrete evidence of solutions and progress demonstrates that achievable actions exist. One of the educational researchers emphasized the importance of acknowledging ocean challenges and their complexity, but at the same time seeing that things can improve. A few of the student teachers replied that they are hopeful, because they have heard about, read about, or otherwise experienced some progress regarding ocean challenges. However, most of them highlighted the importance of receiving more information about ongoing successful efforts and changes toward the positive in order to feel more hopeful. The following quotes from the survey illustrate these views:

> [I feel hopeful] because many people are talking about it, and I occasionally read or hear about measures being taken to improve the condition of the ocean.

> I would feel more hopeful if I saw more tangible results from global agreements and actions, such as less plastic in the ocean and clear improvements in the health of important marine ecosystems.

> Maybe hearing about important steps being taken and decided on to improve the marine environment. Right now, all we hear about are the negative things happening in the ocean.

As presented in the introduction, recent literature highlights the importance of a solutions-oriented focus in ocean narratives. By providing examples of how problems can be solved and what actions can be taken, such narratives can foster evidence-based hope and encourage action to amplify the impact of these solutions (Kelsey 2020; Kelsey and Broadhurst 2024; Knowlton 2021). Ojala (2012a, 2023) emphasizes the importance of positive reappraisal, that is, reframing environmental challenges in ways that highlight opportunities for improvement. It aims to find positive meaning in addressing the challenges by emphasizing how responding to them can lead to positive outcomes and seeing positive possibilities in the changes societies need to make.

5.3.1.6 Awe and Curiosity About the Oceans

According to the guides, marine scientists, and teachers, fostering awe and curiosity toward ocean life is an important way of inspiring hope. Awe and curiosity create emotional connections to ocean life and may reframe challenges as opportunities to care for the ocean. This may include appreciating the strength and adaptability of marine ecosystems. Awe and curiosity were often linked to personal encounters with marine life and the positive feelings they evoke, or to new research and knowledge that awaken curiosity and hope. One of the guides specifically pointed out that when people get to experience special places and have encounters with fascinating marine life, the experience deepens their understanding, motivates them to conserve it, and fosters hope.

As one strategy to help young people cope with environmental change and foster hope, Chawla (2020) encourages building connections with nature. According to her, spending time in natural environments helps individuals become comfortable and competent in nature, while also developing a sense of agency and a feeling of kinship with other living beings. Kelsey (2020) highlights the remarkable capacity of various species to adapt and nature's capacity to recover, as a source of hope. She stresses that this does not diminish human responsibility to care and act, but rather serves as inspiration and a reminder that humans are not alone in responding to change but are part of a reciprocal relationship with the rest of the living world.

5.3.1.7 Visioning Positive Outcomes Through Arts

The teacher educators and educational researchers emphasized the importance of envisioning positive outcomes, particularly through arts, to foster hope for the ocean. Future visions can offer an optimistic picture of what is possible, that a more sustainable future is achievable, and help point the way forward. They may also reinforce the belief in a better world, not only for humans but also for other creatures. One of the educational researchers specifically highlighted fictional stories as a way to engage in envisioning and inspire hope, pointing out that also fiction is, to a certain extent, real, because fiction reflects the world people live in and because fictional characters exist independently in the world.

Stoknes (2014) highlights how people create meaning through stories and narratives, stressing the need for more stories that help them image a renewal of nature, including stories about people who care and act based on their vision. These stories go beyond simply stopping destruction; they describe a richer, regenerated nature that one can look forward to living in. As part of fostering hope, Chawla (2020) suggests engaging young people, through experiential, inquiry-based, and arts-based methods, in envisioning futures they wish to see and identifying steps to move in the desired direction. Likewise, Ojala (2012a) proposes using alternative stories about the future in art and literature as a foundation for critical and creative discussions.

5.3.2 A Proposed Educational Tool

To support an application of the results presented above, an educational tool, aimed at shifting the portrayal of ocean environmental challenges toward hope-inspiring narratives, was developed in the study. The rationale behind the proposed educational tool builds on the relationship between stories of improvement and recovery, evidence-based hope, and environmental engagement and action (Fig. 5.4) as derived from recent research literature reviewed as a part of this study. More specifically, the educational tool aims to contribute to a shift in ocean narratives from being dominated by stories of disaster and loss to greater focus on stories of improvement resulting from environmental action and, through this, to fuel

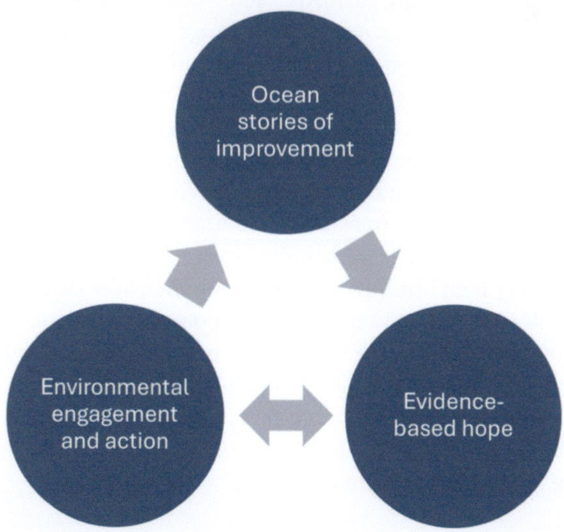

Fig. 5.4 An illustration of the connection between ocean stories, evidence-based hope, and environmental engagement

Fig. 5.5 A schematic flow chart illustration of the proposed educational tool

evidence-based hope. Since feelings of hope can motivate action—particularly when hope is connected to positive reappraisal of problems and finding positive meaning in addressing them—cultivating evidence-based hope is expected to enhance both individual and collective engagement in marine protection and conservation. Increased engagement, in turn, may foster both hope and may generate more stories of improvement and recovery of marine environments.

Applying the proposed tool in educational settings (Fig. 5.5) translates to inviting learners to identify examples of ocean recovery, explore these to develop the necessary understanding of the challenges and solutions involved, create and construct stories of hope, and as a final step, share their stories.

The proposed tool is guided by the seven categories identified through the survey and workshop analyses: *New knowledge and solutions*, *Collective engagement and local agency*, *Personally meaningful issues*, *More policy action*, *Experiencing positive change*, *Awe and curiosity,* and *Visioning positive outcomes* (Fig. 5.3). For instance, the first step of identifying and exploring examples of ocean recovery highlights categories such as *New knowledge and solutions*, *More policy action*, and *Experiencing positive change*, while the process of creating stories from these examples allows for the integration of categories such as *Personally meaningful issues*, *Awe and curiosity*, and *Visioning positive outcomes.* Categories such as *Collective engagement and local agency* inform all three steps. Examples of recovery may be local and stem from collective action, while highlighting local issues and demonstrating that many people make an effort to resolve them can make stories highly engaging. Finally, sharing their stories may help learners feel that they are a part of the collective action, and contribute to raising awareness of the work accomplished and the results achieved.

The tool thus translates the categories from the analyses into practical pedagogical moves, while remaining open to adaptation in various educational settings. Rather than prioritizing some categories over others, the tool is intended to provide a flexible structure that can incorporate all of them, depending on the context. Each of these steps can be shaped according to the local setting, the time available and the age and educational level of the learners involved. It is important to view this proposed tool as open to interpretation, allowing stakeholders to tailor it to their specific contexts. In the following sections, each of the three proposed steps are briefly explored and explained.

5.3.2.1 Identify and Explore Examples of Ocean Improvement and Recovery

Initiating the process of developing stories about ocean improvement and recovery can take many forms and shapes. The examples can span widely, ranging from small individual actions and daily engagement to larger collective efforts over time. They can include examples of recovery, things that are improving, or actions currently being taken that are expected to lead to improvement. These examples might involve improvement of specific marine areas, habitats, or species, or, for instance, a positive shift in people's mindsets and behavior. They can also be either global, regional, or, most preferably local. What unites them is that they all represent initiatives contributing to meaningful change, that is, steps in the positive direction needed to be recognized and built upon.

To be able to construct stories from examples of ocean recovery and improvement, it is essential for the learners to explore the human–ocean interactions behind the improvement, that is, both the cause of the challenge and how the improvement has come about. Such understandings generally require a multidisciplinary approach, encompassing the development of ocean literacy to understand the fundamentals of the ecosystems involved, social and economic information, and the diverse viewpoints and interests of various stakeholders in human–ocean interactions. Identifying different perspectives and scales of the interactions involves collecting information on how humans use and impact the ocean, such as through the fishing industry, ocean transportation, and recreation. The social and economic sustainability dimensions can be considered by examining how communities depend on the ocean for livelihoods and cultural practices. This may include understanding regulatory and policy frameworks that govern these interactions and assessing their effectiveness.

Exploring the differing viewpoints of various stakeholders is useful, as it provides a comprehensive understanding of human–ocean interactions. Stakeholders from organizations and corporations such as schools, fishing industries, government agencies, NGOs, and businesses, have unique interests, value sets, and perspectives, and consequently their own understanding and engagement with the issue. Engaging with the narratives of such groups helps learners identify possible conflicts, collaborative opportunities, and solutions that are sustainable, equitable, and effective.

5.3.2.2 Create and Construct Stories of Hope

In creating and constructing stories of hope based on examples of ocean improvement and recovery, the categories identified through the analyses (Fig. 5.3) can be of guidance. Combinations of knowledge, personally meaningful perspectives, and examples of positive action may help shift the focus toward possibility. Stories that revolve around examples showing that there are solutions, progress is happening, and that engagement at all levels plays a role in creating a more sustainable future for the oceans, both directly and indirectly, can be effective in fostering hope.

Including elements such as emerging knowledge about the ocean and solutions to its challenges, local engagement, or examples of individual, collective, or policy action in the stories, may help learners see that change is possible and that solutions are already being implemented. At the same time, stories that evoke awe, curiosity, or are personally meaningful may encourage emotional connection and a sense of relevance. For instance, a story about a recovery of a local marine species might be even more impactful if it also reflects personal or community involvement in restoration efforts.

Envisioning positive futures through discussion, creative writing, and various arts-based methods may help students process their emotions while imagining constructive ways forward. Such visions can provide space to explore values and to reflect on the role individuals and communities might play in shaping outcomes. It is important to note that depending on the specific human–ocean interaction, the stories can vary greatly in format or genre and can be presented in a multimodal manner, including arts-based methods of storytelling. Who is telling the story, and who it is being told to, will also shape how the story should be constructed.

5.3.2.3 Sharing Ocean Stories of Hope

The final step in the proposed educational tool involves sharing the stories. Sharing helps make the learning process more meaningful for the learners themselves. Knowing their work will be seen and heard by others can increase motivation, deepen engagement, and enhance learners' sense of agency and ownership.

Beyond personal meaning, shared stories can inspire wider audiences. If shared with classmates, families, community members, or local policymakers, they can raise local awareness of ocean recovery efforts and foster a shared sense of purpose. This connection to real-world audiences helps position learners as active contributors to broader conversations around marine conservation.

Importantly, story-sharing can also open space for dialogue and feedback. Inviting others to listen, respond, and reflect through classroom discussions, public events, or digital platforms creates opportunities for meaningful exchange. Engaging with stakeholders who share a commitment to ocean well-being can amplify the impact of the stories and bring new perspectives into the learning process. Furthermore, promoting the stories through partnerships with other stakeholder groups can extend their reach and invite new audiences into the conversation.

Sharing of the stories can take many forms, including oral storytelling, written narratives, posters, performances, and other forms of visual and creative arts. Digital platforms also offer powerful avenues for sharing, provided that ethical guidelines for digital sharing are carefully considered. Ultimately, the format should reflect the interests, resources, and creativity of the learners and educators involved. The emphasis should not lie on the medium itself, but on the opportunity for learners to meaningfully express what they have discovered and to connect with others through their stories.

5.4 Concluding Remarks

A significant finding of this study is the identification of seven distinct categories that illustrate how hope for the ocean may be fostered within educational settings. These categories were established through analyses based on a combination of literature review, survey responses, and workshop discussions with various groups of stakeholders, and offer multiple ways into hope-oriented ocean narratives. They reflect a range of experiences and perspectives, including the importance of new knowledge and solutions, policy action, collective engagement, personally meaningful connections to the ocean, and the emotional resonance of awe, curiosity, and creative envisioning.

While there were some differences in emphasis among the stakeholder groups, the consistency of several of the categories across the data sources is worth noting. This can be interpreted to signify that, despite variation in background and perspective of the stakeholders, these results resonate broadly and may offer common ground for educational practice. They may offer educators a foundation for fostering evidence-based hope in response to the emotional complexity that often surrounds environmental education. This need is underscored by the survey results, which showed a skew toward lower levels of hope among the student teachers. Importantly, participants in the workshops emphasized that evidence-based hope can be fostered without downplaying the seriousness of ocean challenges. They viewed this approach as both meaningful and useful for advancing ocean literacy. The findings suggest that cultivating hope involves reframing complex issues in ways that support agency, relevance, and constructive engagement.

The educational tool developed in this study offers a structured yet adaptable approach to applying these insights in practice. That is, it is intended to support educators and learners in shifting the narrative focus from degradation and loss to stories of improvement and possibility. The tool encourages the identification and exploration of real-world examples of positive change, the creation of stories grounded in these examples, and the sharing of these stories of hope with relevant audiences in meaningful manners. This process may offer a way to approach ocean education that acknowledges complexity while also creating space for constructive hope and engagement.

There is a need for further exploration of how the results of the study, including the proposed educational tool, can support the cultivation of evidence-based hope in diverse educational settings. Future research should investigate how the tool can be adapted across age groups and diverse educational settings, and the various ways it may help learners recognize signs of progress, understand the pathways to positive change, and contribute to broader conversations around ocean protection and conservation. The authors of this chapter have plans to test the tool in practice, which will provide better insight into the issues and offer opportunities to refine the tool. Such testing may include classroom interventions in schools and teacher education, as well as pilot use in nonformal contexts such as museums. The effects on learners' hope and engagement can be assessed through a combination of quantitative and qualitative methods, including questionnaires and interviews.

This study contributes to ongoing discussions about how education, particularly in science and sustainability, can address the emotional and ethical dimensions of environmental problems. By offering a practical tool grounded in research and developed in close cooperation with various stakeholders, the study aims to support educational approaches that help learners engage with complex environmental issues in ways that both promote understanding of ocean challenges, and foster evidence-based hope and environmental engagement.

References

Asikin N, Suwono H, Sumitro SB, Dharmawan A (2025) Teaching Ocean literacy in science education: a systematic review. Environ Educ Res:1–26. https://doi.org/10.1080/13504622.2025.2490227

Burke SEL, Sanson AV, Van Hoorn J (2018) The psychological effects of climate change on children. Curr Psychiatry Rep 20:35. https://doi.org/10.1007/s11920-018-0896-9

Chawla L (2020) Childhood nature connection and constructive hope: a review of research on connecting with nature and coping with environmental loss. People Nat 2(3):619–642. https://doi.org/10.1002/pan3.10128

Elo S, Kyngäs H (2008) The qualitative content analysis process. J Adv Nurs 62(1):107–115. https://doi.org/10.1111/j.1365-2648.2007.04569.x

Hicks D (2014) Educating for hope in troubled times: climate change and the transition to a post-carbon future. Institute of Education Press

Kelsey E (2020) Hope matters: why changing the way we think is critical to solving the environmental crisis. Greystone Books Ltd

Kelsey E, Broadhurst G (2024) Why evidence-based hope is crucial to Salish Sea recovery and beyond. In: Elliser CR (ed) Emerging issues in the Salish Sea (issue 5). Salish Sea Institute, Western Washington University

Knowlton N (2021) Ocean optimism: moving beyond the obituaries in marine conservation. Annu Rev Mar Sci 13(1):479–499

Mokos M, Realdon G, Zubak Čižmek I (2020) How to increase ocean literacy for future ocean sustainability? The influence of non-formal marine science education. Sustainability 12(24):10647. https://doi.org/10.3390/su122410647

Ojala M (2012a) Hope and climate change: the importance of hope for environmental engagement among young people. Environ Educ Res 18(5):625–642. https://doi.org/10.1080/13504622.2011.637157

Ojala M (2012b) How do children cope with global climate change? Coping strategies, engagement, and well-being. J Environ Psychol 32(3):225–233. https://doi.org/10.1016/j.jenvp.2012.02.004

Ojala M (2013) Coping with climate change among adolescents: implications for subjective well-being and environmental engagement. Sustainability 5(5):2191–2209. https://doi.org/10.3390/su5052191

Ojala M (2015) Hope in the face of climate change: associations with environmental engagement and student perceptions of teachers' emotion communication style and future orientation. J Environ Educ 46(3):133–148. https://doi.org/10.1080/00958964.2015.1021662

Ojala M (2023) Hope and climate-change engagement from a psychological perspective. Curr Opin Psychol 49:101514. https://doi.org/10.1016/j.copsyc.2022.101514

Ojala M, Cunsolo A, Ogunbode CA, Middleton J (2021) Anxiety, worry, and grief in a time of environmental and climate crisis: a narrative review. Annu Rev Environ Resour 46(1):35–58. https://doi.org/10.1146/annurev-environ-012220-022716

Plomp T (2013) Educational design research: an introduction. In: Plomp T, Nieveen N (eds) Educational design research Part A: an introduction. SLO, pp 10–51

Ramboll (n.d.) SurveyXact. Ramboll. https://www.ramboll.com

Shellock RJ, Fullbrook L, McKinley E, Beaumont N (2024) The nature and use of ocean literacy in achieving sustainable ocean futures: a systematic map. Ocean Coastal Manag 248:107373. https://doi.org/10.1016/j.ocecoaman.2024.107325

Stevenson RB (2007) Schooling and environmental education: contradictions in purpose and practice. Environ Educ Res 13(2):139–153. https://doi.org/10.1080/13504620701295726

Stoknes PE (2014) Rethinking climate communications and the "psychological climate paradox". Energy Res Soc Sci 1:161–170. https://doi.org/10.1016/j.erss.2014.03.007

Stoll-Kleemann S (2019) Feasible options for behavior change toward more effective ocean literacy: a systematic review. Front Mar Sci 6:273. https://doi.org/10.3389/fmars.2019.00273

Wals AEJ (2010) Between knowing what is right and knowing that it is wrong to tell others what is right: on relativism, uncertainty and democracy in environmental and sustainability education. Environ Educ Res 16(1):143–151. https://doi.org/10.1080/13504620903504099

Tuula H. Sarvas Skarstein is a Professor of science education in the Department of Early Childhood Education at the University of Stavanger (UiS). She also leads the UiS Sustainability Hub, an interdisciplinary learning lab at the Faculty of Arts and Education, and is affiliated with Filiorum—Centre for Research in Early Childhood Education and Care. Her research background includes educational sciences and marine biology. Her current research focuses on early science education in outdoor settings, ocean literacy, and evidence-based hope in sustainability education. She has published her work in several national and international scientific journals.

Frode Skarstein is an Associate Professor of science education at the University of Stavanger, Norway. With a strong background in evolutionary biology and formal training as a science teacher, he serves as both a lecturer and a researcher in teacher education. His research primarily focuses on sustainability education and its integration with science education within teacher and higher education contexts. He has a particular interest in the perceptions and utilizations of science and scientific knowledge in climate and sustainability education, as well as in public discourse. He has published in both natural and educational science journals, and he is also the author of several monographs.

Part II
Interdisciplinary Pathways

Chapter 6
Education for Sustainable Development (ESD) in Teacher Education: Pre-Service Teachers' Perspectives on Interdisciplinary Learning

Anne Bergliot Øyehaug, Jill Tove Buseth, and Robert J. Didham

Abstract Teachers' knowledge about *sustainable development* (SD) and their use of teaching methods in this theme plays a crucial role in *education for sustainable development* (ESD). In this study, our aim was to explore pre-service teachers' pedagogical capacity in ESD, and we interviewed 12 pre-service teachers specialising in science and social studies from one Norwegian teacher education institution. When interpreting the data within an interdisciplinary framework of teachers' pedagogical capacity, two main themes emerged: (1) interdisciplinary and disciplinary content knowledge about sustainable development (SD), and (2) interdisciplinary pedagogical knowledge in ESD. The informants referred to several relevant perspectives from the content of SD, and especially pre-service teachers specialising in social studies showed great insight into the complexity of SD perspectives. Moreover, the pre-service teachers suggested several similar general teaching approaches in ESD, like engaging students in authentic issues and familiarising themselves with special interest groups. Especially the pre-service teachers specialising in science also reflected about the connections between multiple subject perspectives in SD and how to use these connections to develop ESD learning activities.

Keywords Education for Sustainable Development in teacher education · Pre-service teachers' perspectives · Interdisciplinary learning · Interdisciplinary pedagogical content knowledge (IPCK) · Sustainability teaching approaches

---

A. B. Øyehaug (✉) · J. T. Buseth · R. J. Didham
University of Inland Norway, Faculty of Education, Hamar, Norway
e-mail: anne.oyehaug@inn.no

F. Skarstein, L.-A. Wolff (eds.), *Sustainability and Science Education*, Environmental Challenges and Solutions,
https://doi.org/10.1007/978-3-032-24408-6_6

6.1 Introduction

Pre-service teachers will be educating future citizens of the world, and therefore, they play a significant role in *education for sustainable development* (ESD) (Jegstad et al. 2018). However, the complexity of *sustainable development* (SD) as a concept can leave teachers feeling overwhelmed when trying to work with ESD and linking the concept to educational learning outcomes (Kioupi and Voulvoulis 2019). The critical importance of the role of education in helping to achieve SD has been well acknowledged in recent years, especially through the United Nation's decade of ESD 2005–2014 (UNESCO 2014), and thereafter through the international agreement in 2015 of the 17 Sustainable Development Goals (SDGs) (United Nations 2015). The SDGs emphasise the importance of quality education for all enshrined in SDG 4, the recognition of education as a key means of implementation for the achievement of all SDGs, and the call for mainstreaming ESD specifically identified in SDG target 4.7.

To be able to realise ESD, teachers need both relevant SD-knowledge about subject matter (for instance, knowledge about global warming in science or indigenous people in social studies) and knowledge in pedagogy. Over the past three decades, education and learning for sustainable development have achieved significant progress through the further development of concepts and pedagogical approaches, as well as through increased integration in curriculums, implementation in both formal and non-formal education sectors, and inclusion in teacher training. One viewpoint that is widely accepted is that ESD can be considered as a core approach for enhancing the overall quality and relevance of education (e.g. Didham and Ofei-Manu 2018). Based on a study of the impacts of ESD in 18 countries, which actively implemented ESD during the UN decade of ESD, Laurie et al. (2016) report that the pedagogical approaches promoted by ESD were identified as having a stronger transformative impact on primary and secondary education in most countries than the inclusion of specific sustainability content.

In ESD, special attention must be paid to teacher education, as it promises to have multiplier effects that will eventually reach all levels of education. Multiple studies have shown that in both primary and secondary education, as well as teacher education, the most common approach for integrating ESD into education is by working within existing structures and including ESD as topic in traditional subjects (Bagoly-Simó 2013; Didham and Ofei-Manu 2012; Qablan 2018). Based on a systematic literature review on ESD in teacher education, it is reported that ESD is most implemented in the subjects of geography, social studies, and science within teacher education programmes, and that teacher educators regularly state that they find it difficult to make clear connections between their subjects and ESD (Goller and Rieckmann 2022). There are numerous case examples of more integrated approaches to implementing ESD in various educational contexts, and often these are resulting in stronger transformations of learning environments and processes. For instance, transformation can happen through cross-curricular alignment, interdisciplinary projects, and whole school approaches (O'Donoghue et al. 2018), but they are less common in teacher education and far from being mainstreamed in practice.

Recent changes in Norwegian educational policy have created both an increased need and an opportunity to work with both sustainable development and interdisciplinarity within teacher education programmes. The national framework plans on teacher education in Norway have included sustainable development as a topic for teacher education since 2010, first being included as a targeted skill but later being referred to as a general competence (Jónsson et al. 2021). These plans also give guidance that teacher education must respond to the contents of the school curriculum, and in 2020 a curriculum reform was implemented in Norway that lifted SD up to an overarching and interdisciplinary theme. However, there has been little cooperation across subjects in teacher education, and there is a need for restructuring and change of work habits in terms of how interdisciplinary teaching is integrated into teacher education (Biseth et al. 2022).

The study presented here has been collected as part of the BRIDGES (*Bridging the gaps in teacher education and schools through interdisciplinary work*). BRIDGES is a research project in which three well-established Norwegian teacher education institutions work in variegated ways with developing, strengthening, and systematising interdisciplinary work within their teacher education programmes and in collaboration with partner schools. In this chapter, we report from a small scoping study investigating how pre-service teachers reflect on ESD teaching approaches at one teacher education institution. The study is based on qualitative interviews with pre-service teachers specialising in science and social studies. The purpose of the study is to explore the students' pedagogical capacity in ESD, and we aim to address the following two research questions:

- What disciplinary (science and social studies) and interdisciplinary content knowledge about SD do pre-service teachers refer to in interviews?
- What do these pre-service teachers suggest as suitable teaching approaches in ESD?

6.2 Theoretical Framework

The theoretical framework we have used to explore the pedagogical capacity in ESD consists of two main parts. The first part concerns the disciplinary and interdisciplinary content of SD, and the second part concerns pedagogical knowledge in ESD, both general teaching approaches and interdisciplinary teaching approaches. These two main perspectives are then integrated in a model describing teachers' competence by emphasising the importance of both content knowledge and pedagogical knowledge.

6.2.1 *Interdisciplinary and Disciplinary Content Knowledge in the SD Theme*

Sustainable development (SD) is a term that is usually defined as an interaction between ecological, economic, and social conditions (Giddings et al. 2002). These perspectives are also expressed through Agenda 2030 and its SDGs, which address five central aspects (i.e. people, planet, prosperity, peace, and partnership) (United Nations 2015). The Sustainable Development Goals can be seen as a universal call to action to eradicate poverty, protect the planet and ensure peace, prosperity, and well-being for all by 2030. The goals are integrated, and the idea is that action in one area will affect the results in the others, and that development must balance social, economic, and environmental sustainability. Systems thinking is also actively used as a central perspective in SD to examine the interlinked environmental, social, and economic dimensions of sustainable development (Sinakou et al. 2019a). Sachs (2012) argues that the far-reaching goals of sustainable development covering these dimensions represents a broad-reaching, global consensus on the hopes for a good society. Within the 2030 Agenda for Sustainable Development, this holistic perspective is further lifted by highlighting it as an agenda for people, planet, and prosperity which includes peace and partnership as its building blocks, and thus underscoring the strong interlinkages between these dimensions (McKenzie and Abdulkadri 2018). Furthermore, the role of scientific research from different disciplines on pathways to achieve sustainable development has long been a concern of governments and public sectors (Archibald et al. 2016), because findings from research projects related to sustainability issues often lead to debates on how to evaluate and make use of such findings.

Sustainability issues, such as the opposing goals of economic growth and environmental sustainability, are often complex and require examination from multiple perspectives to be fully comprehended. A pluralistic approach aims to utilise diverse perspectives to examine issues related to SD in society, to approach these topics with knowledge from various disciplines, and to incorporate a variety of viewpoints and reflect on the positions of multiple stakeholders (e.g. Rudsberg and Öhman 2018). In addition, SD addresses content knowledge aimed at empowering sustainability action, which includes content about civic participation, political advocacy, and the precautionary principle, as shown in the following quote:

> The concept of citizen participation is unquestionably linked to the concept of sustainable development. Promoting equitable, inclusive, and fair societies involves generating processes for citizen participation that imply shared decision-making and an assumption of responsibilities in favour of global sustainability. (Poza-Vilches et al. 2022, p. 1019)

Along with the aspirations for civic engagement and decision-making, SD also advocates principles for careful consideration and planning of development actions to avoid negative impacts. The 'precautionary principle' emerged as a part of environmental policy in the mid-1970s, and its position in sustainable development was solidified by its inclusion as a principle in the outcomes of the Rio Earth Summit in 1992, that is, Principle 15 of the *Rio Declaration on Environment and Development*

(O'Riordan and Cameron 2013). '[T]he precautionary principle imposes a duty of planetary care on the human will. It seeks to deepen, to widen, and to lengthen all manner of so-called impact assessments. Above all, it provides a basis for public discussion and deliberation over what kind of society and moral accountability we collectively should choose to adopt' (Read and O'Riordan 2017, p. 5). This, however, means that SD itself is not neutral; it contains a particular policy commitment that carries with it ethical and political dimensions that a teacher must grapple with and may be challenged by (Winter 2007).

Content from the disciplines social studies and science is very relevant for the perspectives of SD. For instance, a study of social studies learning designs in an ESD-based learning programme showed that it fulfilled concepts such as human dignity, responsibility for future generations, nature–society relations, social and economic dimensions of SD, justice, and cultural diversity (Nasution et al. 2023). Sinakou et al. (2019a) suggest that the goal of ESD in the environmental dimension (thus mostly in science) aims to raise awareness of the physical environment's resources and fragility, the influence of human activities on the environment, climate change, environmental protection, and biodiversity. Moreover, Cebrián and Junyent (2015) provide evidence that pre-service teachers would prioritise acquisition of knowledge and practical skills related to nature and sciences in a school project related to ESD.

6.2.2 Pedagogical Knowledge in ESD

ESD is noted for actively promoting pluralistic and holistic approaches to education. A pluralistic approach to sustainability issues should present diverse perspectives, drawing on knowledge from multiple disciplines, incorporating a range of viewpoints, and considering the positions of various stakeholders. It should also foster dialogue and deliberation to promote shared understanding and the co-creation of collective meaning (Berglund and Gericke 2016; Boeve-de Pauw et al. 2015; Rudsberg and Öhman 2018). Socio-Scientific Issues (SSIs) concerning sustainability involve the deliberate use of scientific topics that require students to engage in dialogue, discussion, and debate. SSIs are usually controversial in nature but have the added element of requiring the evaluation of ethical concerns in the process of arriving at decisions regarding possible resolution of these issues (Zeidler and Keefer 2003). ESD proponents argue that the inclusion of multiple disciplinary and value-based perspectives in sustainability education simultaneously strengthens the holistic nature of learning outcomes and supports learners to better consider the situational context and relevance for different stakeholders when discussing sustainability solutions (Berglund and Gericke 2016). Holistic approaches in ESD are seen to contribute to learners' overall comprehension and understanding of sustainable development, while pluralistic approaches are found to have a positive effect on learners' sustainability behaviour (Boeve-de Pauw et al. 2015). Holistic, pluralistic as well as action-oriented teaching in ESD are thought to be effective in cultivating students' sustainability behaviour.

Sinakou et al. (2019b) suggest a framework for ESD that consists of five components: (a) action-taking, (b) students' leadership in their learning and teaching, (c) peer interaction, (d) community involvement, and (e) interdisciplinarity. Action-taking means students being active participants in social contexts concerning SD issues, and students' leadership in learning and teaching means that the students take responsibility for their own learning in such issues. Moreover, peer interaction means working in groups in the resolution of an SD issue, participating in collective actions and decision-making processes, and taking an active role in society. This is related to taking part in community-based approaches in which students learn to take part in civic activities. The last perspective is interdisciplinary approaches, and this facilitates students to obtain competencies through the integration of various fields of knowledge to address SD issues.

Integrated interdisciplinary teaching with an emphasis on problem solving, collaboration, and inquiry has led to students learning more than through separated disciplinary teaching (Barr et al. 2015). Drake and Reid (2018) particularly highlight inquiry-based methods as central to integrated interdisciplinary teaching. They point out that such methods can stimulate creativity, collaboration, problem solving, and critical and scientific thinking. Interdisciplinary learning will improve students' abilities to reason by using information from several fields, thus allowing them to respond to the challenges of their time more effectively (Mokski et al. 2023). Klein (1990) defines interdisciplinary learning as neither a subject matter nor a body of content, but as a process for achieving an interpretive synthesis, a process that usually begins with a problem, question, topic, or issue. Boix Mansilla (2005: 15) states the following about interdisciplinary understanding; "the capacity to integrate knowledge and modes of thinking in two or more disciplines to produce a cognitive advancement – e.g., explaining a phenomenon, solving a problem, creating a product, raising a new question – in ways that would have been unlikely through a single disciplinary means… the integration of disciplinary perspectives is a means to a purpose, not an end in itself". Thus, key aspects of interdisciplinary learning and understanding are the abilities to produce synthesis and integration.

6.2.3 Interdisciplinary Pedagogical Content Knowledge in ESD

Teachers' competence in ESD must consist of both Content Knowledge about SD and Pedagogical Knowledge about ESD. Shulman (1986) describes teachers' competence by emphasising the importance of both pedagogical knowledge and content knowledge which together provide pedagogical content knowledge (PCK). The term PCK has later been used by several authors in connection with the education of teachers (Cochran et al. 1993; Gess-Newsome 1999; Olufsen et al. 2017). Kind

(2009) uses the PCK model to describe teachers' various areas of knowledge; thus, demonstrating the need for both insight into the Content Knowledge in a discipline (CK) and a repertoire of suitable teaching methods, or Pedagogical Knowledge (PK), as well as general skills that overlap between PK and CK (i.e. how disciplines are systematised, modified, and taught in the classroom). For example, a science teacher who teaches about the greenhouse effect must apply content knowledge of physics and climate models (CK) and insight into central teaching strategies and principles (PK) to choose relevant teaching methods such as the use of models and dialogues (PCK). However, when the same teacher expands this to work with ESD, they must include content from other disciplines and other teaching approaches to provide interdisciplinary teaching and learning. An (2017) provided an expansion of Shulman's PCK model by introducing a new component across the disciplines—namely interdisciplinary pedagogical content knowledge (IPCK) (see Fig. 6.1).

Figure 6.1 illustrates the PCK framework for interdisciplinary pedagogical content knowledge ($IPCK_{ESD}$) for SD and ESD across the subjects of science and social studies. In the figure, there are three main categories of knowledge: interdisciplinary content knowledge (ICK_{SD}) in SD, content knowledge (CK_{SD}) about SD in social studies, and content knowledge about SD in science (CK_{SD}). A teacher who works with a sustainability issue based on social studies and science must have interdisciplinary content knowledge about SD and content knowledge about SD in both social studies and science. Furthermore, the teacher must have pedagogical knowledge (PK_{ESD}) about teaching approaches in ESD. At the heart of the figure, we see that the overlap between the four categories interdisciplinary content knowledge (ICK_{SD}) in SD, content knowledge (CK_{SD}) about SD in social studies and science, and pedagogical knowledge (PK_{ESD}) about teaching approaches in ESD gives rise to the last category; namely interdisciplinary pedagogical content knowledge ($IPCK_{ESD}$) in ESD. This competence means having a capacity to integrate knowledge and modes

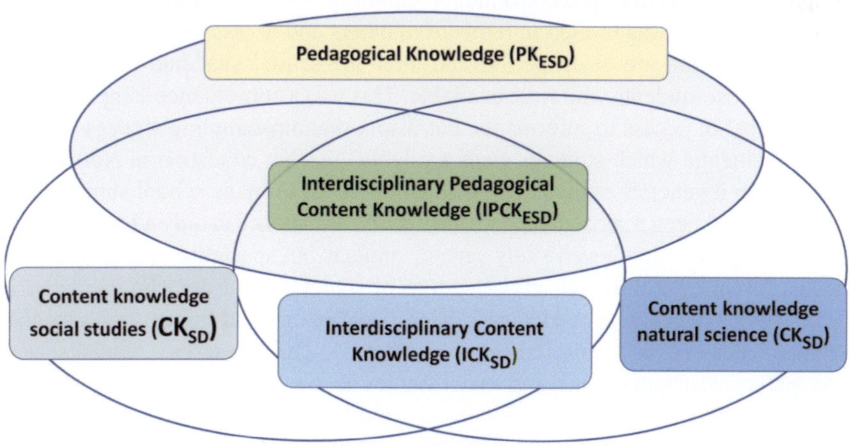

Fig. 6.1 Interdisciplinary pedagogical content knowledge (IPCK) to integrate social studies and science in SD and ESD. (Inspired from An and Tillman (2018))

of thinking in two or more disciplines. An and Tillman (2018) point out that teachers with such competence have insight into representations and concepts from other subject areas (relevant for sustainability), and that they can illuminate SD from several disciplines and perspectives at the same time—drawing connections between these disciplines and perspectives.

6.3 Method

The findings presented in this chapter are based on data collected and analysed through a qualitative methodology approach. We conducted in-depth, semi-structured interviews with pre-service teachers ($N = 12$) enrolled in a teacher education programme at one university college in Norway. This is a so-called initial teacher education programme (Grunnskolelærerutdanning) in which pedagogical courses are integrated together with chosen subjects. The pre-service teachers we interviewed prepared to teach grades 1–7 (ages 6–13) or grades 5–10 (ages 10–16). Below, we outline the methodological choices we made in this research project, as well as reflect upon their validity.

6.3.1 Data Collection

The pre-service teachers in this study were selected because they specialised in either social studies or science, and all of them were between their fourth and fifth (final) year of the study programme. Thus, we carried out a purposeful sampling based on certain criteria (Maxwell 2013: 97). In addition to undertaking purposeful sampling, we also employed convenience sampling. The population from which the sample was taken was limited in terms of numbers and access—the data collection took place during the exam period and after the school year had ended, so we recruited those students who were available. This was a convenience sampling strategy in terms of access to informants, but also a random sampling strategy since it was coincidental which students were available. Teacher education in Norway has earlier been a generalist education (with competence in many school subjects) but has since 2010 been a specialist education (with competence in only a few subjects). The students we interviewed study several subjects but specialise in one subject in the last 2 years. We interviewed two groups of students; students who specialise in social studies ($N = 5$, referred to as S1, S2, S3, and so on), and students who specialise in sciences ($N = 7$, referred to as N1, N2…). The pre-service teachers were selected because both science and social studies have a main responsibility in teaching sustainability according to the current Norwegian school curriculum, and they had to take at least one course in ESD. Thus, they were familiar with the concepts of SD and ESD. It is important to mention that five of the pre-service teachers specialising in science also had taken a minor course in social studies. The five pre-service teachers specialising in social studies did not study science as one of their other subjects.

We conducted individual, semi-structured interviews, which normally are useful for the purpose of obtaining rich, in-depth, and detailed data on a certain topic (Bryman 2004: 320). Such interviews are useful for knowledge production because they allow much more leeway for sliding off track or going a bit back and forth between whichever topics that come up. In-depth interviews make it possible for the informants to elaborate on their experience and views on the topic. However, we did not get insight into the students' actual teaching, as we would if we had observed the pre-service teachers in teaching situations.

The interviews took approximately 30 minutes each and focused on categories of interdisciplinary pedagogical capacity in ESD (see Fig. 6.1). Hence, we asked the pre-service teachers about both disciplinary and interdisciplinary content in SD, as well as general and interdisciplinary pedagogical knowledge in ESD. Finally, we asked the students to elaborate on teaching methods and content particularly related to a well-known case of an interest conflict related to the establishment of a wind power park in Norway (the *Fosen* case), where particularly Sami reindeer herders' rights have been violated. We considered this a relevant case to discuss with the students, because it is highly relevant to complex sustainable development issues and has also been prevalent in the media the last few years. All the pre-service teachers were familiar with the case beforehand. They gave informed consent, and the audio files and transcripts were stored anonymously on a secure platform for storing research data. Moreover, the project was approved by the Norwegian Agency for Shared Services in Education and Research (SIKT). Two of the authors conducted the interviews (separately). The validity of this project lies in the rich descriptions by each informant, which we followed, analysed, and discussed. Furthermore, important to note, we had two groups of informants (pre-service teachers specialising in either social studies or science), but this was not a comparative study. Because of the limited number of participants, we cannot generalise externally nor between the two groups. However, we found interesting patterns that distinguishes the two groups, and we will discuss these further in the analysis.

6.3.2 Data Analysis

After transcribing the interviews, we started the coding process. We used the IPCK model (see Fig. 6.1) as an overall framework for analysing the pre-service teachers' pedagogical capacities in ESD. In the model it is emphasised that teachers must have insight into the SD perspectives in the subject area they teach (CK_{SD}) and have interdisciplinary content knowledge (ICK_{SD}) in SD. In addition, teachers should be able to see the relevant connections between subject areas in SD. Therefore, we identified excerpts in the interviews about relevant content knowledge in SD as an interdisciplinary theme (ICK_{SD}), and excerpts about subject perspectives in social studies and science relevant for SD (CK_{SD}) (research question 1). Teachers must also have a repertoire of suitable teaching methods in ESD (PK_{ESD}), as well as teaching approaches for linking subject perspectives in ESD ($IPCK_{ESD}$). Therefore, we identified excerpts in the interviews about teaching approaches in ESD (PK_{ESD} and $IPCK_{ESD}$) (research question 2). We used the IPCK model to derive the two main categories relevant for the two research questions (interdisciplinary and disciplinary content knowledge in SD, general and interdisciplinary pedagogical knowledge in ESD). To identify sub-categories of these two main categories, we performed an interpretation of the empirical data and in some cases also of relevant existing theories and concepts related to both categories (Alvesson and Sköldberg 2018; Swedberg 2017). We used both deductive and inductive analysis of the data material and used what Kvale and Brinkmann (2009: 223) describe as an 'abductive approach'. In the following sections, we will describe how the analytic process was performed for the two main categories.

6.3.2.1 Interdisciplinary and Disciplinary Content Knowledge in SD

To identify the interdisciplinary content knowledge about SD, we identified excerpts about SD. These excerpts were mainly from pre-service teachers' reflections about what kinds of knowledge are needed to teach about SD, and from this an a priori list of different perspectives was elaborated as a first coding. After a further coding and grouping process, we ended up with sub-categories similar to concepts and theoretical perspectives about SD in the literature. Excerpts about the importance of research and scientific knowledge in SD were categorised as scientific knowledge. Moreover, excerpts about social, environmental, and economic perspectives, and the connections between them were placed in four separate sub-categories (see Table 6.1) (e.g.

Table 6.1 Categories of content knowledge SD

Content knowledge SD (ICK_{ESD})	Category description
Scientific knowledge	Refers to scientific knowledge as relevant for SD
Social perspectives	Refers to social perspectives as society and cooperation between people as relevant for SD
Economic perspectives	Refers to economical perspectives as relevant for SD
Environmental perspectives	Refers to environmental perspectives as climate change and biodiversity loss as relevant for SD
Interconnections	Refers to interconnections between social, economic, and environmental perspectives as relevant for SD
Pluralism	Refers to different perspectives and interest conflicts as relevant for SD
Holism	Refers to seeing the word as a whole system with smaller systems being parts of this as relevant for SD
Precautionary thinking	Refers to precautionary thinking and citizen participation as a part of SD

Based on first coding from interviews and United Nations (2015), Boeve-de Pauw et al. (2015), and Poza-Vilches et al. (2022)

United Nations 2015) to identify what perspectives the pre-service teachers emphasised. We placed excerpts about different perspectives and conflict interests in the category pluralism and excerpts about the importance of seeing the world as an entirety in the category holism (Boeve-de Pauw et al. 2015). Excerpts about precautionary thinking, action, and citizen participation were categorised as precautionary thinking (Poza-Vilches et al. 2022). Table 6.1 shows the description of the sub-categories of SD as a concept, based on first coding and literature above.

In the interviews, the pre-service teachers were not asked to go deeper into or talk more about perspectives of SD. This meant that some of the pre-service teachers gave quite short answers and others gave more extensive answers often in advanced ways. Thus, we categorised the pre-service teachers quotes about SD as either advanced (A, rich, extended with coherent reasoning), intermediate (I, moderate explanation), or basic (B, a short, superficial mention). For instance, this quote about scientific knowledge 'One has to be sceptical and perhaps critical of what you are served (…) Scientific knowledge is important, of course, it is no coincidence that sustainability has become an interdisciplinary topic. It is rooted in something' was categorised as advanced because it was extended with coherent reasoning. On the other hand, this quote 'Teachers need to facilitate the knowledge about climate we know exists' was categorised as basic because it is only a short, superficial mention of scientific knowledge.

Moreover, this quote about the social perspectives in SD: 'One has to have insight in the politics' was categorised as basic, but quotes with extended explanations of social perspectives in SD were categorised as advanced. To get insight into how the individual pre-service teachers from the two groups emphasised the different SD perspectives, we wanted to gain knowledge of the relative frequency of the sub-categories. Thus, we counted the occurrence of quotes in these sub-categories. The length of one quote was 1–3 sentences.

To gain insight into the pre-service teachers' disciplinary content knowledge in SD, we identified excerpts in which they referred to SD perspectives in social studies and science. In this context, we used the a priori list as the results from the two subjects to get an overview of the diversity of findings. The pre-service teachers mentioned similar perspectives (for instance, indigenous people, poverty and different perspectives in social studies and energy, biodiversity, and the greenhouse effect in science). We counted the occurrence of quotes about the two subjects to gain knowledge of their relative frequency. In this context, we were only interested in how the two groups of pre-service teachers (specialising in science and social studies) emphasised the different SD perspectives in science and social studies.

6.3.2.2 General and Interdisciplinary Pedagogical Knowledge in ESD

To identify the pedagogical knowledge in ESD, we identified excerpts about how to work with sustainability issues in school settings, and an initial list was elaborated as a first coding. Some of these excerpts were about general teaching approaches (PK_{ESD}) and others were about interdisciplinary linking strategies ($IPCK_{ESD}$), and we assigned the excerpts in these two categories. We counted the occurrence of quotes in the two categories to gain knowledge of their relative frequency. After a further coding and grouping process, we ended up with sub-categories similar to models of ESD and interdisciplinarity in the literature (Sinakou et al. 2019b; An 2017). Excerpts about bringing in real, complex issues (for instance, 'I will bring in current topics, for instance, what is in the media, to make it more realistic') were placed in the category real issues and complexity, and excerpts about inquiry and social learning were placed in the category participation, inquiry, and social learning. Moreover, we placed excerpts about teaching to give students the insight into different perspectives (for instance, 'I would like students to take other perspectives. Arguing then – that they must stand up for a cause') in the category pluralism and different perspectives, and excerpts about teaching to promote acting or suggesting solutions in the category actions for sustainability. Furthermore, we placed excerpts about the importance of linking subject perspectives for a greater reason (for instance, 'It can be an advantage to draw on more subjects to see the larger picture and students can achieve in-depth learning') in the category linking subject perspectives for a reason, and excerpts with examples of how to link subject perspectives (for instance, 'Gain insight into science and social studies first. Then I use the language to support that debate') in the category linking subject perspectives. Table 6.2 shows the main categories and sub-categories of ESD teaching approaches.

Table 6.2 Main categories and sub-categories of pedagogical knowledge in ESD

Main category	Sub-category	Category description
General teaching approaches (IPKs)	Real issues, complexity	States that teaching approaches should relate to relations and issues of the real world
	Participation, inquiry, and social learning	States that teaching approaches should encourage participation of students with active involvement and inquiry, for instance, by learning in groups
	Pluralism and different perspectives	States that teaching approaches should be based on the culture of complexity, and consider critical analysis through multiple perspectives, reflecting on conflicts of interest
	Actions for sustainability	States explicitly that teaching should result in actions/solutions targeted at effecting real change regarding sustainability issues, the analysis of student lifestyles, behaviour, decision-making and actions
Interdisciplinary teaching approaches (IPCKs)	Linking subject perspectives for a reason	States the importance of or gives examples of how to link subject perspectives for a reason, for instance, to achieve in-depth learning, to understand the interconnections in SD and to obtain sustainability competence
	Linking subject perspectives	Gives examples of how to link subject perspectives in specific teaching cases

6.4 Results

The pre-service teachers reflected about both interdisciplinary (105 quotes) and disciplinary content knowledge (240 quotes) in SD and teaching approaches relevant for ESD (178 quotes). In the following sections, we present important findings from these reflections.

6.4.1 Interdisciplinary Content Knowledge in SD

Table 6.3 presents an overview of the pre-service teachers' specialising in science ($n = 7$) and social studies ($n = 5$) quotes about SD perspectives and to what extent these quotes were on advanced (A), intermediate (I), or basic (B) level. As Table 6.3 shows, the social studies pre-service teachers more frequently referred to all the SD perspectives and seemed to do this in a more advanced way. We also see that all pre-service teachers referred to three or more SD perspectives.

The pre-service teachers most frequently (22 out of 105 quotes) referred to environmental perspectives both in basic and more advanced ways when they mentioned relevant SD content, as reflected in this quote from S5: 'The concept of sustainability is very much linked to the environment, isn't it. It depends a lot on ... if I use less fossil fuel, drive less, save electricity, all such typical environmental measures' (intermediate). However, they often mentioned both economic (10 out of 106 quotes) and social perspectives (17 out of 105 quotes) at the same time. S5 followed up in this way: 'But the concept of sustainability is also within economics and social relations (…) the point is to be able to understand different attitudes and actions towards what people do, whether it is sustainable or not, and that poverty is very central... and sustainable economy (..)' (advanced), thus also referring to interconnections between economic, environmental, and social perspectives. Out of 106 quotes, 12 referred to such interconnections. Some of the pre-service teachers referred to SD in basic ways, as N2 about social perspectives: 'I think that a teacher should have basic knowledge of what sustainable development entails (…) and about why we are an important part of society'. Furthermore, the pre-service teachers also often (16 out of 106 quotes) referred to the holistic perspective of SD, as reflected in this quote from N3: 'You must look to the theory, but also link it to what you can observe. Be able to see big system and to get the whole picture' (intermediate). Moreover, some of the pre-service teachers referred to precautionary thinking (11 out of 106 quotes) as important SD content both in basic ways (for instance, N7: ' (…) teach them to take care of the planet as best as possible'), and in more advanced ways (for instance, S5: ' (..) the whole premise of being sustainable is to help the generations that come after, while at the same time you also get to meet your own needs'). Some of the pre-service teachers suggested scientific knowledge (9 out of 106 quotes) as important, for instance, by referring to temperature measurements in climate research (S3). The pre-service teachers also talked about

Table 6.3 Pre-service teachers' specialising in science (N1, N2, N3, N4, N5, N6, N7) and social studies (S1, S2, S3, S4, S5) number of quotes about SD (ICK_{SD}; advanced (A), intermediate (I), or basic (B))

ICK sustainability	Science pre-service teachers							Social studies pre-service teachers					SUM
	N1	N2	N3	N4	N5	N6	N7	S1	S2	S3	S4	S5	
Scientific knowledge		B (1)	B (1)						I (2)	I (1) A (1)	I (3)		9
Social perspectives		B (1)	B (2)	I (2) A (1)		B (1) I (1)	B (1)	I (1)	I (1)	I (3)	I (1)	A (2)	17
Economic perspectives				I (2) A (1)	B (2)	B (1)	B (1)		I (1)	I (1)		I (1)	10
Environmental perspectives	B (2)	I (2)	A (1)	I (2) A (1)	I (2)	B (1) I (1)	I (2)	B (1)	I (1)	I (3)	I (2)	I (1)	22
Interconnections			B (1) I (2)	I (2) A (1)	B (2)				I (1)	A (1)		A (1)	12
Pluralism	B (1)		I (1)					I (1)			B (1) I (2)	A (1) I (1)	8
Holism	B (1)		I (1)	I (3)			B (1)	I (1)	I (3) A (1)	I (2)	B (1)	A (1) I (1)	16
Precautionary thinking	N (1) I (1)	N (1)				A (1)	B (1)		I (1)	I (1)		A (4)	11

different perspectives and interest conflicts (pluralism) as relevant for SD. Out of 106 quotes, 8 referred to pluralism in sustainability issues as important, as reflected in this quote from S4: 'Insight in and skills to, ability to change opinions and be open to conflicts or other points of view then (...)' (intermediate).

6.4.2 Disciplinary Content Knowledge (Science and Social Studies) in SD

Table 6.4 presents an overview of the relative frequency of the pre-service teachers' reflections involving disciplinary content knowledge (240 quotes). As we see, the science pre-service teachers referred to content in the two subjects just as frequent, but the social studies pre-service teachers referred much more frequent to content in social studies than they did for science.

6.4.2.1 Science Pre-Service Teachers' Reflections About Disciplinary Content in SD

Pre-service teachers specialising in science typically referred to social perspectives when they mentioned relevant SD content from social studies, as reflected in this quote from N4: 'We have social systems in social studies concerning how people cooperate and act together (...)'. Moreover, they referred to human rights and living conditions for indigenous people and living conditions and poverty in other countries. N6 suggested that social studies should have the responsibility for students learning how people live their lives in other countries. Most of the pre-service teachers referred to terms such as politics and democracy as relevant SD content from social studies. In addition, they referred to communities, knowledge about cities, infrastructure, different industries (for instance, the clothing industry), and economy. A couple of them mentioned how people can have a great impact on nature, and how climate change can influence other places in the world, thus referring to the interconnections between social and environmental perspectives. Different perspectives and roles in sustainability issues were also mentioned as relevant for social studies. Furthermore, students typically referred to environmental perspectives when they discussed relevant SD content in science. When they talked about these

Table 6.4 Number of reflections from pre-service teachers about SD content in science and social studies

	Science pre-service teachers ($n = 7$)		Social studies pre-service teachers ($n = 5$)	
Content knowledge SD	Number of quotes about social studies	Number of quotes about science	Number of quotes about social studies	Number of quotes about science
	Around 60	Around 60	Around 100	Around 20

environmental perspectives in science, they mentioned ecosystem, species, biodiversity, how ecosystems react to influences, how a wind turbine works, energy production, emissions, greenhouse gases, greenhouse effect, renewable energy, carbon cycle, global warming, climate crisis, vaccines, and systems thinking in the natural world.

6.4.2.2 Social Studies Pre-Service Teachers' Reflections About Disciplinary Content in SD

Pre-service teachers specialising in social studies typically referred to social perspectives (not economical or environmental) when they mentioned relevant SD content in social studies. When they talked about these social perspectives in social studies, they mentioned historical contexts, timelines, indigenous people, the rule of law and trials, politics, democracy, big society, conflicts of interests, living conditions, livelihoods, how people socialise, humanity, poverty and hunger, consequences and actions, demonstrations, and action competence. However, a few of them referred to economic perspectives as relevant content for sustainability in social studies. Moreover, pre-service teachers in social studies referred to all categories of sustainability (see Table 6.3) as relevant for social studies. In fact, there is an overlap between what they said about the sustainability content in social studies and what they said about sustainability in general, as reflected in the quote below from S2: 'A great deal of the topics in social studies could be linked to sustainability without necessarily being directly mentioned. So, my strengths as a social studies teacher will be that I hopefully have more knowledge and overall understanding.' Moreover, these pre-service teachers rarely referred to relevant content for sustainability in science (20 out of 120 quotes). Those who did referred to environmental perspectives (for instance, energy, how nature can be destroyed by wind turbines, climate change, research in science, and actions for the environment).

6.4.3 Pedagogical Knowledge in ESD

The pre-service teachers suggested several ways of working with ESD and mentioned both general teaching approaches and more interdisciplinary perspectives. They referred to general teaching approaches more frequently (124 out of around 178 quotes) than interdisciplinary perspectives (54 out of 178 quotes).

6.4.3.1 General Pedagogical Knowledge in ESD

Both pre-service teachers specialising in science and social studies suggested several general teaching approaches in ESD, and there were little differences between the two groups. They referred to perspectives from all categories in Table 6.2 (Real issues, complexity; Participation, inquiry and social learning; Pluralism and different perspectives; and Actions for sustainability).

Real Issues, Complexity The pre-service teachers typically suggested that sustainability teaching should engage students in relevant or authentic issues, as reflected in this quote from N7: 'I will bring in current themes, for instance what is in the media, to make it more realistic.' Some of the pre-service teachers suggested that the teaching and learning should occur in authentic contexts, for instance, by making agreements with farms or industries (N4) or by having excursions to observe a spot in nature and to discuss what would happen if you built something there (for instance, a nuclear power plant) (S1). Moreover, two of them suggested that ESD projects should be long lasting, for example, by planting and following a school garden over time (N1) or by rearranging the timetable and spending quite a few hours on the project (N7).

Participation, Inquiry, and Social Learning The pre-service teachers typically suggested that students should learn about sustainability in groups working actively together in a project (for instance, N6: 'It is important that the students must do something themselves. Maybe both outdoors and indoors'). N2 suggested that the students' active process in groups could be assessed. Some of the pre-service teachers suggested that students should use inquiry-based methods (for instance, N1: 'I think students can do inquiry-based methods' and N7: 'I want students to do inquiry – collect data – to be active, preferably in a group'). Other pre-service teachers explicitly suggested that students should seek for information on the Internet.

Pluralism and Different Perspectives The pre-service teachers typically suggested that students should familiarise themselves with the perspectives of a special interest group (for instance, S1: 'Map the conflict of interest. I like to create a large joint mind map for the students'). All the pre-service teachers specialising in science and one specialising in social studies suggested that students could participate in a debate or write an appeal with the perspectives of special interest groups, as reflected in the following quote from N2: 'Then the students have to try to extract the arguments both for and against, and then you group students preparing arguments from one interest group before they participate in the debate.'

Actions for Sustainability The pre-service teachers also typically suggested that ESD should promote actions in SD issues as reflected in the quote from N5: 'I would like to increase their action competence – that they want to do something. Engage them.' A couple of them suggested that students should be involved in action projects, as reflected in this quote from N7: 'Let the students participate actively. (…) let them see that it has an impact. Let them sort plastic and rubbish at home, for example. Get them to go out and pick up trash.' A few of them suggested that teachers should promote sustainability action (for instance, S3: 'It is difficult to talk about skills, but the teacher must be a good role model at least through action and values').

6.4.3.2 Interdisciplinary Pedagogical Knowledge in ESD

The pre-service teachers suggested several ways to combine perspectives when they work with the sustainability theme. Pre-service teachers specialising in science referred more often to the linking of subject perspectives (41 quotes out of 54) than those specialising in social studies (13 quotes out of 54).

Linking Subject Perspectives for a Reason The pre-service teachers frequently stated the importance of linking subject perspectives for a reason, as reflected in this quote from N2: 'It can be an advantage to draw in more subjects to see the larger picture, and students can achieve in-depth learning.' Several of the pre-service teachers gave examples of how different disciplinary perspectives can contribute to understanding of the interconnections between various SD perspectives (for instance, N4: '… that students understand that these are in fact systems and that they are connected, and that it is not only separate parts and that the different subjects are connected') or to obtain sustainability competence (for instance, S5: '… giving the students a boost in the ability to think critically – if the students must think from two perspectives and different content perspectives this will give them this ability').

Linking Subject Perspectives The pre-service teachers (especially those who specialised in science) gave examples of how relevant SD perspectives in subjects can be linked in specific teaching cases. For instance, they gave examples of how social studies, science, and language can be linked in a debate on the Fosen case, as reflected in this quote from N5: 'Gain insight into science and social studies first. Then I use language to support that debate.' Furthermore, some of them also gave other examples of how other subjects than social studies, science, and language subjects could be linked to the Fosen case. N7 and S3 suggested bringing in religion and ethics related to the case of the Sami people, and N7 suggested bringing in art by building models of wind turbines. S5 also suggested bringing in mathematics, for instance, by using statistics. Moreover, two pre-service teachers gave examples of how to assess products by linking perspectives from different subjects, for instance, that teachers in social studies, science, and language could assess different subject perspectives in a written appeal.

6.4.4 Interdisciplinary Pedagogical Content Knowledge ($IPCK_{ESD}$)

Figure 6.2 illustrates the quotes and thus the interdisciplinary pedagogical content knowledge for one of the pre-service teachers in this study (N7, specialising in science). Even though this informant mainly demonstrated basic understanding of SD (ICK_{SD}), N7 referred to relevant sustainability content both from social studies and science (CK_{SD}) (see Fig. 6.2). Moreover, this pre-service teacher referred to and

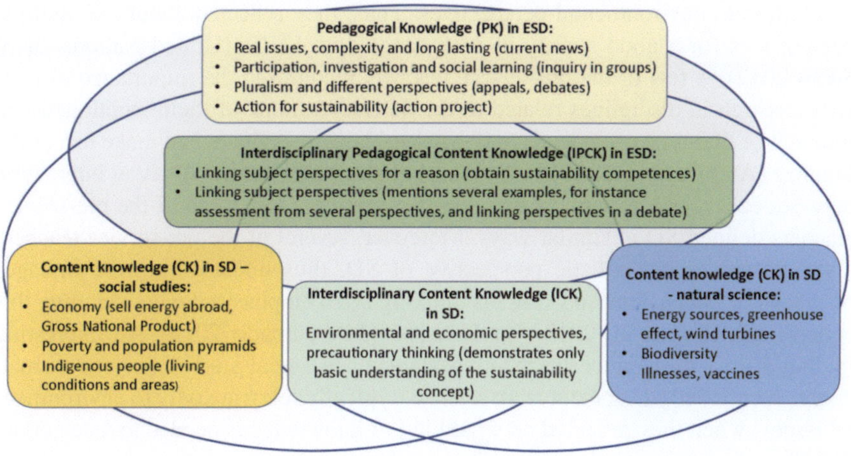

Fig. 6.2 Interdisciplinary pedagogical content knowledge (IPCK) to integrate social studies and science in sustainability for one of the pre-service teachers (N7)

gave examples from all the categories of general teaching approaches (PK_{ESD}) and gave several examples of how to link subject perspectives ($IPCK_{ESD}$) for a reason and in specific teaching cases. This student case is representative of pre-service teachers specialising in science. Pre-service teachers specialising in social studies show an opposite tendency, they demonstrated advanced to intermediate understanding of SD, but they referred to few SD perspectives from science and suggested few examples of linking subject perspectives.

6.5 Discussion

In the following section, we discuss how SD is understood generally and in subjects by pre-service teachers, and what they suggest as suitable ways to work with ESD. Finally, we consider what they suggest as ways of linking subject perspectives when teaching about SD.

6.5.1 Content Knowledge in Sustainability (CK and ICK)

The pre-service teachers in this study frequently referred to environmental perspectives when they mentioned relevant content for sustainable development. This is similar to pre-service teachers in a study from Spain, in which the students prioritised acquisition of knowledge and practical skills related to nature and sciences in a school project related to ESD (Cebrián and Junyent 2015). However, most of the pre-service teachers in our study referred to both economic and social perspectives in addition to environmental perspectives. Thus, they reflected about the multiple perspectives for a good society in accordance with SDGs (United Nations 2015; Sachs 2012). A few of the pre-service teachers recognised the importance of relevant research in disciplines related to the SDGs, but none of them mentioned that research of that type often leads to debates on how to evaluate and make use of the findings (Archibald et al. 2016). Giddings et al. (2002) define SD as an interaction between ecological, economic, and social conditions, and some of the pre-service teachers defined SD in similar ways. Moreover, several of the pre-service teachers also referred to the holistic perspective of SD, thus underscoring the linkages between the dimensions and systems thinking often emphasised when defining this concept (Sinakou et al. 2019a; McKenzie and Abdulkadri 2018). An interesting finding is that pre-service teachers specialising in social studies only to a small degree mentioned SD content from science (typically referring only to environmental issues) when they reflected on what kind of knowledge is needed to teach about SD. Thus, there is a risk that they will put little emphasis on environmental perspectives when teaching about SD perspectives in the future.

Pluralism is another facet of SD, and it is characterised by the attempt to recognise and partake in a dialogue, inclusive of different viewpoints, on issues related to sustainable development (Boeve-de Pauw et al. 2015; Rudsberg and Öhman 2010). Several of the pre-service teachers in this study considered different perspectives in SD issues as important for a teacher to recognise. Moreover, some of them suggested precautionary thinking as important content related to SD. According to Poza-Vilches et al. (2019), the concept of citizen participation is unquestionably linked to the concept of sustainable development by promoting equitable, inclusive, and fair societies, and it involves generating processes for citizen participation. In line with Poza-Vilches et al., these pre-service teachers acknowledged that SD itself is not neutral, and that it contains a particular policy commitment with a view of a society and moral accountability it is possible to adopt to (Winter 2007; Read and O'Riordan 2017, p. 5).

Kioupi and Voulvoulis (2019) note that the complexity of sustainability as a concept makes it difficult to relate the Sustainable Development Goals (SDGs) to educational learning outcomes. In this study, it seems like pre-service teachers specialising in social studies reflect on a wider range of sustainability perspectives than pre-service teachers specialising in science, and it seems like they have better insight into the conceptual complexity of sustainable development. There are overlaps between how the pre-service teachers specialising in social studies reflected on their own subject and the concept of SD, and this could be linked to the fact that they recently had taken a course in social studies named Education for Sustainable Development. ESD-based learning programmes in social studies often include several relevant SD concepts (Nasution et al. 2023). However, it could also be due to a stronger conceptual aggregation between the aspects of environmental sustainability and social/human development from a cross-disciplinary perspective.

6.5.2 *Pedagogical Knowledge in ESD*

Both pre-service teachers specialising in science and social studies suggested several general teaching approaches in ESD. All informants most frequently suggested that sustainability teaching should engage students in relevant or authentic issues, thus acknowledging the fact that socio-scientific issues often are controversial in nature (Zeidler and Keefer 2003). Moreover, they suggested teaching approaches using the Fosen case with great enthusiasm, demonstrating their willingness to use such authentic cases in ESD. Sinakou et al. (2019b) point out precisely that students should be active participants in social contexts concerning SD issues, but also that peer interaction is important. Students should work in groups in resolution of the SD issue, taking an active role in society. The pre-service teachers frequently suggested that students should learn about SD in groups working actively together, investigating in a project (for instance, to seek information and do inquiry-based methods). This is in line with Drake and Reid (2018) and Barr et al. (2015), who highlight collaboration and inquiry-based methods as central to integrated interdisciplinary teaching.

Moreover, pre-service teachers frequently suggested that students should familiarise themselves with the perspectives of a special interest group, and even participate in the debates with these roles. This shows an appreciation of pluralism in ESD, and it is characterised by recognising different viewpoints in issues related to sustainable development (Boeve-de Pauw et al. 2015). For instance, several of the informants suggested that students should familiarise themselves with the different interest groups in the Fosen case and learn about the politics in that case. Sinakou et al. (2019a) point out something similar, namely offering students collective actions to experience decision-making processes. Moreover, Sinakou et al. (2019a, b) suggest that ESD should be based on community-based approaches, for students to take part in civic activities. Some of the pre-service teachers in this study suggested action projects where students actively do things in the local community or in society as an ESD teaching approach. A couple of the pre-service teachers suggested that the teacher should be a role model when it comes to SD. Cheung (2020) and Vukic (2019) argue that teachers serve as models for their students; hence, teachers' perspectives and how these are enacted will influence their students.

Several of the pre-service teachers in this study pointed to the importance of linking subject perspectives in ESD, for instance, by giving examples of how different subject perspectives can contribute to sustainability competence, in-depth learning and understanding of interconnections in SD. This shows an understanding of the connections between different subject areas (An and Tillman 2018), and that interdisciplinary learning is a process for achieving an interpretive *synthesis*, a process that usually begins with a problem, question, topic, or issue (Klein 1990). Moreover, mostly the pre-service teachers specialising in science gave examples of how different subject perspectives can be linked in specific teaching cases. This could be due to that these pre-service teachers were trained in doing this both in assignments and orally in class in some of the science courses. For instance, they gave several examples of how social studies and science could be linked. Thus, they demonstrated their competence in the middle of the IPCK model (An and Tillman 2018) illuminating the sustainability theme from several subject areas at the same time and seeing connections between different subjects. The Fosen case is a typical example of an SD issue that cannot be solved or discussed through a single disciplinary means, and the pre-service teachers showed a capacity to integrate knowledge and modes of thinking in two or more disciplines to produce a cognitive advancement (Boix Mansilla 2005). Moreover, this is in line with the holistic approach to ESD, according to which SD is considered an integrated concept of three dimensions, namely environment, economy, and society (Sinakou et al. 2019a).

6.5.3 Interdisciplinary Pedagogical Content Knowledge in ESD

In Fig. 6.2, we see one example of a pre-service teacher's (N7) competence in ESD, consisting of both content knowledge (CK_{SD} an ICK_{SD}), pedagogical knowledge (PK_{ESD}), and interdisciplinary pedagogical content knowledge ($IPCK_{ESD}$). Despite demonstrating a basic insight into the content knowledge of SD (ICK_{SD}), N7 mentions several examples of general teaching approaches (PK_{ESD}) and examples of how to link subject perspectives ($IPCK_{ESD}$). This can be due to the knowledge of relevant SD perspectives from social studies and science (CK_{SD}), but it also seems like linking relevant subject perspectives (Klein 1990) and brings forward content knowledge perspectives from SD (ICK_{SD}). Generally, the pre-service teachers referred to several relevant perspectives in SD, some of them showing great insight into the complexity of these perspectives (ICK_{SD}). Moreover, they all seemed to agree that ESD should be interdisciplinary, and that SD as viewed in the Fosen case is a complex issue. They suggested several similar general teaching approaches, like engaging students in authentic issues and familiarising themselves with a special interest group (PK_{ESD}). However, not all of them suggested ways of linking perspectives ($IPCK_{ESD}$), and the pre-service teachers specialising in social studies rarely referred to SD perspectives from the view of science (CK_{SD}).

6.6 Conclusion and Implications

This chapter has presented the results from interviews with 12 pre-service teachers in a Norwegian teacher education context. Even if the sample is quite small, the rich, qualitative nature of these in-depth interviews has been valuable to analyse through the lens of the IPCK model to understand how well current pre-service teachers perceive ESD from an interdisciplinary perspective. Reporting the number of quotes in this study may appear somewhat misleading since similar wordings by the same students could be counted more than once. However, this was only applied for interdisciplinary SD perspectives and in the subject's science and social studies, and it did provide useful knowledge about a possible overlap between the general SD perspectives and the subject perspectives.

The IPCK model appeared useful in analysing and describing the pre-service teachers' competence in ESD by emphasising the importance of both content knowledge in SD and pedagogical and interdisciplinary pedagogical knowledge in ESD (Shulman 1986), and it could be a valuable tool/framework when improving teacher education towards ESD. In the future, all pre-service teachers should get insight into ESD in more of their teacher education courses. Even though teacher educators regularly report that they find it difficult to make clear connections between their subjects and ESD (Goller and Rieckmann 2022), and there is little cooperation across subjects in teacher education (Biseth et al. 2022), ESD is often viewed as a core approach for enhancing the overall quality and relevance of education (Didham and Ofei-Manu 2018). Pre-service teachers should be given situations where they practice addressing content from multiple subjects simultaneously, highlighting connections among different disciplines and assessing students' learning of content from multiple subjects. This is supported by Laurie et al. (2016), who report that pedagogical approaches promoted by ESD have a strong transformative impact. The results from this study highlight a need for a more detailed review of what constitutes specific interdisciplinary pedagogical knowledge in general and specifically in its application in ESD. Follow-up research, using quantitative data and interviews or focus groups, could provide insights on how to better use, rethink, redesign, and combine pedagogical approaches to provide a more sustainability-oriented education.

References

Alvesson M, Sköldberg K (2018) Reflexive methodology: new vistas for qualitative research, 3rd edn. Sage

An SA (2017) Preservice teachers' knowledge of interdisciplinary pedagogy: the case of elementary mathematics-science integrated lessons. ZDM 49(2):237–248. https://doi.org/10.1007/s11858-016-0821-9

An SA, Tillman DA (2018) Preservice teachers' pedagogical use of 'Gerrymandering' to integrate social studies and mathematics. J Math Educ 11(3):33–53. https://doi.org/10.26711/007577152790031

Archibald T, Sharrock G, Buckley J, N C (2016) Assumptions, conjectures, and other miracles: the application of evaluative thinking to theory of change models in community development. Eval Program Plann 59:19–127. https://doi.org/10.1016/j.evalprogplan.2016.05.015

Bagoly-Simó P (2013) Tracing sustainability: an international comparison of ESD implementation into lower secondary education. J Educ Sustain Dev 7(1):95–112. https://doi.org/10.1177/0973408213495610

Barr DJ, Boulay B, Selman RL, McCormick R, Lowenstein E, Gamse B, Fine M, Brielle Leonard M (2015) A randomized controlled trial of professional development for interdisciplinary civic education: impacts on humanities teachers and their students. Teach Coll Rec Voice Sch Educ 117(2). https://doi.org/10.1177/016146811511700202

Berglund T, Gericke N (2016) Separated and integrated perspectives on environmental, economic, and social dimensions – an investigation of student views on sustainable development. Environ Educ Res 22(8):1115–1138. https://doi.org/10.1080/13504622.2015.1063589
Biseth H, Svenkerud SW, Magerøy SM, Rubilar KH (2022) Relevant transformative teacher education for future generations. Front Educ 7:806495. https://doi.org/10.3389/feduc.2022.806495
Boeve-de Pauw J, Gericke N, Olsson D, Berglund T (2015) The effectiveness of education for sustainable development. Sustainability 7(11):15693–15717. https://doi.org/10.3390/SU71115693
Boix Mansilla V (2005) Assessing student work at disciplinary crossroads. Chang Mag High Learn 37(1):14–21. https://doi.org/10.3200/CHNG.37.1.14-21
Bryman A (2004) Social research methods, 2nd edn. Oxford University Press, Oxford
Cebrián G, Junyent M (2015) Competencies in education for sustainable development: exploring the student teachers' views. Sustainability 7(3):2768–2786. https://doi.org/10.3390/su7032768
Cheung P (2020) Teachers as role models for physical activity: are preschool children more active when their teachers are active? Eur Phys Educ Rev 26(1):101–110. https://doi.org/10.1177/1356336X19835240
Cochran KF, Deruiter JA, King RA (1993) Pedagogical content knowing: an integrative model for teacher preparation. J Teach Educ 44(4):263–272. https://doi.org/10.1177/0022487193044004004
Didham RJ, Ofei-Manu P (2018) Advancing policy to achieve education for sustainable development. In: Leicht A, Heiss J, Byun W (eds) Issues and trends in education for sustainable development. UNESCO Publishing, pp 87–110
Didham R, Ofei-Manu P (2012) Monitoring and evaluation of education for sustainable development: A framework of factors and leverage points in the implementation of ESD in the Asia-Pacific region. Institute for Global Environmental Strategies. https://www.iges.or.jp/en/pub/monitoring-and-evaluationeducation/en
Drake SM, Reid JL (2018) Integrated curriculum as an effective way to teach 21st century capabilities. Asia Pac J Educ Res 1(1):31–50. https://doi.org/10.30777/APJER.2018.1.1.03
Gess-Newsome J (1999) Pedagogical content knowledge: An introduction and orientation. In: Gess-Newsome J, Lederman NG (eds) Examining pedagogical content knowledge. Science & technology education library, vol 6. Springer, Dordrecht. https://doi.org/10.1007/0-306-47217-1_1
Giddings B, Hopwood B, O'Brien G (2002) Environment, economy and society: fitting them together into sustainable development. Sustain Dev 10(4):187–196. https://doi.org/10.1002/sd.199
Goller A, Rieckmann M (2022) What do we know about teacher educators' perceptions of education for sustainable development? A systematic literature review. J Teach Educ Sustain 24(1):19–34. https://doi.org/10.2478/JTES-2022-0003
Jegstad KM, Sinnes AT, Gjøtterud SM (2018) Science teacher education for sustainable development: from intensions to realisation. NorDiNa *14*(4):1–18. https://doi.org/10.5617/nordina.3263
Jónsson ÓP, Guðmundsson B, Øyehaug AB, Didham RJ, Wolff L-A, Bengtsson S, Andreasen Lysgaard J, Sóley Gunnarsdóttir B, Árnadóttir SM, Rømoen J, Sund M, Cockerell E, Plummer P, Brückner M (2021) Mapping education for sustainability in the Nordic countries. Nordic Council of Ministers. https://www.norden.org/en/publication/mapping-education-sustainability-nordic-countries
Kind V (2009) Pedagogical content knowledge in science education: perspectives and potential for progress. Stud Sci Educ 45(2):169–204. https://doi.org/10.1080/03057260903142285
Kioupi V, Voulvoulis N (2019) Education for sustainable development: a systemic framework for connecting the SDGs to educational outcomes. Sustainability 11(21):6104. https://doi.org/10.3390/SU11216104
Kvale S, Brinkmann S (2009) InterViews: learning the craft of qualitative research interviewing (2nd ed.). Gyldendal Akademisk.
Klein JT (1990) Interdisciplinarity: history, theory, and practice. Wayne State University Press

Laurie R, Nonoyama-Tarumi Y, McKeown R, Hopkins CA (2016) Contributions of education for sustainable development (ESD) to quality education: a synthesis of research. J Educ Sustain Dev 10(2):1–17. https://doi.org/10.1177/0973408216661442

Maxwell JA (2013) Qualitative research design: An interactive approach, 3rd edn. Sage, Los Angeles

McKenzie S, Abdulkadri A (2018) Mechanisms to accelerate the implementation of the Sustainable Development Goals in the Caribbean. (ECLAC) Economic Commission for Latin America and the Caribbean. https://www.cepal.org/en/publications/43362-mechanisms-accelerate-implementation-sustainable-development-goals-caribbean

Mokski E, Filho WL, Sehnem S, Baltazar J, Os S, De O, Guerra A (2023) Education for sustainable development in higher education institutions: an approach for effective interdisciplinarity. Int J Sustain High Educ 24(1):96–117. https://doi.org/10.1108/IJSHE-07-2021-0306

Nasution N, Harti H, Witjaksono AD, Purnomo NH, Sadewo S, Wasino M, Mansor WM, Lailiyah NF (2023) The development of social studies lesson plan based on ESD: theme of the rise of local entrepreneurs. In: Proceedings of the international joint conference on arts and humanities 2022 (IJCAH 2022). https://doi.org/10.2991/978-2-38476-008-4_167

O'Donoghue R, Taylor J, Venter V (2018) How are learning and training environments transforming with ESD? In: Leicht A, Heiss J, Byun WJ (eds) Issues and trends in education for sustainable development. UNESCO, pp 111–131

O'Riordan T, Cameron J (2013) Interpreting the precautionary principle. Routledge

Olufsen M, Karlsen S, Ødegaard M (2017) Endringer i lærerstudenters kompetanser? En casestudie fra en ny lærerutdanning ved UiT Norges arktiske universitet. Nordic Stud Sci Educ 13(2):117–133. https://doi.org/10.5617/NORDINA.3140

Poza-Vilches M, López-Alcarria A, Mazuecos-Ciarra N (2019) A professional competences' diagnosis in education for sustainability: A case study from the standpoint of the Education Guidance Service (EGS) in the Spanish context. Sustainability 11(6):1568. https://doi.org/10.3390/su11061568

Poza-Vilches F, García-González E, Solís-Espallargas C, Velasco-Martínez LC, López-Alcarria A, Estrada-Vidal LI, Jiménez-Fontana R, Rodríguez-Marín F, Puig-Gutiérrez M, Tójar Hurtado JC, Gutiérrez-Pérez J (2022) Greening of the syllabus in faculties of education sciences through sustainable development goals: the case of public Andalusian universities (Spain). Int J Sustain High Educ 23(5):1019–1044. https://doi.org/10.1108/IJSHE-02-2021-0046/FULL/XML

Quablan A (2018) Building capacities of educators and trainers. In A. Leicht, J. Heiss, & W. J. Byun (Eds.), Issues and trends in education for sustainable development (pp. 133–156). UNESCO

Read R, O'Riordan (2017) The precautionary principle under fire. Environ Sci Policy Sustain Dev 59(5):4–15. https://doi.org/10.1080/00139157.2017.1350005

Rudsberg K, Öhman J (2010) Pluralism in practice: Experiences from Swedish evaluation, school development and research. Environ Educ Res 16(1):95–111. https://doi.org/10.1080/13504620903504073

Rudsberg K, Öhman J (2018) Pluralism in practice – experiences from Swedish evaluation, school development and research. Environ Sustain Educ Policy:193–209. https://doi.org/10.4324/9780203732359-13

Sachs JD (2012) From millennium development goals to sustainable development goals. Lancet 379(9832):2206–2211. https://doi.org/10.1016/S0140-6736(12)60685-0

Shulman LS (1986) Those who understand: knowledge growth in teaching. Educ Res 15(2):4–14. https://doi.org/10.3102/0013189X015002004

Sinakou E, Boeve-de Pauw J, Van Petegem P (2019a) Exploring the concept of sustainable development within education for sustainable development: implications for ESD research and practice. Environ Dev Sustain 21(1):1–10. https://doi.org/10.1007/S10668-017-0032-8/METRICS

Sinakou E, Boeve-de Pauw J, Van Petegem P (2019b) Designing powerful learning environments in education for sustainable development: a conceptual framework. Sustainability 11(21):2–23. https://doi.org/10.3390/su11215994

Swedberg R (2017) Theorizing in sociological research: a new perspective, a new departure? Annu Rev Sociol 43:189–206

UNESCO (2014) Shaping the future we want: UN decade of education for sustainable development (2005–2014) – final report. UNESCO. https://unesdoc.unesco.org/ark:/48223/pf0000230171

United Nations General Assembly (2015) Resolution A/RES/70/1: transforming our world: the 2030 agenda for sustainable development. United Nations. https://undocs.org/A/RES/70/1

Vukic T (2019) Sustainable development from high school teacher's perspective. Facta Univ Ser Philos Sociol Psychol Hist 18(3):131–148. https://doi.org/10.22190/FUPSPH1903131V

Winter C (2007) Education for sustainable development and the secondary curriculum in English schools: rhetoric or reality? Camb J Educ 37(3):337–354. https://doi.org/10.1080/03057640701546656

Zeidler DL, Keefer M (2003) The role of moral reasoning and the status of socioscientific issues in science education. In: The role of moral reasoning on socioscientific issues and discourse in science education, pp 7–38. https://doi.org/10.1007/1-4020-4996-X_2

Anne Bergliot Øyehaug, PhD, is a Professor at the University of Inland Norway. She has years of teaching experience in science and science education and has a PhD in science education. She serves as both a lecturer and a researcher in teacher education and has extensive experience in both national and international collaboration. Her research interest is within nature of science, inquiry-based teaching, and education for sustainable development (ESD). She has a particular interest in the perceptions and utilizations of scientific knowledge and processes in ESD. She has published in several educational science journals and has been one of the editors in a book about in-depth learning in science.

Jill Tove Buseth, PhD, is an Associate Professor in human geography at the University of Inland Norway. Her teaching and research areas cover education for sustainable development, green economy, political ecology, policies and discourses, internationalization in teacher education, and global development studies. She holds a PhD in international environment and development studies (NMBU, 2019). She has extensive experience and a particular interest in education for sustainable development in a social science context. She has published within the field of development programs and policy, green transitions in agriculture and environment, particularly in the context of Eastern Africa, and she has done research in the field of interdisciplinary learning and education for sustainable development.

Robert J. Didham, PhD, holds the UNESCO Chair on Education for Sustainable Lifestyles. He is director of the Centre for Collaborative Learning for Sustainable Development and an Associate Professor at the University of Inland Norway. With a background in community-based sustainable development, public participation, and educational leadership, he has extensive experience in interdisciplinary research and capacity building projects. He has coordinated multi-country research on education for sustainable development, sustainable consumption, and adaptation planning, and he has led policy support projects for several national governments. He is also involved in development of pedagogical approaches and learning resources for promoting active learning on sustainable lifestyles, and the Centre disseminates thousands of copies of their active learning toolkits to teachers internationally every year.

Chapter 7
Systemic Challenges and Specific Solutions: Analysis of Sustainability in Finnish Engineering Education Curricula

Noora Jaakkola

Abstract Science and technology are essential levers of transformation for sustainability. Consequently, engineering education plays a key role to educate graduates with a capacity to create solutions for sustainability crises. However, the sustainability challenges are not only technical, and it is essential that engineering graduates understand how technical and social change interact and identify their own role in the transition processes. This chapter presents a document analysis of Finnish engineering education curricula through a sustainability lens. The aim is to understand how sustainability is approached in the curricula, what sustainability-related educational goals are included in the curricula and what differences can be identified between separate engineering disciplines when engaging sustainability. Five ways of approaching sustainability were identified. The approaches varied between non-existent integration with no sustainability-related learning objectives to integration that adapted sustainability to the norms and practices of engineering disciplines and, eventually, to integration of sustainability in a holistic way, embedding both technological and non-technological aspects of sustainability challenges into the core of engineering discipline. Disciplinary differences were identified, indicating that sustainability is not yet always perceived as a topic that should be embedded across all engineering disciplines. The approaches for integrating sustainability can be used as a tool for engineering programmes to develop their own way of approaching sustainability.

Keywords Sustainability · Higher education · Engineering education · Curriculum · Educational goals

---

N. Jaakkola (✉)
University of Helsinki, Helsinki, Finland
e-mail: noora.jaakkola@helsinki.fi

F. Skarstein, L.-A. Wolff (eds.), *Sustainability and Science Education*, Environmental Challenges and Solutions,
https://doi.org/10.1007/978-3-032-24408-6_7

7.1 Introduction

The ongoing climate crisis and declining biodiversity call for quick transitions of societies. Science and technology have been recognised as essential levers of transformation for sustainability (e.g. Global Sustainable Development Report 2019). In addition to sustainability challenges, engineering industries are also affected by rapid technological development. Higher education institutions face significant pressure to ensure that future engineering graduates have the capabilities to develop new technologies and to put them into effect in society (Gumaleius and Kolmos 2019). However, the graduates should also be able to prevent and mitigate the negative effects that technologies, whether old or new, might cause (Bergen and Robaey 2022).

In this study, sustainability is understood as a goal to fulfil the 'needs of the present while safeguarding Earth's life-support system, on which the welfare of current and future generations depends' (Griggs et al. 2013, p. 306). The changes that sustainability crises cause in the ecological environment as well as in societies are so fundamental that academics in engineering education, like in other disciplines, must rethink their curriculum and perhaps even the purpose of education (Paulsen 2021). The urge to drive sustainability transitions has created a demand for engineering education institutions to educate professionals who are able to analyse the ecological and social impact of technological innovations and design engineering solutions that not only decrease the harm but also restore ecological systems (Desha and Hargroves 2014a).

Researchers studying sustainability in higher education have made several suggestions to improve the situation. A large body of research points towards the need for addressing transdisciplinarity in education (Barth 2015; Sandri 2022; Tejedor et al. 2018). Various studies have suggested the most relevant competencies for sustainability transition processes (Wiek et al. 2011; Rieckmann 2012; Wiek et al. 2016; Brundiers et al. 2021; Redman and Wiek 2021). These competency frameworks have received responses highlighting the importance of including inner capabilities and ethical considerations (Jaakkola et al. 2022; Giangrande et al. 2019; Anderson 2013) and suggestions to align the framework further with sustainability science (Salovaara et al. 2020). Suggestions have been made about teaching methods to support sustainability-relevant learning, including active learning pedagogies (Quelhas et al. 2019) such as transdisciplinary problem-based pedagogies (Mulder et al. 2012; Barth 2015; Guerra 2017) and experiential learning emphasising students' self-reflection (Frank and Stanszus 2019).

There is strong pressure to rapidly renew higher education, and the pace of traditional curriculum renewal processes has not been enough to respond to this need for change (Mulder et al. 2012; Desha and Hargroves 2014b). Embedding sustainability into traditional disciplines remains a challenge (Desha and Hargroves 2014b; Barth 2015). Barth (2015) suggests three ways to approach sustainability in the curriculum. First, sustainability may be discussed in the disciplinary setting. Second, it may be introduced as an interdisciplinary issue, whereby the students from various disciplinary backgrounds provide their expertise and methods of their discipline to solve the challenge at hand. Third, according to Barth (2015) the approach can be transdisciplinary. In this case, a variety of academic actors and societal stakeholders are involved to create sustainable contributions.

Academic disciplines in higher education vary in their ontological and epistemological groundings, and some disciplines may be more ready than others to apply a transdisciplinary approach when integrating sustainability (Christie et al. 2015). Kolmos et al. (2016) introduced three response strategies related to embedding sustainability into engineering education. The first strategy, the 'add-on' strategy, means adding some sustainability-related content or a separate course, but the rest of the curriculum remains intact. The integration strategy, according to Kolmos et al., means embedding sustainability-related aspects across the programme, but in a way, that the educational paradigm is not affected. Finally, the 're-building' strategy means building connections between academia and the societal context and redefining the goals of the programme in relation to sustainability challenges. Regarding the integration of sustainability into disciplinary teaching contexts, Sandri (2021) has noted that educators make conscious and unconscious decisions on how to make sustainability 'fit' with the topic at hand. According to Sandri, a tension may exist between the disciplinary norms and sustainability, and in line with Barth (2015) and Kolmos et al. (2016) Sandri suggests that educators need to make decisions on whether to make sustainability fit into the disciplinary norms or to approach sustainability in a way that may broaden or even transform the norms of the discipline.

This chapter presents a document analysis of engineering education curricula in Finnish universities and explores how sustainability is made to fit into these curricula. The analysis aimed to answer the following research question: In what ways is sustainability approached in engineering education curricula?

This research question is addressed from several directions in the analysis, aiming to create understanding of how sustainability is understood and addressed in engineering education. Attention is given to sustainability-related educational goals and the differences that can be identified across various engineering education disciplines.

7.2 Theoretical Framework

Traditionally, engineering education has focused on creating scientific solutions to technical challenges (Mulder 2004). Embedding sustainability in engineering education has been called for, for a long time (e.g. Mulder 2004; Mulder et al. 2012). In addition to researchers, engineering education accreditation agencies have acknowledged that engineering education must respond to sustainability crises. The European Network for Accreditation of Engineering Education (ENAEE) updated their guidelines on engineering education learning outcomes in 2021 with an aim to consider 'the challenges facing our planet and society and the ever increasing need for engineers to ethically challenge their work for the benefit of society' (ENAEE 2021, p. 2).

Despite this, the implementation of such change remains slow. According to Nicolaou and Conlon (2012), sustainability is often narrowed down to environmental issues and students have little knowledge about the social aspects of sustainability. Mulder (2017) suggests that currently the change is viewed from the perspective of technological innovations while the political processes related to these innovations are often left with little attention. Guerra (2017) and Gutierrez-Bucheli et al. (2022) note in their studies that in engineering education, sustainability is often approached in a technocratic manner, leaving out the contextual elements that make sustainability-related challenges so complex. However, there are also positive signs of change. For example, Sigahi et al. (2023) present in their study a process of developing an undergraduate programme where the starting point for development was the complexity of sustainability challenges.

7.2.1 Suggested Sustainability Competencies for Engineers

To be able to solve and manage sustainability challenges, engineering graduates need to be able to frame the challenges they are solving as more than technical (Takala and Korhonen-Yrjänheikki 2013; Mulder 2017; Guerra 2017). According to Mulder et al. (2012), understanding sustainability challenges requires transdisciplinary approach and capabilities to analyse the challenges from various scales and viewpoints. Takala and Korhonen-Yrjänheikki (2013) emphasise the need to involve multiple stakeholders in development processes to be able to view the challenge from a range of perspectives. According to Mulder et al. (2012), solving these challenges requires both knowledge about the criteria and principles of sustainable technologies and, eventually, the ability to develop these technologies. Mulder (2017) underlines the connection between technological and social changes: technological changes do not happen in isolation but are intertwined with societal processes, and this sets requirements for engineering graduates' competencies and capabilities.

In research about sustainability in higher education, frameworks of sustainability competencies have become a popular way of conceptualising graduate capabilities for solving and managing sustainability challenges. The framework developed by Wiek et al. (2011, 2016) and further developed by, for example, Brundiers et al. (2021) and Redman and Wiek (2021) is based on a problem-solving framework and includes several competencies that are needed to develop sustainable solutions. Originally, Wiek et al. (2011) included five competencies in their framework: systems thinking competency, strategic competency, anticipatory competency, normative competency and interpersonal competency. The framework has been further assessed and developed to include implementation competency and intrapersonal competency (Brundiers et al. 2021).

Recent studies by Quelhas et al. (2019) and Beagon et al. (2022) confirm the relevance of sustainability competencies in engineering education. Beagon et al.

(2022) studied the views of engineering education stakeholders by comparing sustainability competencies suggested by students, academics and employers with UNESCO (2017) competencies for the Sustainable Development Goals (SDGs) as well as five other competencies they had identified as being relevant according to previous research. As a result, they found high resonance with systems thinking, normative competency, strategic competency, interpersonal competency, critical thinking competency and integrated problem-solving competency, and some resonance with self-awareness competency. According to the results of their study, the focus groups also highlighted fundamental disciplinary competencies and interdisciplinarity and some also continuous learning as essential competencies for solving the SDGs. According to the researchers, the only UNESCO competency that was not brought up in the focus group discussions was anticipatory competency.

In their study on Finnish engineering education, Takala and Korhonen-Yrjänheikki (2013) identified that some general competencies, such as ethics, systemic thinking and life-cycle thinking would be relevant across various engineering disciplines, but field-specific differences also need to be considered. Moreover, beyond single competencies they call for a systems-level change in engineering education towards learning for sustainability. According to them, engineering challenges should be viewed broadly. Non-technical aspects of the challenges should be considered, and engineers should have a role in not only solving but also defining the challenges with multiple stakeholders.

Thus, to build a connection between sustainability and engineering education sustainability challenges should be addressed on two levels: first, as challenges that require professional expertise to solve them, and second, by considering the societal context and systems perspective (Guerra 2017). To be able to approach sustainability challenges as multifaceted and complex and to be able to connect them with engineering practice, both are needed: expertise of the technologies and solutions that can support sustainability but also widening the perspective from problem-solvers to problem-definers and including to problem-defining and problem-solving also non-technological, social, environmental and ethical aspects.

Finally, it is important to consider the development of personal capabilities (Anderson 2013; Quelhas et al. 2019; Jaakkola et al. 2022). The participants in the survey by Quelhas et al. (2019) valued self-knowledge as the most important competency of engineering graduates, indicating that the graduates should have 'the ability to reflect on one's own role in the local community and society, as well as continually evaluate and increasingly motivate one's own actions while dealing with one's own feelings and desires' (Quelhas et al. 2019, 625). Also, Guerra (2017) highlights the importance of supporting the development of students' personal capabilities. According to Guerra, openness of learning objectives as well as the learning process is needed to support the development of students' agency.

7.2.2 Status of Embedding Sustainability in Engineering Education Curricula

Despite the confirmed relevance of sustainability challenges for the engineering professions, as well as the relevance of sustainability competencies for engineering graduates, existing studies show that embedding sustainability as a cross-cutting theme across all engineering disciplines is still underway. Institution-wide analyses of the implementation of the SDGs (Pálsdóttir and Jóhannsdóttir 2021a) and sustainability competencies (Pálsdóttir and Jóhannsdóttir 2021b) at the University of Iceland show that the integration of sustainability in the engineering discipline was relatively low compared to other disciplines at the same university. Studies on the integration of sustainability into specific engineering disciplines have come up with similar results. For example, Ruiz-Rivas et al. (2020) analysed the energy engineering curricula of ten higher education institutions across the world and identified that integration of sustainability was weak and focused too much on the environmental dimension. According to Kolmos et al. (2016), many of the change initiatives regarding embedding sustainability into engineering curricula remain at a level of making course-level changes that do not lead to wider institutional change or organisational learning.

According to Vehmaa et al. (2018), many of the engineering education learning outcomes support the sustainability competencies even though sustainability might not be at the core of the curriculum. However, when sustainability is not systematically introduced, graduates may find it difficult to make connections between their discipline and sustainability, which highlights the importance of building the connection during the studies (Vehmaa et al. 2018; Brundiers et al. 2021). A study by Gutierrez-Bucheli et al. (2022) shows a worrying gap between the expected and actual learning outcomes of engineering graduates in relation to sustainability, especially in the area of intrapersonal capabilities such as identity, values and self-awareness, as well as intersubjective learning outcomes, such as learning about cultural and political perspectives of change.

To conclude, the relevance of sustainability is agreed upon by many, but the engineering disciplines still struggle to build connections between the disciplinary expertise and challenges and solutions of sustainability. In the process of embedding sustainability into curriculum the broad topic might be framed as an environmental challenge, since resource efficiency and optimisation of production processes are at the heart of many engineering disciplines. The importance of finding ways to embed broad sustainability topics into often tightly technology-focused programmes has been identified as a crucial issue (Guerra 2017), but research on sustainability in engineering education has not yet succeeded in developing frameworks that would support curriculum developers in building bridges between sustainability and their specific engineering disciplines.

7.3 Methods

The concept of the curriculum is a central concept in this study. As a theoretical concept, curriculum has several conceptualisations (e.g. Annala et al. 2016; Lindén et al. 2017). In this study, curriculum is understood as a dynamic entity. The understanding of society, the role of education in society and the educational goals are constantly negotiated in formal and informal curriculum processes among academic staff, students and societal stakeholders (Annala et al. 2016). Analysis of curriculum documents sheds light on what is emphasised in these processes of negotiating the purposes and the goals of education. In this study, written curricula are viewed as a dataset that provides a view to one moment of these negotiations.

The higher education curricula in Finland are defined by higher education institutions but also affected by international education policies, such as the Bologna process. A central aim of the Bologna process has been to develop standardised and comparable education across the European higher education area. The aim has been that all countries would adopt, at least to some extent, outcome-based education and the practice of defining graduate learning outcomes (Allais 2014). These education policy standards also guide the Finnish engineering education curricula and how they are documented. Usually, curriculum documents state explicit learning outcomes of graduates and follow a typical learning objective formulation of what a graduate is expected to know or be able to do after completing their degree (Biggs and Tang 2011). However, in this study, the interest is beyond specific learning objectives. Therefore, the curriculum texts defining broader educational goals such as the purpose of the programme or role of the graduates in the society have been included in the analysis. In addition, goals related to students' learning are understood broadly. Thus, in the analysis, the interest is on how the three domains of curriculum –knowing, acting and being (Barnett and Coate 2005)—are included in the documents.

7.3.1 Data Selected for the Study

In Finland, engineering education is provided via two types of higher education institutions: universities and universities of applied sciences. In total, 27 educational institutions have the right to grant degrees in the field of technology, 9 of them universities and 18 universities of applied sciences. In this study, the focus is on the programmes offered by universities leading to a master's degree in science (technology). The master's degree in science (technology) is equal to Level 7 in the European Qualifications Framework (Government Decree on the National Framework, 120/2017).

This study identified in total 112 engineering education degree programmes offered by eight Finnish universities. One of the institutions with a right to offer education in the field of technology had not yet started their programmes at the time of retrieving the data. The notion of a degree programme varied both between and within universities, as sometimes the programme included several specialisation areas, sometimes only one.

The curriculum data were collected in December 2022 from public curriculum databases published on the universities' websites. The descriptions of specialisation areas were also collected and included in the analysis. The curricula that were valid in the academic year 2022–2023 were selected for analysis. Out of the 112 programmes, 3 did not include any description in their curriculum, thus the final number of curriculum documents analysed was 109. Additionally, in three of the curriculum documents some of the areas of specialisation did not have any descriptions or learning objectives. These documents were included in the analysis, but they may present an incomplete view of the programme.

The texts describing the educational programmes and areas of specialisation were included in the analysis, thus excluding the degree structures, course names and descriptions of individual courses. This set of data was chosen because it enabled to create an overview to Finnish engineering education curricula and its goals. The data was rich to serve the purpose of this study (i.e. developing an understanding of how sustainability is approached in engineering education). However, it should be noted that the data does not cover the details of each curriculum and its implementation nor information on how the curriculum is experienced by the students.

7.3.2 Analysis Methods

The analysis was conducted by the author in two stages. As there is not yet consensus about a framework that would best support embedding sustainability into engineering education, an inductive thematic analysis (Byrne 2022) was chosen as the first step of the analysis. The data were read several times and mentions of sustainability or its dimensions and educational goals as well as specific learning objectives framed as sustainability-relevant were coded using the ATLAS.ti (2023) software. The coding was developed through numerous iterations with an aim to identify recurring patterns regarding how sustainability was approached in the curriculum documents. Preliminary themes were identified already during the coding process and finally all codes were organised under four themes representing separate ways of approaching sustainability integration.

During the second stage of the analysis, the curriculum documents were organised according to the approaches of sustainability integration identified in the previous stage. The ATLAS.ti network tool was used to identify the presence of the themes in each curriculum document and the documents were categorised according to those identified themes. When forming the categories, a decision had to be made on how to categorise the documents which included several themes. In these cases, the documents were categorised according to the theme that represented a more holistic view to sustainability. For example, the programmes that viewed sustainability as a complex challenge that requires systemic changes (theme 1) would usually also describe sustainability as a technical challenge (theme 2). One additional category was also formed for those curriculum documents that excluded sustainability. The results of this analysis form a view of how sustainability was approached in the curricula, what sustainability-related educational goals were typical in each category and how various engineering disciplines approached sustainability.

7.4 Results

This section discusses the results of the analysis. First, the ways how sustainability was understood in the analysed curricula are presented regarding the whole data. Then, findings regarding the identified ways of approaching curricula are presented, along with the educational goals associated to these approaches. Also, the presence of each approach across various engineering education disciplines is discussed while presenting each approach.

7.4.1 Understanding of Sustainability in Engineering Education Curricula

The analysis revealed sustainability-related educational goals in 77 out of the total of 109 curriculum documents. In addition, one of the documents mentioned sustainability without linking it to educational goals, but instead mentioned the energy efficiency of learning environments as a sustainability measure taken in the programme in question. In addition to direct mentions of sustainability, sustainable development or the SDGs, the documents included concepts such as green transition, the circular economy, climate change, carbon neutrality, energy transition and food security. The ecological dimension was emphasised in comparison with social and economic sustainability, although economic viability as such was emphasised in the educational goals. Global perspectives on social sustainability were only rarely addressed. Most often the social dimension of sustainability was mentioned as a requirement to balance environmental harm and human well-being.

In the analysed data, the approach to sustainability was often technocratic, limiting the perspective to technological aspects of sustainability challenges such as the development of resource and energy efficiency, innovation of sustainable materials and avoidance of environmental harm in production processes. The natural environment was reduced to natural resources and linked to the responsibility for avoiding environmental harm. The following quotation is the only one mentioning animals or plants, but it also names oceans as important international trade passageways and as production sites of energy, raw materials and food.

> The oceans are the main highways of international trade. Moreover, they provide us with many vital resources: the oceans have vast living resources in the form of fish and plankton and there are gigantic resources of raw materials below the seas such as oil, gas and minerals. [...] Only by being better at utilizing the resources of the seas will it be possible to feed the growing population of the earth. And only by improving our activities at sea can we reduce our harmful impact on the environment and in this way make the living conditions better, for both humans and animals. (Programme in maritime engineering, U1)

The sustainability-related solutions presented as relevant in the analysed curricula focused on optimising the current technologies and developing new ones. Sometimes the sustainability challenge was approached as more complex, and framed as societal, not only technological. In these cases, the educational goals were framed with wider societal goals, for example, carbon neutrality, and sustainability challenges were identified as challenges that require transdisciplinary collaboration. The goal was not to optimise current technologies but to develop solutions to sustainability challenges by utilising a range of technologies and to create completely new, sustainable technologies.

7.4.2 *Five Ways of Approaching Sustainability in Engineering Education Curricula*

The aim of the analysis was to understand what ways of approaching sustainability can be identified in the curricula. During the first stage of analysis, four themes on approaching sustainability were formed, and in the second stage the curriculum documents were categorised based on these themes (Fig. 7.1). Based on the categorisation, five ways of approaching sustainability were identified.

In total, 31 curriculum documents across the engineering disciplines did not connect sustainability to any educational goals (category A). Sustainability was often excluded from the curriculum in programmes representing disciplines such as electrical engineering, computer science engineering and biomedical engineering programmes. Next, the categories which represent the various approaches of including sustainability into curricula (categories B, C, D and E) are presented (Table 7.1).

7.4.2.1 Engineering Discipline as an Enabler of Sustainability

The vaguest way to take sustainability into consideration in the curricula was to name the technological development related to the discipline an enabler of sustainability. Sustainability was approached from the perspective of the environment and energy efficiency. Social sustainability was mentioned as development of electronic communication was suggested to support 'social equity and social sustainability around the world' (Programme in computer science, U3). The following quote from the field of mathematics and operations research names the environment as an application area into which the methods of the discipline could be applied.

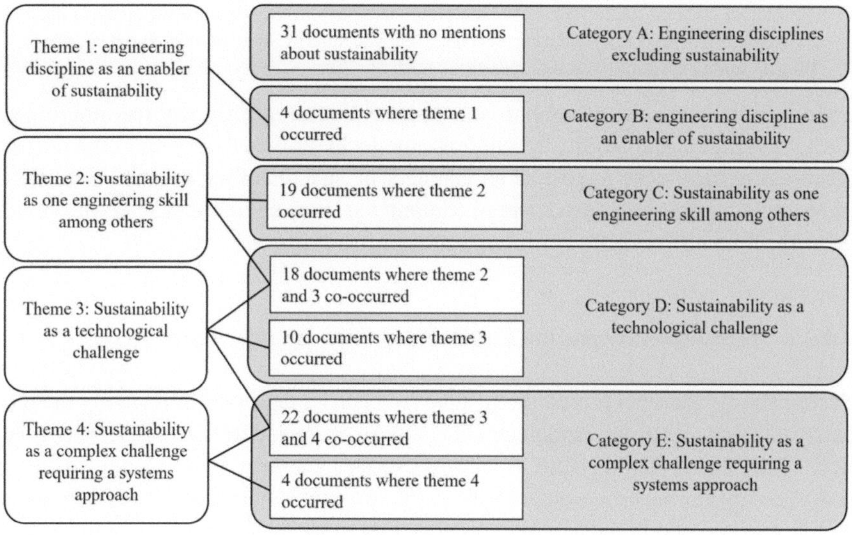

Fig. 7.1 Themes, occurrence of themes in the dataset and the formed categorisation

Table 7.1 Identified curriculum approaches

Approach to sustainability	Description of the approach
Category A: Engineering disciplines excluding sustainability	No mentions of sustainability in relation to the educational goals of the programme.
Category B: Engineering discipline as an enabler of sustainability	The discipline, or its applications, are presented as enablers of sustainability. The curricula do not include any sustainability-related learning objectives.
Category C: Sustainability as one engineering skill among others	Sustainability is mentioned as part of learning objectives, usually as an ability to take sustainability into account as part of multiple planning parameters.
Category D: Sustainability as a technological challenge	Sustainability is presented as a core engineering task, but viewed from the perspective of technological innovations. Curricula may include several sustainability-related learning objectives, but these are focused on discipline-specific knowledge and skills.
Category E: Sustainability as a complex challenge requiring a systems approach	Sustainability is viewed as a complex challenge requiring systems approach and engineers' contribution. Learning objectives cover all three domains of learning and include discipline-specific knowledge and skills, knowledge of sustainability, skills related to solving complex sustainability challenges and goals related to supporting graduates' action competency.

> Methods of systems and operations research are used extensively in application areas such as finance, logistics, manufacturing, services, defence, energy, environment and healthcare. (Programme in mathematics and operations research U1)

In this category, the curriculum documents did not specify any learning objectives related to sustainability. In contrast, the technologies that were in the core of the programmes were presented as enablers of sustainability as such. The disciplines utilising this approach covered computer science engineering, electrical engineering and mathematics.

7.4.2.2 Sustainability as One Engineering Skill among Many

Sustainability was also present in the analysed curricula as one skill among the others, usually as one of the parameters that should be considered in engineering practice. These parameters defined aspects of product viability, for example, customer needs, economic viability and efficiency. In addition to the ability to take sustainability into account in the planning process, sustainability was named as one area in which the graduate would gain expertise, like in the following quote:

> During their studies, students will further acquire expertise in other areas, such as project planning and management, teamwork and coordination, entrepreneurship, critical thinking and sustainability. (Programme in communications and data science, U6)

Typically, sustainability was mentioned among a long list of many things a graduate should consider. Common in this theme was that sustainability was either listed without any further definitions of the concept or narrowed down to cover only the environmental dimension of sustainability.

Regarding the learning objectives, it was common that they remained ambiguous. The typical learning objective in this category was the ability to consider sustainability among several other planning parameters. In addition, knowledge of discipline-specific sustainability issues and understanding the importance of sustainability were mentioned. Curriculum documents thus addressed all three domains of curriculum, but only one at a time, leaving it open what knowing, acting and being would actually support graduates' capability to act sustainably as an engineering professional.

Programmes that approached sustainability as one engineering skill among many represent a wide range of engineering disciplines. These included industrial engineering, mechanical engineering, materials and process engineering, computer science engineering and energy engineering.

7.4.2.3 Sustainability Challenge as a Technological Challenge

A stronger, and well-reflected, approach to sustainability can be identified in curriculum documents that identified sustainability as a core engineering challenge. In these curricula, the focus was on a certain sustainability challenge, and the technological solutions that are needed to cope with it. The documents cover topics such as climate change, energy transition, sustainable industrial processes and sustainable production of materials. In an energy engineering programme, environmental issues and climate change were set as the context of educational goals, but the graduates' learning focused on technologies:

> Programme gives extensive and future-proof capabilities for students in order to make the ongoing energy transition into reality. Programme provides good analytical skills and extensive knowledge about key smart energy system and components related technologies (like battery storages, microgrids, big data analytics/artificial intelligence) as well as future market and business models. (Programme in energy engineering, U7)

The challenges that the graduates should meet were connected to grand sustainability challenges. Sustainability was still approached from the perspective of the disciplines, but this time sustainability challenges were considered when defining the technological challenges engineers should solve. Programmes approaching sustainability as a technological challenge covered a wide range of engineering disciplines. This approach was especially highlighted in chemical engineering, industrial engineering and energy engineering programmes.

Curricula in this category emphasised skills related to developing technological solutions to tackle sustainability-related challenges. These skills included analysis of the environmental or sustainability impacts of a range of solutions and development of discipline-specific sustainable solutions, for example, solutions in energy engineering, software development, manufacturing or chemical engineering. The curricula also highlighted disciplinary knowledge connected to sustainability. The goals stated that students would understand the carbon footprint of software solutions, the environmental effects of new materials or sustainable solutions and technologies. Additionally, knowledge about sustainability beyond the disciplinary context and sustainability competencies was mentioned in some documents. In comparison to the previous category, the documents often connected knowledge and skills. Learning objectives related to the domain of being were emphasised in two documents, but in most documents the domain of being was absent.

7.4.2.4 Sustainability as a Complex Challenge Requiring a Systems Approach

The curricula approaching sustainability as a grand challenge requiring a systems approach used concepts such as sustainability crisis, climate change and biodiversity loss and green cities. The social dimension of sustainability was discussed from the perspective of the effects of human activities to the environment. The sustainability challenge was defined as complex, including several dimensions that needed to be balanced. The programmes combined the broad view to sustainability challenges with specific engineering topics, emphasising the need to connect the wide ecological and societal context with the development of engineering solutions. To be able to deal with complex sustainability challenges a transdisciplinary approach was needed. The graduates should have the ability to work with diverse people and to deal with competing interests, like in the following quote:

> Complex sustainability challenges, such as climate change, loss of biodiversity, or poverty and inequality, call for new mindsets and skills. These mindsets should incorporate a systems-based view of the planet and the ability to work with professionals from different fields and cultures. Sustainability challenges require multi-level solutions that involve a wide range of stakeholders with often competing interests. (Sustainability-focused engineering programme, U1)

Curricula in this category emphasised knowledge of sustainability and sustainability competencies. In line with the previous category, skills related to developing engineering solutions to tackle sustainability-related challenges were present in most of the documents. Regarding the domain of being, programmes emphasised graduates' action competence, more specifically graduates' capacity and courage to act.

Approaching sustainability as a complex challenge was a dominant approach in environmental engineering and in engineering programmes with a specific sustainability focus. Additionally, more than half of the programmes in the area of civil engineering were identified to belong to this category. In most of the disciplines, there was at least one programme that approached sustainability as a complex challenge, thereby showing that it is possible to embed sustainability across engineering disciplines in a manner that approaches sustainability as a complex challenge requiring a systems approach.

7.5 Discussion

Of the 109 education programmes that were analysed, 77 linked sustainability to the educational goals of the programme. However, what is meant by sustainability, and what sustainability-related learning the students should gain often remained open, indicating that sustainability-related educational goals might not always be well-reflected, as suggested previously by Guerra (2017). In line with previous research (e.g. Weisser 2017; Guerra 2017), the results confirm that sustainability was conceived as human-centred and often approached in a technocratic manner. The concept was viewed from the ecological perspective, as noted earlier by Nicolaou and Conlon (2012) with an emphasis on optimising the use of natural resources and resource efficiency. When sustainability challenges were at the core of the programme, the challenges were viewed as being more complex, and other than ecological dimensions of sustainability were also included. Some of these programmes emphasised the non-technological, social aspects linked to technological challenges, in line with what Mulder (2017) and Takala and Korhonen-Yrjänheikki (2013) have previously suggested.

The five ways of approaching sustainability evolved from non-existent integration with no sustainability-related learning objectives to integration that adapted sustainability to the norms and practices of engineering education and, eventually, to integration of sustainability in a holistic way, embedding both technological and non-technological aspects of sustainability challenges into the core of engineering disciplines. When sustainability was not placed in the core of the programme, an important way to embed sustainability was to see it as part of engineering practice, that is, as part of the skills required for analysing a challenge and designing technological solutions. Only when sustainability was approached as a complex challenge was it evident that sustainability was widening the disciplinary norms. These findings add to those of Sandri (2021), who has identified that academics approach sustainability either by adapting it to disciplinary norms or by challenging disciplinary norms and taking sustainability-related questions into the core of the discipline. According to the results of this analysis, it seems that in engineering education it is still far more usual to adapt sustainability to the disciplinary norms than to challenge them and take sustainability-related challenges into the core of the discipline.

Moreover, the results imply that delving into complex sustainability challenges is still addressed as a task of only some engineering programmes, and not as a topic that should be embedded across all engineering education disciplines. Viewing sustainability as a complex challenge appeared frequently in environmental engineering programmes and programmes with specific sustainability focus, but it was rare in other areas of engineering education. When comparing these results to the integration strategies suggested by Barth (2015), Sandri (2021) and Kolmos et al. (2016), it is evident that the change has mostly remained on a level that does not renew the educational paradigm or address sustainability as a transdisciplinary issue. However, the analysis also shows that across engineering education disciplines it is possible to challenge the existing disciplinary norms and renew of even rebuild the curriculum so that sustainability with all its complexity is at the core of the programme. Although not significant in terms of quantity, across almost all engineering education disciplines there were signs of approaching sustainability as a complex topic that requires not only technological expertise, but ability to also consider the societal context and systems perspective.

The analysis of educational goals and specific learning objectives sheds light on different options for embedding sustainability into engineering disciplines. Analysis shows that all three domains of curriculum—knowing, acting and being—were utilised when embedding sustainability into curricula. Specific learning objectives usually covered the domains of knowing and acting and were usually tightly connected to discipline-specific knowledge and engineering practice. Despite the previous findings regarding the relevance of sustainability competencies in engineering education (Beagon et al. 2022; Quelhas et al. 2019), these competencies were seldom applied in the analysed curricula. When the domain of being was addressed, the learning objectives highlighted appreciation of the importance of sustainability and development of mindsets and courage to drive sustainable solutions. However, most curricula did not address educational goals of this type. These results are aligned with the findings of Gutierrez-Bucheli et al. (2022), who identified in their study that interior approaches were lacking in the sustainability-related learning outcomes of engineering programmes.

Integration of sustainability across higher education curricula requires negotiating and balancing diverse educational goals, especially those in the area of disciplinary expertise, professional competencies, digitalisation and sustainability-related capabilities (Gumaleius and Kolmos 2019). The approaches to integrate sustainability identified in this study highlight the various strategies academics use to balance sustainability with other educational goals. According to the results, several strategies for integrating sustainability into curricula are possible. To further the integration, most relevant are those approaches in which sustainability was embedded at the level of learning objectives (categories C, D and E). These approaches can be used as a starting point for engineering programmes to discuss and consciously develop their own way of approaching sustainability.

As Guerra (2017) has suggested, defining explicit sustainability-related learning objectives supports systematic integration of sustainability into curriculum. Even though the analysed curricula highlight disciplinary knowledge and skills and give less attention to knowledge of sustainability, sustainability competencies and capabilities related to students' personal development, the results are useful for future development. The results show that across the engineering disciplines and selected ways to approach sustainability it is possible to attach sustainability-related learning objectives related to all domains of curriculum: knowing, acting and being. Since the findings are based on existing disciplinary curricula, they may support academics and curriculum developers to develop meaningful learning objectives and to identify ways to connect sustainability to various engineering disciplines.

Based on this study, at least two topics for future research can be identified. First, the framework of approaches to sustainability integration should be further developed with research design that goes beyond written curricula and enables analysing curriculum as a process of continuous negotiations or as experienced by students. Second, further research on sustainability-related learning objectives in engineering curricula is needed. To reach the pressing general educational goal of embedding sustainability across engineering disciplines, it is necessary to identify and develop ways for balancing the various educational goals and learning objectives in a way that supports both disciplinary development but recognises and cherishes the complexity of sustainability challenges.

References

Allais S (2014) Selling out education. Sense Publishers, National qualifications frameworks and the neglect of knowledge. https://doi.org/10.1007/978-94-6209-578-6

Anderson MD (2013) Higher education revisited: sustainability science and teaching for sustainable food systems. In: Albrecht S, Braun R (eds) Future of food: STATE of the art, challenges and options for action. UIT, Cambridge, pp 179–188

Annala J, Lindén J, Mäkinen M (2016) Curriculum in higher education research. In: Case J, Huisman J (eds) Researching higher education. International perspectives on theory, policy and practice. Society for Research into Higher Education & Routledge, pp 171–189. https://doi.org/10.4324/9781315675404

ATLAS.ti Scientific Software Development GmbH (2023) ATLAS.ti Mac (version 23.2.1) [Qualitative data analysis software]. https://atlasti.com

Barnett R, Coate K (2005) Engaging the curriculum in higher education. Society for Research into Higher Education

Barth M (2015) Implementing sustainability in higher education. In: Learning in an age of transformation. Routledge. https://doi.org/10.4324/9780203488355

Beagon U, Kövesi K, Tabas B, Nørgaard B, Lehtinen R, Bowe B (2022) Preparing engineering students for the challenges of the SDGs: what competences are required? Europ J Eng Educ 48(1):1–23. https://doi.org/10.1080/03043797.2022.2033955

Bergen JP, Robaey Z (2022) Designing in times of uncertainty: what virtue ethics can bring to engineering ethics in the twenty-first century. In: Dennis MJ, Ishmael G, Umbrello S, van den Hoven J (eds) Values for a post-pandemic future, philosophy of engineering and technology, vol 40, pp 163–184. https://doi.org/10.1007/978-3-031-08424-9_9

Biggs J, Tang C (2011) Teaching for quality learning at university, 4th edn. Open University Press

Brundiers K, Barth M, Cebrián G, Cohen M, Diaz L, Doucette-Remington S, Dripps W, Habron G, Harré N, Jarchow M, Losch K, Michel J, Mochizuki Y, Rieckmann M, Parnell R, Walker P, Zint M (2021) Key competencies in sustainability in higher education–toward an agreed-upon reference framework. Sustain Sci 16(1):13–29. https://doi.org/10.1007/s11625-020-00838-2

Byrne D (2022) A worked example of Braun and Clarke's approach to reflexive thematic analysis. Qual Quant 56(3):1391–1412. https://doi.org/10.1007/s11135-021-01182-y

Christie BA, Miller KK, Cooke R, White JG (2015) Environmental sustainability in higher education: what do academics think? Environ Educ Res 21(5):655–686. https://doi.org/10.1080/13504622.2013.879697

Desha C, Hargroves K (2014a) Higher education and sustainable development. A model for curriculum renewal. Routledge

Desha C, Hargroves K (2014b) A peaking and tailing approach to education and curriculum renewal for sustainable development. Sustainability 6(7):4181–4199. https://doi.org/10.3390/su6074181

ENAEE (2021) EUR-ACE® framework standards and guidelines. European network for accreditation of engineering. Education. Retrieved on September 18 2023, from https://www.enaee.eu/wp-content/uploads/2022/03/EAFSG-04112021-English-1-1.pdf

Frank P, Stanszus LS (2019) Transforming consumer behavior: introducing self-inquiry-based and self-experience-based learning for building personal competencies for sustainable consumption. Sustainability 11(9):2550. https://doi.org/10.3390/su11092550

Giangrande N, White RM, East M, Jackson R, Clarke T, Saloff Coste M, Penha-Lopes G (2019) A competency framework to assess and activate education for sustainable development: addressing the UN sustainable development goals 4.7 challenge. Sustainability 11(10):2832. https://doi.org/10.3390/su11102832

Global Sustainable Development Report (2019) The future is now—science for achieving sustainable development. United Nations, New York

Government Decree on the National Framework (120/2017) Retrieved on September 22 2023, from https://www.finlex.fi/fi/laki/ajantasa/2017/20170120

Griggs D, Stafford-Smith M, Gaffney O, Rockström J, Öhman MC, Shyamsundar P, Steffen W, Glaser G, Kanie N, Noble I (2013) Policy: sustainable development goals for people and planet. Nature 495(7441):305–307. https://doi.org/10.1038/495305a

Guerra A (2017) Integration of sustainability in engineering education. Why is PBL an answer? Int J Sustain High Educ 18(3):436–454. https://doi.org/10.1108/IJSHE-02-2016-0022

Gumaleius L, Kolmos A (2019) The future of engineering education: where are we heading? In: Proceedings of the SEFI 47th annual conference, Budapest, Hungary. 16–20 Sept 2019

Gutierrez-Bucheli L, Kidman G, Reid A (2022) Sustainability in engineering education: a review of learning outcomes. J Clean Prod 330:129734. https://doi.org/10.1016/j.jclepro.2021.129734

Jaakkola N, Karvinen M, Hakio K, Wolff L-A, Mattelmäki T, Friman M (2022) Becoming self-aware–how do self-awareness and transformative learning fit in the sustainability competency discourse? Front Educ 7:2022. https://doi.org/10.3389/feduc.2022.855583

Kolmos A, Hadgraft RG, Holgaard JE (2016) Response strategies for curriculum change in engineering. Int J Technol Des Educ 26(3):391–411. https://doi.org/10.1007/s10798-015-9319-y

Lindén J, Annala J, Coate K (2017) The role of curriculum theory in contemporary higher education research and practice. In: Huisman J, Tight M (eds) Theory and method in higher education research, vol 3. Emerald Publishing Limited, Bingley, pp 137–154. https://doi.org/10.1108/S2056-375220170000003008

Mulder K (2004) Engineering education in sustainable development: sustainability as a tool to open up the windows of engineering institutions. Bus Strateg Environ 13(4):275–285. https://doi.org/10.1002/bse.407

Mulder KF (2017) Strategic competences for concrete action towards sustainability: an oxymoron? Engineering education for a sustainable future. Renew Sust Energ Rev 68(2):1106–1111. https://doi.org/10.1016/j.rser.2016.03.038

Mulder KF, Segalàs J, Ferrer-Balas D (2012) How to educate engineers for/in sustainable development: ten years of discussion, remaining challenges. Int J Sustain High Educ 13(3):211–218. https://doi.org/10.1108/14676371211242535

Nicolaou I, Conlon E (2012) What do final year engineering students know about sustainable development? Eur J Eng Educ 37(3):267–277. https://doi.org/10.1080/03043797.2012.681863

Pálsdóttir A, Jóhannsdóttir L (2021a) Signs of the United Nations SDGs in university curriculum: the Case of the University of Iceland. Sustainability 13(16):8958. https://doi.org/10.3390/su13168958

Pálsdóttir A, Jóhannsdóttir L (2021b) Key competencies for sustainability in University of Iceland curriculum. Sustainability 13(16):8945. https://doi.org/10.3390/su13168945

Paulsen M (2021) From late Holosene to early Anthropocene educational thinking (humanism revisited). In: Petersen KB, Brömssen K, Jacobsen GH, Garsdal J, Paulsen M, Koefoed O (eds) Rethinking education in light of global challenges: Scandinavian perspectives on culture, society, and the Anthropocene. Routledge, pp 204–219. https://doi.org/10.4324/9781003217213

Quelhas OLG, Lima GBA, Ludolf NV-E, Meiriño MJ, Abreu C, Anholon R, Vieira Neto J, Rodrigues LSG (2019) Engineering education and the development of competencies for sustainability. Int J Sustain High Educ 20(4):614–629. https://doi.org/10.1108/IJSHE-07-2018-0125

Redman A, Wiek A (2021) Competencies for advancing transformations towards sustainability. Front Higher Educ 6. https://doi.org/10.3389/feduc.2021.785163

Rieckmann M (2012) Future-oriented higher education: which key competencies should be fostered through university teaching and learning. Futures 44(2):127–135. https://doi.org/10.1016/j.futures.2011.09.005

Ruiz-Rivas U, Martinez-Crespo J, Venegas M, Chinchilla-Sanchez M (2020) Energy engineering curricula for sustainable development, considering underserved areas. J Clean Prod 258:120960. https://doi.org/10.1016/j.jclepro.2020.120960

Salovaara JJ, Soini K, Pietikäinen J (2020) Sustainability science in education: analysis of master's programmes' curricula. Sustain Sci 15(3):901–915. https://doi.org/10.1007/s11625-019-00745-1

Sandri O (2021) Providing a 'point of entry': approaches to framing sustainability in curriculum design in higher education. Aust J Environ Educ 37(1):56–68. https://doi.org/10.1017/aee.2020.19

Sandri O (2022) What do we mean by 'pedagogy' in sustainability education? Teach High Educ 27(1):114–129. https://doi.org/10.1080/13562517.2019.1699528

Sigahi TFAC, Rampasso IS, Anholon R, Sznelwar LI (2023) Classical paradigms versus complexity thinking in engineering education: an essential discussion in the education for sustainable development. Int J Sustain High Educ 24(1):179–192. https://doi.org/10.1108/IJSHE-11-2021-0472

Takala A, Korhonen-Yrjänheikki K (2013) A National Collaboration Process: Finnish Engineering Education for the benefit of people and environment. Sci Eng Ethics 19:1557–1569. https://doi.org/10.1007/s11948-011-9330-y

Tejedor G, Segalàs J, Rosas-Casals M (2018) Transdisciplinarity in higher education for sustainability: how discourses are approached in engineering education. J Clean Prod 175:29–37. https://doi.org/10.1016/j.jclepro.2017.11.085

UNESCO (2017) Education for sustainable development goals: learning objectives. UNESCO

Vehmaa A, Karvinen M, Keskinen M (2018) Building a more sustainable society? A case study on the role of sustainable development in the education and early career of water and environmental engineers. Sustainability 10(8):2605. https://doi.org/10.3390/su10082605

Weisser CR (2017) Defining sustainability in higher education: a rhetorical analysis. Int J Sustain High Educ 18(7):1076–1089. https://doi.org/10.1108/IJSHE-12-2015-0215

Wiek A, Withycombe L, Redman CL (2011) Key competencies in sustainability: a reference framework for academic program development. Sustain Sci 6(2):203–218. https://doi.org/10.1007/s11625-011-0132-6

Wiek A, Bernstein M, Foley R, Cohen M, Forrest N, Kuzdas C, Kay B, Withycombe Keeler L (2016) Operationalising competencies in higher education for sustainable development. In: Barth M, Michelsen G, Rieckmann M, Thomas I (eds) Handbook of higher education for sustainable development. Routledge, pp 241–260

Noora Jaakkola is an expert in curriculum development in Aalto University and a PhD candidate at the University of Helsinki's Faculty of Education, located both in Finland. She is deeply invested in sustainability education and the integration of sustainability across higher education disciplines. Her research focuses on understanding how sustainability goals are affected when integrated into educational contexts across various fields, especially in engineering education. She, working at the intersection of research and university education development, is especially interested in how research can support the sustainability transition within the education sector.

Chapter 8
Transformative Teaching and Learning in Biology: Supporting Biodiversity Education for Promoting Sustainability

Eija Yli-Panula, Isabel Banos-González, Eila Jeronen, and Sandra Bueno-Férez

Abstract The planetary crisis requires an educational approach that seeks to comprehensively change people's knowledge, attitudes and values to support sustainable behaviour. Transformative education plays an important role in this process. When teaching is connected to students' everyday lives and values, as well as to the meanings of the community and includes critical discussion and reflection on personal assumptions and issues, this kind of teaching and learning opens new perspectives that can change the way they think and act. In this study, a systematic review was conducted to investigate the extent to which transformative teaching and learning have been conceptualised and put into practice in biology, specifically in biodiversity education (BE). The study was guided by two research questions: (i) How are the learning processes concerning transformative education supporting BE? (ii) What are the transformative teaching and learning methods used to promote BE? Our main finding was that prior learning is important for promoting transformative learning and disorienting dilemmas. Various teaching methods and strategies have been used, such as action learning, project and problem-based learning, teamwork, field trips and use of technological tools and discourses. This chapter also presents findings concerning learning outcomes, such as improvement of practical skills, positive long-term impact of technology on authentic problem-solving, change of attitude towards and positive assessment of biodiversity, impact of curricula and need for interdisciplinarity. The findings are discussed in connection with biodiversity, sustainability and transformative teaching and learning.

E. Yli-Panula (✉) · I. Banos-González
University of Turku, Turku, Finland

University of Murcia, Murcia, Spain
e-mail: eijyli@utu.fi

E. Jeronen
University of Oulu, Oulu, Finland

S. Bueno-Férez
University of Murcia, Murcia, Spain

F. Skarstein, L.-A. Wolff (eds.), *Sustainability and Science Education*, Environmental Challenges and Solutions,
https://doi.org/10.1007/978-3-032-24408-6_8

Keywords Biodiversity education • Biology • Future teacher • Primary school • Secondary school • Science education • Sustainability • Transformative learning • Transformative teaching

8.1 Introduction

Effective and timely actions are needed to curb climate change, biodiversity loss and environmental degradation, and to foster people's environmental awareness for a sustainable future (Mochizuki 2016). *Transformative education* has an important role to play in this, as the planetary crisis requires an educational approach that aims comprehensively to change knowledge, attitudes and values to support sustainable behaviour (Carr et al. 2018). According to the transformative paradigm, it is important to maintain and realise human potential in relation to the need to achieve and maintain social, economic and ecological well-being and recognising that they are part of the same dynamic (Sterling 2001). The entire education system, its members and society could benefit from teaching and learning that is consistent with transformative education, for example, in promoting sustainability transformation (Sterling et al. 2018).

In the last half-decade, educational researchers have offered numerous conceptions of what transformative education might look like. In one context, for example, transformative education is closely linked to the project of eradicating oppression and injustice, while in another, it is derived from phenomenological experience. Sometimes, transformative education refers to deep forms of learning in which students experience an expansion of value and meaning in relation to new scientific concepts and, sometimes, it is primarily concerned with learning how to learn (Yacek et al. 2020). In this study, transformative education is understood as an approach that offers students the opportunity to reflect on who they are, how they relate to the social world and who they want to be (Paul 2020; Yacek 2021). This endeavour is supported through transformative teaching and learning.

Transformative teaching is student- and competence-oriented, is rooted in transformative learning theory and engages with the normativity of research for sustainable development, as well as with experiences and inter- and transdisciplinary methods (Förster et al. 2019). In this study, it is also seen to be connected to students' everyday lives and values and community meanings (Osguthorpe and Jensen 2023). It promotes choices and develops expression and collaboration between teachers and students. Transformative teaching also requires teachers to reflect on and make decisions about what is defensible in relation to what is morally acceptable, just and caring (Osguthorpe and Jensen 2023). It emphasises teacher self--awareness and awareness of teaching goals and methods. Transformative teaching focuses on student-centred learning and uses active learning techniques and offers students, as individuals and as part of society, transformative learning situations (Osguthorpe and Jensen 2023).

Many educators refer to *transformative learning* as events that promote reflective learning in an environment that is relatively free from social and political constraints. Many researchers, on their part, describe transformative learning as a pedagogical philosophy. It is seen as a pedagogy that is consistent with transformative education, but distinct from critical pedagogy, progressive education and adult education. Some people use transformative learning and transformative education interchangeably (McWhinney and Markos 2003). In this study, transformative learning refers to events where students can examine their values and interests regarding the issues to be learned and possible to notice contradictions in their preconceptions (Osguthorpe and Jensen 2023). Critical discussion and reflection on personal assumptions and issues can lead to the transformation and adaptation of frames of reference or systems of meaning (Mezirow 1990, 1997; Fraser 2015). Learning experiences thus become a formative tool, providing new perspectives that can change the way students think and act (Mezirow 2009).

Transformative pedagogy is an approach that helps teachers and students develop their identity as whole individuals. It aims to develop cooperative social skills and knowledge of living together as a member of society, as well as individual socio-economic development (Farren 2019). Transformative pedagogy is based on moral-ethical concepts that guide stakeholders to work together to understand their own role and meaning in society (Farren 2019). It emphasises personal awareness and social change through action and critical reflection. It is important for educators to develop pedagogical skills that transform students into critical thinkers, problem solvers and innovators capable of applying learned information in the workplace and society as a whole. The study of Farren (2019) highlights the critical importance of transformational pedagogies in educating students to face the difficult problems of the twenty-first century.

According to Sterling (2010), 'environmental and sustainability education' is seen as a process by which the development of more sustainable lifestyles can be achieved. It combines theories of education and learning that, in turn, combine the best of liberal education with new ideas about transformative education, capacity building, creativity and adaptive leadership, which are considered crucial for sustainable development (Sterling 2008).

Sterling (2004, 2008, 2010, 2011, 2013) discusses transformative learning in the context of sustainability education, emphasising ecological and systems thinking to address sustainability challenges. According to both Sterling (2004) and Mezirow (1981), transformative learning is essential to achieving profound change in ways of thinking and acting and critical reflection. Both acknowledge the importance of social and cultural context in transformative learning. Transformative teaching focusing on student-centred learning uses active learning techniques and promotes critical reflection (Merzel 2023).

According to the Organisation for Economic Co-operation and Development (OECD 2018), education plays a fundamental role in developing the knowledge, skills, attitudes and values that enable people to contribute to and benefit from an inclusive and sustainable future. Nevertheless, O'Brien et al. (2013, p. 48) question whether it would be possible to achieve transformative learning without an axial revolution to address 'the complex transdisciplinary challenges posed by global environmental change' and which allows individuals to question current assumptions and beliefs. In this sense, Sjöström and Eilks (2018) claim that the main goal of science education at all levels is to support the development of critical, socially aware and participatory citizens that can apply scientific knowledge and skills when solving complex sustainability challenges. Preventing the loss of biodiversity is one of these challenges that should be addressed, and pro-nature approach supported (UN 2022).

Loss of biodiversity at all levels, including species extinction and functional and phylogenetic levels of diversity erosion, can lead to the breakdown of ecosystems (Van Weelie and Wals 2002). However, this risk is hardly understood by the general public because of its complexity and ambiguity (Renn 2008). This situation has arisen through both the lack of empathy or connection with nature (Pyle 2003), and the lack of knowledge about biological diversity and conservation (Heberlein 2013). Educators can contribute to the solution to this problem and its prevention by teaching students about biodiversity and biodiversity protection in an engaging way. Previous studies have shown that many factors, such as peoples' experiences of species, are related to species conservation motivation (Eyster et al. 2022). Teaching about biodiversity, for example, at the regional level, can motivate both teachers and students towards conservation efforts (Lavorel and Morel-Deville 2010).

Based on an environmental and sustainability education approach, biodiversity education could guide students into the understanding and analysing of the many meanings and dimensions of biodiversity and encourage the construction of knowledge applied to solving problems in various contexts (Navarro-Perez and Tidball 2012). Students benefit also of having the opportunity to develop critical skills and increase their awareness of scientific and non-scientific aspects of biodiversity. In addition, education should emphasise the importance of biological diversity, responsibility towards nature and action for the protection of biodiversity (Kollmuss and Agyeman 2002). The goal is to understand the various factors, interpretations and uses of biodiversity that are relevant, to make students aware of and to understand their own ecological, economic, ethical, spiritual and cultural values, and to make them recognise the relationship between biodiversity and human well-being (Lindemann-Matthies et al. 2009). Biodiversity education (BE) is thus considered a useful educational approach when teaching issues on ecosystems and sustainable development. It emphasises the importance of understanding the values of biodiversity and the responsible attitude towards living nature and the urgency of action for biodiversity conservation (UNESCO 2017; Wolff 2022). In this context, the goal of transformative education is to promote the skills needed to implement a sustainability breakthrough (Odell et al. 2019). The document 'Education for the Sustainable Development Goals' highlights transformative teaching and it calls for 'a transformative, action-oriented pedagogy that supports self-directed learning, participation and collaboration, problem orientation, interdisciplinarity and transdisciplinarity'

(UNESCO 2017, p. 2). This transformative approach is closely related to education for sustainability (EfS), which represents a comprehensive educational vision and emphasises the cognitive, social, emotional and functional learning of sustainability at all educational levels (UNESCO 2017). In this sense, for Sterling (2010), 'transformative learning' means a quality of learning that is deeply engaging, and touches and changes deep levels of values and believes through a process of realisation and recognition. For developing active citizenship and enable people's participation it is necessary to continue research on BE and EfS from the perspective of a transformative approach at all educational levels.

In these times, when climate change has caused widespread adverse impacts and related losses and damages to nature and people (IPCC 2023), and biodiversity across the world is deteriorating at rates unprecedented in human history (CBD 2022, p. 1), the key question for education is how it can support sustainable development regarding biodiversity. Although a fair amount of research has been published on transformative education and EfS, only a few studies have examined and reflected on the kind of teaching and learning (T/L) strategies that can be used to support BE in promoting sustainability. Therefore, we conducted a systematic review to analyse the state of the matter, considering that systematic reviews offer the opportunity to explore, survey, question and improve the published evidence in the field of interest (Monroe et al. 2019).

8.2 Theoretical Framework

In this theoretical framework, the focus is on transformative pedagogy. In the school culture, transformative learning plays a key role in promoting social processes among students and in the entire school community (Bivens et al. 2009; Mogren 2019). Transformative learning means developing students' participation, agency and solutions (Stevenson et al. 2012). Participation and agency are related to joint planning and visioning of the future, as well as critical thinking and reflection (Tilbury 2007). Transformative learning from the view of biodiversity can mean, for example, how everyone as an individual and community member can actively work to protect biodiversity and promote a sustainable use of it. This is shown, for example, in the study of Wickenberg et al. (2022), in which the interaction between committed individuals and stakeholder groups, as well as participation in collaborative learning and integrative and reflexive knowledge production, led to doing better things concerning nature-based solutions. In terms of teaching and other activities, this means a strong commitment to sustainable practices and sustainability education and the adoption of sustainability competencies (Collins-Figueroa 2012). At the learner level, it is essential to critically examine one's own assumptions and beliefs, to take emotions into account and to be aware of and tolerate contradictions.

8.2.1 The Need for Transformative T/L

T/L plays a key role in developing society members that support sustainability. For this, it is necessary that T/L processes are transformative. However, what is meant by transformative T/L? Already in the 1980s, the American social scientist Jack Mezirow developed an adult learning theory on transformation and defined 'transformative learning' as 'learning that transforms problematic frames of reference—sets of fixed assumptions and expectations (mental habits, meaning perspectives, mindsets)—to make them more inclusive, discriminating, open, reflective and emotionally capable of change' (Mezirow 2003, p. 58). Furthermore, in one of his early papers, Mezirow (1981) specified some key features of transformative learning, focusing on the following:

(i) Learning processes (how people learn): These encompass a number of elements that are evolutionary, recursive and spiral in nature (Taylor 1997) and aimed at changing entrenched beliefs that might impede learning, leading individuals to critical reflection and a commitment to action.
(ii) The results (what they learned): These are what students are able to do and think at the end of the learning period after being integrated into disciplinary and professional contexts (Mezirow 1981).
(iii) The conditions (how to improve the students' learning): These are external and internal conditions and factors that influence the students' ability to participate in a learning situation.

Conceptualisations of transformative learning encompass various individual and social purposes such as empowerment, ecological awareness, social action, citizenship and democracy (Rodríguez and Barth 2020). For some authors, contributing to a more significant social change (or transformation) through education makes transformative learning attractive and necessary for EfS (Rodríguez and Barth 2020). Mezirow argued that through transformative learning, individuals may learn to reflect on hidden experiences, preconceptions, values and assumptions, to think critically and freely engage in rational discourses (Mezirow 1990, 1991).

According to UNESCO (2017), there is a need for a 'transformative pedagogy' that awakens students' critical awareness and enables them to become active transformers of their own reality. Based on previous research on transformative T/L, we recognise that this process involves creating dynamic relationships between teachers, students and also shared knowledge in ways that promote student learning and personal growth (Blake et al. 2013). In contrast to traditional education, which involves the transfer of information from teachers to students, transformative education involves teachers as agents who guide students in the process of development as learners and individuals (Slavich and Zimbardo 2012). Therefore, it is important to examine teachers' strategies and students' responsibilities in shaping their own and their peers' learning experiences. For all these reasons, it can be said that *transformation* has become a central objective in educational policies in the twenty-first century (UNESCO 2017).

8.2.2 EfS and Education for Biodiversity

EfS aims to teach people to be aware of current socio-ecological problems and to make appropriate actions towards environmental responsibility (Aarnio-Linnanvuori 2019). In order to achieve this aim, students should be motivated and aware of values, as well as able to put sustainability ideas into practice (Banos-González et al. 2021). Although many educational efforts have already been undertaken, EfS still remains a challenge. Researchers have recommended alternative approaches to traditional teaching that emphasise the personal relevance of content, acknowledge students' emotions and integrate the socio-economic and scientific dimensions of the problem (Chawla and Cushing 2007; Trott 2019). The teaching of biology is closely related to the content of EfS. This content exists especially in the fields of biology, ecology, biodiversity and conservation (Jeronen et al. 2017). These authors recommend that teaching about biological phenomena related to socio-scientific issues, such as biodiversity or climate change, has an integrative and interdisciplinary approach to ensure that the topic is thoroughly taught and learned.

In their study, Van Weelie and Wals (2002) reflected on the complexity of the concept of biodiversity and the lack of an explicit definition of it and pointed out that it is important to contextualise the term in teaching situations. They highlighted that it is beneficial for learner's understanding if biodiversity is linked with real contexts such as political, scientific and media sources when these contexts are used as educational resources. Through this contextualisation, students become aware of the conflict of interest underlying the conservation issue. Contextualisation facilitates critical reflection on the normative aspects of socio-scientific issues typical of transformative T/L.

8.2.3 Principles and Teaching Methods in EfS and Biodiversity

To achieve the goals of EfS, several researchers recommend active, memorable and experiential teaching methods, through which students learn to discuss their own choice of values and critically evaluate phenomena and information sources (Maina 2004). These methods include problem-based learning and outdoor education (Jeronen et al. 2017), collaborative learning (Volet et al. 2013) and inquiry-based learning (Hakkarainen and Sintonen 2002).

Problem- or project-based learning (PBL) is a set of instructional strategies that empower students to conduct research, integrate theory and practice and apply knowledge (Savery 2006). Students start from a problem setting to define their own learning objectives. For Chin and Chia (2004), 'learning in this way [PBL] is useful and motivating as students learn while looking for solutions to problems. Students actively participate and learn in the context in which the knowledge will be used' (p. 708). In PBL, students collect and work with information with the purpose to solve a problem; therefore, problems are presented at the beginning of the intervention. Thus, the order is reversed from traditional teaching methods, which introduce problems only after that the students have learned the necessary body of knowledge.

Outdoor education (OE) takes place when T/L activities 'are organised in authentic, often natural environments, where students can connect their theoretical thinking with experiences of various kinds of real nature' (Jeronen et al. 2017, p. 5). It includes general interdisciplinary aspects of the world outside the classroom, including experiential education and outdoor environmental education (Jeronen et al. 2009). Borsos et al. (2022) argued that this methodology is successful and effective only when the teacher can guide students through an active, transformative learning process. In addition, various learning environments, which could be meaningful to the student, and the current and contextualised tasks used in PBL and OE support self-efficacy, autonomy, engagement and meaningful learning, and encourage creativity and flexibility (Turner and Fulmer 2013).

Collaborative learning is considered an active process that results in knowledge of higher quality than that produced by an individual. The increasing importance of solving complex problems and building knowledge in modern society emphasises the need for collaborative activities and environments in schools to foster learning (Volet et al. 2013). This is especially the case for socio-environmental issues such as biodiversity conservation.

Inquiry-based learning, in its part, is well suited to the context of collaborative learning, as students' mutual interactions make it possible to achieve a deeper and higher level understanding. This action is important in solving complex socio-environmental problems when working towards a more sustainable lifestyle (Hakkarainen and Sintonen 2002).

Therefore, in this chapter, we set out to clarify whether these transformative T/L methods are used to promote BE in primary and secondary education and teacher education.

8.3 Objective and Research Questions

An important question concerning T/L in biology is how transformative T/L approaches can support BE for promoting sustainability. Therefore, in this study, we conducted a systematic review to investigate the extent to which transformative T/L have been conceptualised and put into practice in biology and, specifically, in BE. In addition, we intended to collect evidence from previous studies on how to promote transformative learning in relation to the studied content.

This study was guided by the following research questions:

1. How do the learning processes concerning transformative education support the BE presented in primary and secondary education and in teacher education?
2. What transformative T/L methods are used to promote BE in primary and secondary education and teacher education?

To answer these research questions, we analysed the characteristics of the transformative teaching methods (TTM) used in research literature and the implications of TTM for the teaching of biology.

8.4 Material and Methods

In this section, the data collection, processing and analyses of the selected articles are described in detail in order to answer the research questions. A systematic literature review (SLR) was conducted to understand the relationships between transformative T/L, sustainability education and BE. The SLR was conducted on all available peer-reviewed articles published in English and Spanish (also in Finnish, but none was found) that explicitly focused on transformative learning, sustainability and BE. SLR represents a typical method for mapping the field and tracking recent developments in both the educational sciences and the science of sustainability (Spangenberg 2011).

In this work, we followed the SLR approach described by Sánchez-Meca and Botella (2010), and the PRISMA 2020 statement (Page et al. 2021). First, the research objectives and questions were established, followed by the locations of the empirical studies. In the identification phase, the inclusion and exclusion criteria of the studies were defined. The following aspects were included in SLR: (a) an identification of the study designs admissible for review; (b) a definition of the types of programmes, treatments or interventions intended to be investigated; (c) a definition of the study participants' characteristics; (d) a determination of the statistical data that the studies provide to calculate the effect sizes; and (e) an identification of how the outcome variables were to be measured (duly graded performance tests, self-report measures, etc.). Finally, in the SLR, the language in which the study was written and the time range of the articles were considered. An SLR is about providing a systematic and replicable search and analysis strategy that is fully documented and transparent (Sánchez-Meca and Botella 2010).

The SLR conducted in this study followed three steps: (1) data collection, (2) data processing and coding and (3) data analysis, which yielded a bibliometric view that combines quantitative and qualitative analyses of the learning process, results and conditions.

8.4.1 Data Collection

To provide a complete sample universe, searches were performed in electronic, scientific databases such as Scopus, Web of Science (WOS) and ERIC, which are considered among the most enriched and popular academic databases. The search strategy was based on a systematic organisation, categorisation and selection of keywords related to transformative T/L in biology, sustainability and BE. We used a single broad search term instead of a series of discrete search terms. The keyword terms used in the literature search were as follows: TITLE-ABS-KEY (Transformative) AND (teach* OR learn*) AND (biodiversity OR biology) AND (education).

Based on these keywords, a common search strategy was developed for the consulted databases, adapting it to the characteristics of the given platform. Depending on the requirements of each database, the search fields were basically limited to the titles, abstracts and keywords of the articles. The following criteria were used to select the articles for in-depth analyses: (a) scope: national and international research, (b) period: 2012–2022, (c) target groups: primary and secondary school students and education of future biology teachers and (d) languages: English, Finnish and Spanish.

The review was limited to scholarly papers published in peer-reviewed journals because these have undergone rigorous review and are therefore higher quality documents. The authors were aware of the existence of other types of documents that could have been analysed, such as theses, research reports, books and book chapters and conference proceedings, but they were not included in the final SLR.

8.4.2 Data Processing and Coding

This study focuses on T/L methods to promote transformative learning, sustainability education and BE in primary and secondary school and teacher education. A raw database was created from the sample universe ($n = 130$), which included all available bibliographic data, the abstract and the full text. Additional variables were added to better capture the nature of the contributions (e.g. target groups, date of study and language). Coding was performed iteratively by selecting first the abstracts and then the full articles using an inductive coding approach. Then, 31 documents that did not meet the study date criteria (2012–2022) were excluded. After this filtering, 35 documents were excluded on the basis of the type of ineligible document (e.g. book chapters, notes and editorials), along with three documents that did not meet the language criteria (Fig. 8.1).

After an in-depth analysis of the thematic relevance of the remaining 61 articles, they were classified into four large distinctive research areas in the sample universe, and then 52 articles were excluded in areas (i–iv), as follows:

(i) Higher education in general: University-level contributions in undergraduate or graduate programmes ($n = 6$ were deleted).
(ii) Non-formal or informal learning: Encompasses research on learning environments that are not directly linked to formal educational institutions and programmes ($n = 16$).
(iii) Policy and guidelines: Mainly cover conceptual elaborations on policies and general discourses on education and their links with sustainability and BE ($n = 30$).

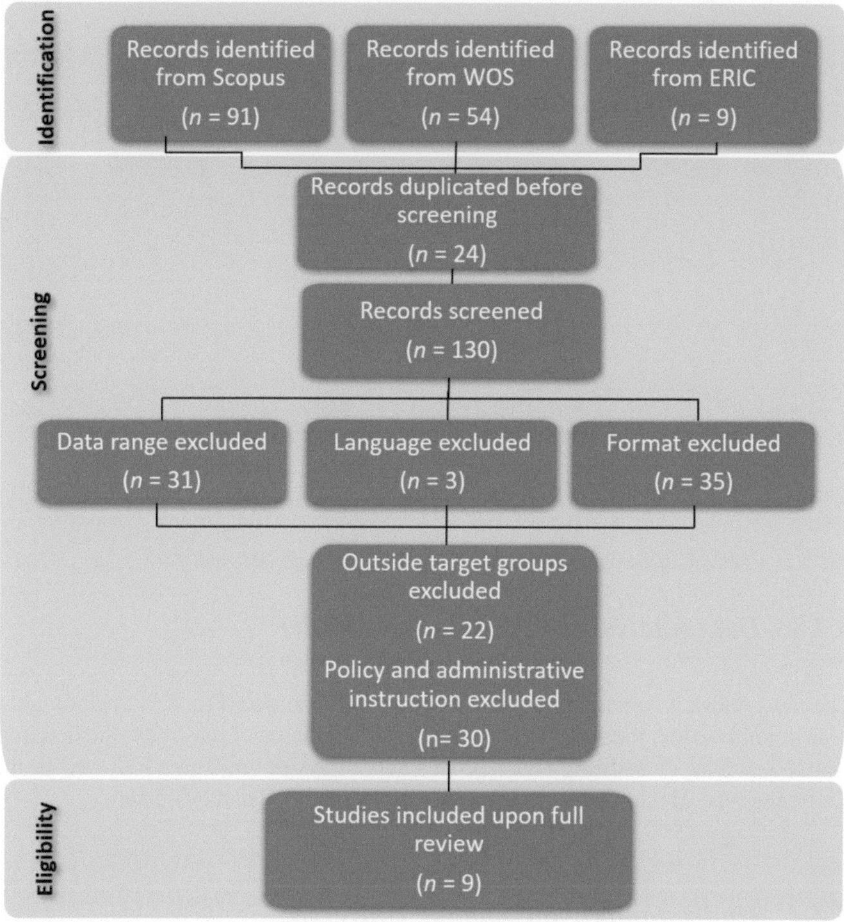

Fig. 8.1 Flow diagram of the review selection process. (Modified after PRISMA, 2020)

(iv) Educational programmes: Programmes for future teachers and students in primary and secondary education (n = 9), which represent the core sample of our study.

To ensure the reliability of the SLR, all three co-authors were involved in the article screening. Two authors reviewed the title, abstract and keywords of each article, coding the article as 'accept', 'reject', or 'further review' based on the inclusion criteria. A third author reviewed any articles coded 'further review' or whose coding by the other co-authors did not match. Articles that were still considered for inclusion underwent full-text analysis by two of the co-authors, with the third co-author again reviewing any cases of disagreement. Finally, the evaluation of inter-rater reliability was completed with adequate agreement among the three authors ($\kappa = 0.7$) (Landis and Koch 1997).

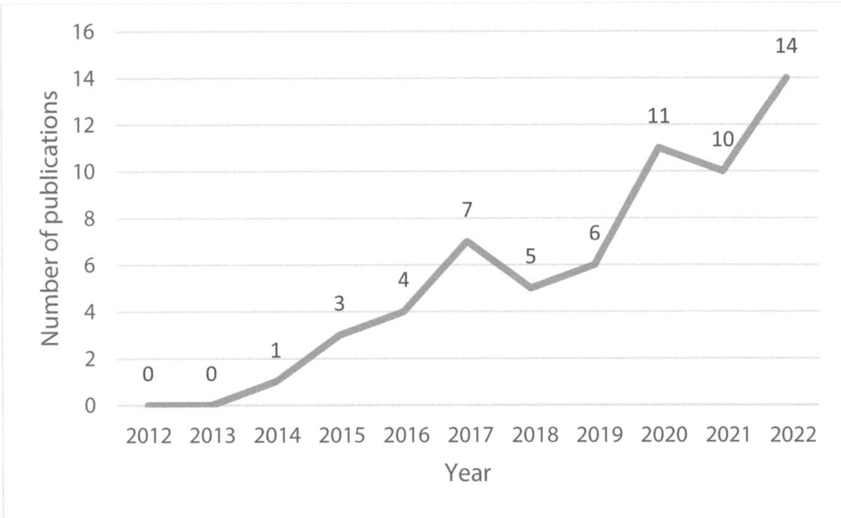

Fig. 8.2 Evolution in the number of publications found between 2012 and 2022

8.4.3 Data Analyses of the Selected Articles

The most relevant content of the nine studies were recorded (Fig. 8.2), including the year of publication, location, target population and theme of the study; the transformative teaching or learning and type of intervention or strategies followed in the study; the type of data collection and the results obtained and evaluated.

8.4.4 Understanding the Sample Universe

A raw data set, which included the identified articles, was obtained from each database consisting of 154 publications (91 from Scopus, 54 from WOS and nine from ERIC). Article screening revealed duplicates because the databases presented overlaps. After removing the duplicates (n = 24), we retrieved a raw sample of 130 publications (Fig. 8.1). Then, the raw sample was initially screened using the predefined inclusion and exclusion criteria.

In this study, the focus was on articles on transformative T/L and BE for promoting sustainability and dedicated to primary and secondary students and future teachers. It represents 14.8% of the found contributions (9/61 articles; Fig. 8.1).

During the data processing and coding, the following observations were made. In the study period, the number of articles found in relation to transformative T/L and education for biodiversity was increasing, with the highest number of publications found in 2022 (Fig. 8.2). The annual average for the studied period 2012–2022 was 5.4 articles.

Most of the articles included secondary school students (A2, A4, A5 and A8) and future biology or science teachers (A3, A7 and A9), respectively, and A1 included both groups as a collaborative team (Table 8.1). Transformative learning in BE played a lesser role in learning and research than in EfS. Hence, only few studies have addressed BE in any of the theoretical uses of transformative learning.

8.5 Findings

In this section, findings of transformative learning processes are presented according to unintentional and foreseen situations and critical reflection. Also, the methods and strategies connected to transformative teaching and learning outcomes are related to sustainability and biodiversity.

8.5.1 Findings Concerning Transformative Learning Processes

The first finding was the importance of prior conceptions. Four articles explicitly investigated prior perspectives and worldviews (A1, A3, A8 and A9). The rest of the articles mentioned prior learning as part of participants' general background and experiences related to biodiversity. Among the former four, the A1 article argued that learning is an active and dynamic process that occurs through experience. On the other hand, the A3 article explored the problem of not considering emotional and motivational variables to advance in professional development programmes for future biology teachers when they address conservation issues such as the natural remigration of the grey wolf in Germany. In A9, the preconceived ideas of anthropocentrism as domination over nature in the emulation of the subject were dismantled. This means that environment preservation cannot be achieved from the prevailing selfish perspective of the human species but, according to A9, it should be based on the improvement of the conditions of human life. Finally, A8, by means of an interpretive phenomenological analysis based on a qualitative research approach, aimed to understand the lived experiences of secondary school students during their first year of A-level study through the theoretical lens of the threshold concept (TC) framework. For A8, TC is likely to be transformative, occasioning ontological and epistemological shifts in the learner's view of the world and, potentially, their own identity.

The second finding was the occurrence of disorienting dilemmas. The reviewed articles showed various situations in which such disorienting dilemmas can occur (Rodríguez and Barth 2020).

(i) *Structured and unintentional situations*: These refer to planned learning activities in which educators do not deliberately trigger a dilemma; rather, owing to

Table 8.1 The analysed articles (*n* = 9)

Article code	Authors (year)	Study location	Participants	Data collection	Study theme
A1	Harnisch et al. (2014)	Illinois (USA)	Collaborative teams composed of graduate teaching scholars in education, secondary students and high school teachers	Unstructured interviews Surveys at the beginning and end of the service	Science and mathematics education
A2	Ortega--Torres and Moncholí (2021)	Valencia (Spain)	96 ninth-grade students	Group research dossier Expert experimentation dossier Exhibition products with co-evaluation	Albufera biodiversity study
A3	Büssing et al. (2018)	Germany	120 pre-service biology teachers	Questionnaire on contextual factors and their motivation to teach on the subject	Biodiversity conservation of sustainable development
A4	Zoller (2015)	Israel	Tenth-grade high school, undergraduate and graduate students	Pre-post questionnaires Observations in class Interviews	Sustainability
A5	Salami (2015)	Nigeria	274 upper secondary education students	Observation verification Rating scale of the activities Achievement test	Study of plant and animal nutrition
A6	Rüdisser et al. (2017)	Austria	548 students, aged between 6 and 20 years	Identification template Project database	Watching and monitoring of butterfly species
A7	Bezeljak et al. (2020)	Slovenia and Austria	60 biology students and 60 pre-service biology teachers	Survey with open and Likert-type questions	Sustainable development and education for sustainability
A8	Dunn (2019)	United Kingdom	Six high school students from a biology class	Class diaries Semi-structured interviews	Biology education
A9	Vasquez et al. (2020)	Venezuela	Future teachers from various disciplines	Stories and reflections on an educational process that adopted the science, technology and society (CTS) approach	Education for sustainable development and biodiversity of the semi-arid territory

the nature of the activity, the student experiences such a dilemma. Examples include new learning environments such as interdisciplinary environments at work (A1). In A4, students found situations in which they could participate in change processes and establish new social relationships.

(ii) *Structured and foreseen situations*: These refer to cases in which the disorienting dilemma is planned and deliberately induced during educational interventions to challenge students' frames of reference by asking critical questions such as those in A5. In this work, a modified laboratory learning environment was used, which had a significant positive impact on students' use of science process skills.

(iii) *Critical reflection:* Three types of reflections were identified:

- Reflection on content encompasses the analysis of information, concepts, values and norms that promote critical thinking; they are collaborative and create knowledge, such as those in A1, which was aimed at creating engaging collaborative learning environments. In addition, A7 was aimed at identifying knowledge and understanding of sustainable development and EfS among future biology teachers from Slovenia and Austria, and comparing the results between the two countries.
- Reflection on the process refers to assessing the way in which learning experiences are developed. In the process, it is vital to provide participants with frequent opportunities for feedback, reflection and idea modification. This is the case of A2, in which one main finding was from the reflection and teamwork of four secondary school teachers involved in the project 'We explain the Albufera' and students' reflections about the change produced by the methodology, with challenge-based learning.
- Reflection on premises occurs when learners assess the assumptions that underlie their thoughts, beliefs, attitudes, actions and behaviours. In the context of teacher education, A3 highlighted the importance of educating future teachers to reflect on values and transform them into more sustainable ways of living and teaching.

8.5.2 Findings on Teaching Methods and Strategies

Among the teaching methods or types of intervention, eight were differentiated. One was *action learning*. For Rodríguez and Barth (2020), the transformation process is sometimes followed by the adoption of sustainable behaviours as part of the designed experiments. In A9, teachers were urged to get rid of traditional educational standards and assume commitment to doing and relearning in a flexible process of high creativity. It is 'a requirement of social commitment, not a promise, where ethics, honesty and transparency are put ahead, that the embryos take their steps and follow their course towards sustainability' (A9, p. 4).

Project- or problem-based learning is another method that occurs in the analysed articles. In A2, the didactic sequence designed for the interdisciplinary project about Albufera was structured in accordance with challenge-based learning methods. This is linked to a goal that promotes students' active participation through the need to use higher order cognitive skills (HOCS) to achieve greater depth in learning. Moreover, A4 developed methodologies for the promotion of HOCS; for example, students, not teachers, ask the questions, solve problems that require their systems and evaluative thinking and decision-making skills in high schools and colleges in Israel.

Teamwork was also present in the included articles. Several references are available to this type of work. For example, in A1, various collaborative T/L teams were formed among National Science Foundation graduate teaching fellows in the K-12 education programme, which 'not only provided space for the exchange of ideas, but also fostered collaboration, community and mutual learning and teaching between educators and students' (A1, p. 7). In A2, the cooperative technique of the Aronson puzzle was developed (Aronson and Bridgeman 1996). Following this, each student belonging to the initial core group was assigned various roles to form five groups of experts according to their affinities and interests.

Use of technological tools was observed in A1, A4 and A5. Through GK-12, meaningful technologies and collaborative teaching models are integrated for students to learn science. A *modified laboratory learning environment* was demonstrated in A5. The study of this article examined the impact of a modified laboratory learning environment, particularly on secondary school students' science process skills. The students designed and conducted experiments on two research topics, animal and plant nutrition, with the help of trained research assistants under a researcher's supervision. *Field trips* were represented in two articles. Both A2 and A6 used field trips to create transformative learning contexts or environments that support the students' learning and motivation.

The *integrative learning of science, technology and society (STS) approach* was presented in A9. In this case, the approach was used for transformative and critical environmental EfS to achieve meaningful and collaborative learning among the participants.

Discourse was the eighth finding. Some authors referred to the stage of discourse as a process of sharing knowledge and practices related to biodiversity conservation (A6) or experiences of participation in a community activity (A9).

8.5.3 Findings of the Learning Outcomes

Five issues regarding learning outcomes were presented in the articles included in this review. One issue is the *improvement of practical skills, and the increase of new knowledge related to sustainability and biodiversity* (Table 8.2). In the study reported in A6, authors investigated about systematically collected data by trained students, together with their teachers, on the occurrence of diurnal butterflies. The

Table 8.2 Summary of the findings about learning outcomes in the SLR

Learning outcomes	Citation
Improvement of practical skills and increase of new knowledge related to sustainability and biodiversity	A5. Salami (2015) A6. Rüdisser et al. (2017) A7. Bezeljak et al. (2020)
Positive long-term impact of technology when used for authentic problem solving	A1. Harnisch et al. (2014)
Change of attitude and positive assessment of biodiversity issues to improve teaching and learning	A2. Ortega-Torres and Moncholí (2021) A3. Büssing et al. (2018) A6. Rüdisser et al. (2017)
Impact of the curricula on the teaching and learning about sustainability and biodiversity education	A4. Zoller (2015) A7. Bezeljak et al. (2020)
The need for interdisciplinarity	A7. Bezeljak et al. (2020) A9. Vasquez et al. (2020)

participants gradually gained experience, 'sometimes even turn[ed] into experts' (p. 685). According to the authors, this kind of participation could substantially contribute to the successful implementation of biodiversity monitoring programmes. In A5, a modified laboratory learning environment had a significant positive impact on students' use of science process skills. Scientific process skills were learned and acquired by students in environments that encourage thinking and enthusiasm for working and discovering new things. The results also showed that the process skills of low-achieving students improved significantly when they were taught in a modified laboratory setting. Meanwhile, A7 revealed that less than half of the future biology teachers who answered the questionnaire had a good understanding of the environmental aspects of sustainability, but most lacked an understanding of the interconnections between the environmental, economic and social dimensions.

The *positive long-term impact of technology when used for authentic problem solving* (A1) was presented to create engaging experiences for both students and teachers. The A1 article demonstrated the fundamental integration of meaningful technologies (e.g. informatics and visualisation tools) and collaborative teaching models for student-learners to develop critical scientific literacy in primary and secondary school classrooms.

Change of attitude and positive assessment of biodiversity issues to improve T/L were the key findings reported in A2 and A3. The connection between the selection of adequate contexts (e.g. the returning wolves to the region) and the attitudes, enjoyment and perceived behavioural control of future teachers to teach about the issue was demonstrated in A3. This positive attitude and increased motivation of future teachers might be a stepping stone to strengthening BE in schools, leading to a more sustainable future for everyone. A2 described changes in students' interest in science and attitudes due to the transformation of a field trip into an interdisciplinary project. This positive assessment by the students was considered a good predisposition towards new projects, as described by A2. On the other hand, A6 found that 87% of the students reported a high degree of interest and enjoyment in participating in the butterfly observation project. Thus, the authors concluded that the project could enthuse more individuals for nature observation and biodiversity issues.

The *impacts of the curricula on T/L about sustainability and BE* were differentiated in A7. The article described that future biology teachers had little awareness of the complexity of sustainability issues and even less awareness of social and intergenerational equity. For this reason, the authors argued to convert biodiversity and sustainability into cross-cutting issues, which could benefit both the natural and social sciences education and also in humanities. A4 criticised the fact that contemporary science education in secondary and high school education is mainly disciplinary. This article also described a slow shift from disciplinary teaching at the undergraduate level to interdisciplinary learning oriented to higher order cognitive skills.

The *need for interdisciplinarity* was also identified. A9 showed that integrative learning of STS for the theme of biodiversity allowed for achieving the transversality of environmental education, and meaningful and collaborative learning between students and teachers. This integration is achieved in multidisciplinary and transversal teaching to build alternatives for sustainable local development. In addition, A7 reported that future biology teachers, despite having a good understanding of the environmental aspects of sustainability, lacked an understanding of the interconnections between the environmental, economic and social dimensions, which are essential for active T/L approaches in the fields of sustainability and BE.

8.6 Discussion

In this section, we first discuss the sample universe and, second, the main RQs. Third, we discuss the main learning outcomes found in the literature and, finally, we summarise what we learnt from the review.

8.6.1 Understanding the Sample Universe

The transformative learning theory in EfS and BE has become an emerging field of research, as evidenced by the increasing trend in the number of publications over the last few years. In this study, the focus was on articles referring to transformative T/L and BE in promoting sustainability.

Rodríguez and Barth (2020) suggested that transformative learning has become an attractive theory used in the field of EfS, although it often lacks deep enough analyses. However, in the field of education for biodiversity, the results obtained in this study do not confirm this idea, apparently because of the limited number of papers. Nevertheless, this idea seems to be proven for other fields such as social learning and sustainability (Reed et al. 2010; Taylor 1997). The complexity of this problem could lie in the unpredictability and subjectivity of the results of experiences related to transformative learning and in the methodological and ethical implications when evaluating those results.

8.6.2 *Transformative T/L and BE*

Dewey (1938) suggested that learning is based on continuous reflection on lived experiences. This statement is supported by A1, which emphasised the importance of active and dynamic processing of experiences in learning situations. The authors of A1 used the threshold concept framework as a transformative approach to learning, which is based on the idea that in the process of change, learners are awakened to new ways of seeing and become ready to let go of their prevailing perceptions (Land and Rattray 2017). The learning of the concepts, such as photosynthesis or biodiversity, can bring about ontological and epistemological changes in learners' worldviews and, possibly, their own identities.

Understanding the importance of biodiversity is a prerequisite for nature conservation. In this sense and from the perspective of EfS, it is important to consider students' personal experiences, emotions and motivation (Chawla and Cushing 2007). According to A3's conclusions, the professional development programmes for future biology teachers have neglected this view.

On the other hand, the preconceived notions about anthropocentrism are dismantled in A9. The authors of A9 suggested that nature and biodiversity conservation cannot be achieved from an anthropocentric point of view; but could be based on personal experiences and conducted as collective activities that acknowledge unsustainability and are aimed at creating a just and equal society and viable ecological conditions. Weinberg et al. (2020) argued that education is a critical element in a global transformation towards sustainability and is urgently needed to stabilise socio-ecological systems worldwide.

8.6.2.1 Transformative Learning Processes

According to the SLR performed by Rodríguez and Barth (2020) about transformative learning in the field of sustainability, structured and unintentional situations, structured and foreseen situations and critical reflection are areas of concern, since they can contain confusing dilemmas. Through various learning environments, students' self-efficacy, autonomy, commitment and meaningful learning can be fostered and their creativity and flexibility are encouraged (Turner and Fulmer 2013). Social relationships and collaborative learning can result in jointly processed knowledge, which is important when discussing environmental issues (Huang and Lajoie 2023).

Structured and foreseen situations of disorienting dilemma such as a modified laboratory learning environment were evident from the data reported in A5. This dilemma seemed to have a significant positive impact on the students' use of science process skills (SPS) such as thinking and research skills, including basic and integrated process skills (Elmas et al. 2018). Basic process skills create a foundation for learning integrated (more complex) skills. They consist of observing, guessing, measuring, communicating, classifying and predicting. Integrated process skills include variable control, operationalisation, formulating hypotheses, interpreting data, conducting experiments and designing models (Elmas et al. 2018). SPS is used in problem solving, critical thinking, understanding concepts (Candrasekaran 2014), and in supporting students to understand and solve everyday life problems and cope with global challenges (Abungu et al. 2014), such as biodiversity loss and climate change.

In our SLR, we identified three types of critical reflections: reflection on learning content and results, reflection on the learning processes and reflection on premises of students' assumptions. Reflection related to learning conditions, learning processes, content and results is typical for transformative learning (Mezirow 1981). Thus, education aims to increase individual empowerment, ecological awareness and social activity (Rodríguez and Barth 2020). Reflecting on premises concerns students' assumptions that underlie their thoughts, beliefs, attitudes, actions and behaviours. The results in A3 emphasised that in teacher education, it is important that future teachers reflect on their values to become aware of norms and to change their practice (including teaching activities) to be more sustainable. This transformative pedagogy can also awaken students' critical awareness and enable them to become active transformers of their own reality according to the goal of the UNESCO (2017).

8.6.3 Teaching Methods and Strategies

Teaching methods are important because they affect all types of learning in cognitive, affective and psychomotor areas (Karami et al. 2012). In this SLR, we identified the following types of learning: action learning, problem-based learning, teamwork, information and communication technology, laboratory work, field trips and integrative learning of STS and discourse.

Approaches to problem-solving involve acting and reflecting upon the results (Reynolds 2011), which may facilitate the development of problem-solving processes and simplifying the solutions developed by the team. In A9, teachers were encouraged to abandon traditional teaching methods and to apply flexible action learning that developed creativity. The results of A6 also suggested that citizen science projects related to biodiversity conservation could promote public engagement and provide transformative learning situations.

Problem-based learning and project-based learning are active learning methods, by which students solve questions related to, for example, their own lives, as in the interdisciplinary Albufera project (A2). The goal in A2 was to support the students' HOCS from the perspective of deep learning. A4 had the same goal, but students, not teachers, presented questions and solved problems that required systems thinking and decision-making skills. According to Fuglei (2014), problem-based learning is an especially useful method of transformational education. As an active-learning method, it can support structural changes in students' thoughts, feelings, actions and consciousness, which can permanently change the way they view and behave in the world.

In the articles analysed in our SLR, teamwork was found among the T/L methods. Teamwork includes the collaborative effort of a group to achieve a common goal or to complete a task in the most effective and efficient way (Salas et al. 2008). In A1, various collaborative T/L teams were formed among the National Science Foundation graduate teaching fellows in the K-12 education programme, which fostered collaboration community and mutual learning and teaching between educators and students. In A2, the collaborative technique of the Aronson puzzle was used.

Regarding laboratory work in science and biology education, Gericke et al.'s (2023) systematic review showed that one of its main purposes is to provide students with conceptual and theoretical knowledge to support them in learning scientific concepts and, through scientific methods, in understanding the nature of science. In A5, authors stated that a modified laboratory learning environment might enhance this kind of learning, as was the case in the animal and plant nutrition study. Laboratory work also provides students with the opportunity to experience science by using scientific research procedures. To achieve meaningful learning, Ottander and Grelsson (2006) stated that scientific theories and their application methods should be experienced by students. Moreover, laboratory work should encourage the development of the students' analytical and critical thinking skills and their interest in science.

The purpose of field trips is usually to provide students with experiences outside their everyday school issues, as the aim is to observe things in their natural state and possibly collect samples (Jeronen et al. 2017). According to our SLR findings, a field trip is meant to create transformative learning environments that serve to support students' learning and motivation. Liefländer and Bogner (2018) argued that when a field trip includes problem-based activities, learning by doing, exploring real-life phenomena and sensory engagement, it significantly improves students' interest, knowledge, attitudes and behaviour in relation to sustainable development. According to Borsos et al. (2022), field trips should be an integral part of education at all educational levels. The integrative learning of STS approach was used in A9 for transformative and critical environmental EfS to achieve meaningful and collaborative learning among the participants.

As defined by Van de Walle et al. (2014), a classroom discourse includes 'the interactions between all the participants that occur throughout a lesson' (p. 20). When discussing the protection of diversity (A6) and participation in community activities (A9), information and practices related to these topics can be shared. This is important from the point of view of understanding scientific knowledge and the development of scientific thinking. Knowledge of species and species identification are fundamental prerequisites of understanding biodiversity, ecosystems and ecological sustainability, according to Palmberg et al. (2018) and Randler (2008). The studies of these authors showed that teachers' and student teachers' knowledge of species and species identification and their interest in nature in an ecological context have significantly declined. One possible reason is that students' scientific vocabulary varies considerably. Some students have a good command of the terminology of biology and the complex meanings associated with biological terms, whereas others struggle with these terms and do not master even the threshold concepts enough to explain the relationships and meanings between the phenomena (Mouton and Rootman-Le Grange 2020).

It can be concluded that after the in-depth analysis of the nine selected articles, the most common methodologies used when considering transformative learning were based on problem-based learning and outdoor activities.

8.6.4 Learning Outcomes

The key findings to enhance transformative learning in relation to biodiversity and EfS in this SLR study were as follows: improvement of practical skills and increase of new knowledge related to sustainability and biodiversity, positive long-term impact of technology when used for authentic problem solving, change in attitude and positive assessment of biodiversity issues to improve T/L, impact of the curricula on T/L about sustainability and BE and the usefulness of interdisciplinarity.

Developing practical skills and adding new knowledge related to sustainable development and biodiversity enhanced transformative learning. For example, students' ability and participation in gathering systematically information about the occurrence of diurnal butterflies significantly contributed to the success of diversity monitoring programmes (A6). For its part, the modified laboratory learning environment had a significant positive effect on the development and use of students' natural science process skills (A5). Active T/L processes appeared to develop higher order learning and practical skills (Elmas et al. 2018). On the other hand, our findings show that most student teachers lacked an understanding of the connections between the environmental, economic and social dimensions of sustainability. Without knowledge of the relationship between human beings and nature, it seems difficult to understand the structure and function of ecosystems (life-supporting systems on Earth) and the principles of biodiversity and its role in maintaining sustainability (Randler 2008).

Learning occurs when people solve problems in authentic contexts and participate in community activities. Thus, authentic learning can be considered closely related to problem solving and other community practices. Consequently, authentic learning can support students to become members of the culture and participate in meaningful practices (Cho et al. 2015), especially when the fundamental integration of meaningful technologies (e.g. informatics and visualisation tools) takes place and when student learners develop critical scientific literacy in primary and secondary school classrooms. The positive long-term impact of technology can lead to transformative learning when used for authentic problem solving.

The findings of this study show that the choice of context has meaning for the relationships between teachers' attitudes, enjoyment and perceived behaviour and students' increased interest in science through their participation in an interdisciplinary field project, such as the butterfly observation project, which fostered transformative T/L. According to previous studies (Duerden and Witt 2010; Palmberg et al. 2018), people's knowledge about nature and interest and experiences in nature influenced their values and emotions regarding nature. Transformative learning touches and changes deep levels of values and attitudes through realisation and recognition (Sterling 2011) and could therefore motivate teachers in more value-based T/L about biodiversity.

Another finding of this study was the impact of curricula on T/L about sustainability and biodiversity. Some education experts have noted that most twenty-first-century students are still being taught by teachers using twentieth-century pedagogical practices in nineteenth-century school organisations (OECD 2019). The transition from traditional teaching methods to interdisciplinary learning and teaching methods has been slow. This may be partly because the concepts of sustainability and sustainability education are difficult for both teachers and students to understand (Pathan et al. 2013). In addition, the need for thinking of issues as an entity is fundamental, and the ability to systems thinking is essential. Systems thinking is a way of understanding the world as a complex system, in which everything is connected to everything else (Palmberg et al. 2018). It supports the identification of the interrelationships of various biophysical and social factors in a given environment (Sterling 2003). It is a way of thinking that enables a person to understand their role from a holistic perspective. Learning situations offer encounters in which teachers and students can learn from each other and build paths to a sustainable future together through open discourse.

Finally, it is important to point out some of the shortcomings of this review. The reduced number of papers included is considered a limitation of this study. It might be due, on the one hand, to the difficulty of implementing the theory of transformative learning and, transforming results may become visible much later or not at all. On the other hand, to the specificity of the content, education for biodiversity conservation, we argue that for future research on biodiversity transformative T/L, it would be prudent to note that more empirical research is needed in the areas of teacher education and formal learning other than higher education. This review can be considered a starting point from which potential articles that we might have missed (and future articles) can be integrated into the sample universe. Finally, this study warrants the need to develop good methods and tools that can provide insights into transformative learning processes and outcomes.

8.6.5 What We Can Learn from this Review

A change in 'conventional' pedagogical practices towards practices that seek impactful learning results both individually and socially is necessary. Various pedagogical endeavours such as problem- or project-based learning have been made in interdisciplinary and intercultural settings. However, introducing these methods and settings can cause unexpected dilemmas. Therefore, before embarking on these processes, it is essential to consider both the disorienting dilemmas that can arise during learning interventions and the individuals' prior learning experiences. In addition, the social norms and emotional charge inherent in environmental issues can lead to problematic situations.

Social norms can reduce the discussion of environmental issues and affect environmental behaviour, such as electricity consumption and recycling (Jachimowicz et al. 2018). If the students think that their peers do not support their environmental views, they will avoid discussing the topic. A recent systematic review of social determinants of climate change research highlights the need to better understand the effects of social network members on an individual's climate change behaviour (Tam et al. 2021).

Research on emotions has in its part shown how personal emotional attitudes differently affect the decision-making of individuals (Peters et al. 2006). Today, information is lacking on the effects of emotions on environmental behaviour due to the complex nature of environmental issues such as climate change. According to Latkin et al. (2021), avoiding the topic of climate change is negatively related to citizens' willingness to participate in climate change actions. Possibly the reason is that people avoid discussing topics that they do not feel they can influence. Understanding the role of emotions is very important in the current situation, when environmental decisions will be made, for example, about curbing climate change (Xie et al. 2019) and environmental policy (Smith and Leiserowitz 2014). Decisions related to these topics may have long-term consequences, and people's emotional reactions can strongly influence their attitudes and actions (Davidson and Kecinski 2022). One the one hand, positive feelings such as empathy and concern for nature can motivate individuals to participate in, for example, recycling or in supporting conservation efforts (Berenguer 2007). On the other hand, negative emotions such as fear or denial can prevent environmental actions or leads to unsustainable practices (Büssing et al. 2018). The findings of the study reveal the importance of planning educational interventions to create supportive learning conditions as a space for reflection and discourse, and social and collaborative interactions. The need for interdisciplinarity was one important finding of this study. Biodiversity can be taught and learned meaningfully in, for example, environmental education using collaborative learning. However, such active teaching methods require a good understanding of the connections between the environmental, economic and social dimensions of sustainability. For this reason, topics related to sustainability should be added to all teacher education subjects. Moreover, time for practical training and teaching experiences should be increased in teacher education programmes to develop teachers' pedagogical content knowledge concerning biodiversity and sustainability (Großschedl et al. 2015) to support transformative teaching and learning.

Acknowledgements **This work was supported by the Grant PID2019-105705RA-I00 funded by MCIN/AEI/10.13039/501100011033 and by the Moving Minds Program of the University of Murcia.**

References

Aarnio-Linnanvuori E (2019) How do teachers perceive environmental responsibility? Environ Educ Res 25(1):46–61. https://doi.org/10.1080/13504622.2018.1509300

Abungu HE, Okere MIO, Wachanga SW (2014) The effect of science process skills teaching approach on secondary school students' achievement in chemistry in Nyando District, Kenya. J Educ Soc Res 4(6):359–372. https://doi.org/10.5901/jesr.2014.v4n6p359

Aronson E, Bridgeman D (1996) The jigsaw technique. In: Banyard P, Grayson A (eds) Introducing psychological research. Palgrave, pp 65–68

Banos-González I, Esteve-Guirao P, Jaén M (2021) Future teachers facing the problem of climate change: meat consumption, perceived responsibility, and willingness to act. Environ Educ Res 27(11):1618–1637. https://doi.org/10.1080/13504622.2021.1926433

Berenguer J (2007) The effect of empathy in proenvironmental attitudes and behaviors. Environ Behav 39:269–283. https://doi.org/10.1177/0013916506292937

Bezeljak P, Scheuch M, Torkar G (2020) Understanding of sustainability and education for sustainable development among pre-service biology teachers. Sustainability 12(17):6892. https://doi.org/10.3390/su12176892

Bivens F, Moriarty K, Taylor P (2009) Transformative education and its potential for changing the lives of children in disempowering contexts. IDS Bull 40(1):97–108. https://doi.org/10.1111/j.1759-5436.2009.tb00395.x

Blake J, Sterling S, Goodson I (2013) Transformative learning for a sustainable future: an exploration of pedagogies for change at an alternative college. Sustainability 5(12):5347–5372. https://doi.org/10.3390/su5125347

Borsos É, Banos-González I, Boric E, Lyngved Staberg R, Fekete AB (2022) Trainee teachers' perceptions of outdoor education. Environ Educ Res 28(10):1490–1509. https://doi.org/10.1080/13504622.2022.2031901

Büssing AG, Schleper M, Menzel S (2018) Do pre-service teachers dance with wolves? Subject--specific teacher professional development in a recent biodiversity conservation issue. Sustainability 11(1):47. https://doi.org/10.3390/su11010047

Candrasekaran S (2014) Developing scientific attitude, critical thinking and creative intelligence of higher secondary school biology students by applying Synectics techniques. Int J Humanit Soc Sci Invention 3(6):1–8

Carr P, Rivas E, Molano N, Thésée G (2018) Pedagogías contra el despojo: Principios de una eco/demopedagogía Transformativa como vehículo para la justicia social y ambiental. Rev Int Educ Justicia Soc 7(1):69–93. https://doi.org/10.15366/riejs2018.7.1.004

CBD (Convention on Biological Diversity) (2022) Post-2020 global biodiversity framework. Open--ended working group on the post-2020 global biodiversity framework. CBD/WG2020/5/L.2. UN Environment Programme. 2020. https://www.cbd.int/doc/c/409e/19ae/369752b245f05e88f760aeb3/wg2020-05-l-02-en.pdf

Chawla L, Cushing DF (2007) Education for strategic environmental behavior. Environ Educ Res 13(4):437–452. https://doi.org/10.1080/13504620701581539

Chin C, Chia LG (2004) Implementing project work in biology through problem-based learning. J Biol Educ 38(2):69–75. https://doi.org/10.1080/00219266.2004.9655906

Cho YH, Caleon IS, Kapur M (2015) Authentic problem solving and learning for twenty-first century learners. In: Cho YH, Caleon IS, Kapur M (eds) Authentic problem solving and learning in the 21st century perspectives from Singapore and beyond. Springer Science + Business Media, Singapore, pp 3–16

Collins-Figueroa M (2012) Biodiversity and education for sustainable development in teacher education programmes of four Jamaican educational institutions. J Educ Sustain Dev 6(2):253–267. https://doi.org/10.1177/0973408212475257

Davidson DJ, Kecinski M (2022) Emotional pathways to climate change responses. Wiley Interdiscip Rev Clim Chang 13:e751. https://doi.org/10.1002/wcc.751De

Dewey J (1938) The philosophy of the arts. In: Dewey J (ed) The later works 1925–1953, vol 13. Southern Illinois University Press, Carbondale, pp 357–368

Duerden M-D, Witt PA (2010) The impact of direct and indirect experiences on the development of environmental knowledge, attitudes, and behavior. J Environ Psychol 30:379–392. https://doi.org/10.1016/j.jenvp.2010.03.007

Dunn MJ (2019) Threshold concepts and the troublesome transition from GCSE to A-level: exploring students' experiences in secondary school biology. Curric J 30(4):375–391. https://doi.org/10.1080/09585176.2019.1646664

Elmas R, Bodner GM, Aydogdu B, Saban Y (2018) The inclusion of science process skills in multiple choice questions: are we getting any better? Eur J Sci Math Educ 6(1):13–23. https://doi.org/10.30935/scimath/9519

Eyster HN, Olmsted P, Naidoo R, Chan KMA (2022) Motivating conservation even for widespread species using genetic uniqueness and relational values. Biol Conserv 266:109438. https://doi.org/10.1016/j.biocon.2021.109438

Farren P (2019) 'Transformative pedagogy' in language teacher education. In: Savvidou C (ed) Second language acquisition: pedagogies, practices, and perspectives. IntechOpen, pp 1–23. https://doi.org/10.5772/intechopen.89470

Förster R, Zimmermann AB, Mader C (2019) Transformative teaching in higher education for sustainable development: facing the challenges. Gaia 28(3):324–326. https://doi.org/10.14512/gaia.28.3.18

Fraser SP (2015) Transformative science teaching in higher education. J Transform Educ 13(2):140–160. https://doi.org/10.1177/1541344615571417

Fuglei M (2014) Start your own learning revolution with transformational teaching. Concordia University. https://education.cu-portland.edu/blog/classroom-resources/learningrevolution-transformational-teaching/

Gericke N, Högström P, Wallin J (2023) A systematic review of research on laboratory work in secondary school. Stud Sci Educ 59(2):245–285. https://doi.org/10.1080/03057267.2022.2090125

Großschedl J, Harms U, Kleickmann T, Glowinski I (2015) Preservice biology teachers' professional knowledge: structure and learning opportunities. J Sci Teach Educ 26:291–318. https://doi.org/10.1007/s10972-015-9423-6

Hakkarainen K, Sintonen M (2002) The interrogative model of inquiry and computer-supported collaborative learning. Sci Educ 11:25–43

Harnisch DL, Comstock SL, Bruce BC (2014) Collaborative inquiry with technology in secondary science classrooms: professional learning community development at work. E-learning Digit Media 11(5):495–505. https://doi.org/10.2304/elea.2014.11.5.495

Heberlein TA (2013) Navigating environmental attitudes. Oxford Academic. https://doi.org/10.1093/acprof:oso/9780199773329.001.0001

Huang X, Lajoie SP (2023) Social emotional interaction in collaborative learning: why it matters and how can we measure it? Soc Sci Humanit Open 7:100447. https://doi.org/10.1016/j.ssaho.2023.100447

IPCC (Intergovernmental Panel on Climate Change) (2023) Sections. In: Lee H, Romero J (eds) Climate change 2023: synthesis report. Contribution of Working Groups I, II and III to the sixth assessment report of the Intergovernmental Panel on Climate Change. IPCC, pp 35–115. https://doi.org/10.59327/IPCC/AR6-9789291691647

Jachimowicz JM, Hauser OP, O'Brien JD, Sherman E, Galinsky AD (2018) The critical role of second-order normative beliefs in predicting energy conservation. Nat Hum Behav 2:757–764. https://doi.org/10.1038/s41562-018-0434-0

Jeronen E, Jeronen J, Raustia H (2009) Environmental education in Finland: a case study of environmental education in nature schools. Int J Environ Sci Educ 4(1):1–23

Jeronen E, Palmberg I, Yli-Panula E (2017) Teaching methods in biology education and sustainability education including outdoor education for promoting sustainability: a literature review. Educ Sci 7(1):1. https://doi.org/10.3390/educsci7010001

Karami M, Pakmehr H, Aghili A (2012) Another view to importance of teaching methods in curriculum: collaborative learning and students' critical thinking disposition. Procedia Soc Behav Sci 46:3266–3270. https://doi.org/10.1016/j.sbspro.2012.06.048

Kollmuss A, Agyeman J (2002) Mind the gap: why do people act environmentally and what are the barriers to pro-environmental behaviour? Environ Educ Res 8:239–260. https://doi.org/10.1080/13504620220145401

Land R, Rattray J (2017) Threshold concepts and conceptual difficulty. Pract Evid Scholarsh Teach Learn Higher Educ 12(2):63–80

Landis JR, Koch GG (1997) Measurement of observer agreement for categorical data. Biometrics 33(1):159–174

Latkin C, Dayton L, Coyle C, Yi G, Lee D-I, Winiker A (2021) The relationship between social norms, avoidance, future orientation, and willingness to engage in climate change advocacy communications. Int J Environ Res Public Health 18:13037. https://doi.org/10.3390/ijerph182413037

Lavorel S, Morel-Deville F (2010) Study of biodiversity at the regional scale based upon the achievements of the Santo2600Expedition. In: Ulbrich K, Settele J, Benedict FF (eds) Biodiversity in education for sustainable development: reflection on school-research cooperation. Pensoft Publishers, pp 113–118

Liefländer AK, Bogner FX (2018) Educational impact on the relationship of environmental knowledge and attitudes. Environ Educ Res 24:611–624. https://doi.org/10.1080/13504622.2016.1188265

Lindemann-Matthies P, Constantinou C, Junge X, Köhler K, Mayer J, Nagel U, Raper G, Schüle D, Kadji-Beltran C (2009) The integration of biodiversity education in the initial education of primary school teachers: four comparative case studies from Europe. Environ Educ Res 15(1):17–37. https://doi.org/10.1080/13504620802613496

Maina FW (2004) Authentic learning: perspectives from contemporary educators. J Authentic Learn 1:1–8

McWhinney W, Markos L (2003) Transformative education: across the threshold. J Transform Educ 1(1):16–37. https://doi.org/10.1177/1541344603252098

Merzel C (2023) Pedagogy for transformative teaching and learning. Pedagogy Health Promotion 9(4):231–233. https://doi.org/10.1177/23733799231208392

Mezirow J (1981) A critical theory of adult learning and education. Adult Educ 32(1):3–24

Mezirow J (1990) Fostering critical reflection in adulthood: a guide to transformative and emancipatory learning. Jossey-Bass Publishers, pp 1–20

Mezirow J (1991) Transformative dimensions of adult learning. Jossey-Bass Publishers

Mezirow J (1997) Transformative learning: theory to practice. New Dir Adult Continuing Educ 74:5–12. https://doi.org/10.1002/ace.7401

Mezirow J (2003) Transformative learning as discourse. J Transform Educ 1(1):58–63. https://doi.org/10.1177/1541344603252172

Mezirow J (2009) An overview on transformation learning. In: Illeris K (ed) Contemporary theories of learning. Routledge, pp 90–105

Mochizuki Y (2016) Educating for transforming our world: revisiting international debates surrounding education for sustainable development. Curr Issues Comp Educ 19(1):109–124. https://doi.org/10.52214/cice.v19i1.11538

Mogren A (2019) Guiding principles of transformative education for sustainable development in local school organisations: investigating whole school approaches through a school improvement lens. Doctoral thesis, Karlstad University Studies, 33. Karlstad University, Karlstad, Sweden

Monroe MC, Plate RR, Oxarart A, Bowers A, Chaves WA (2019) Identifying effective climate change education strategies: a systematic review of the research. Environ Educ Res 25(6):791–812. https://doi.org/10.1080/13504622.2017.1360842

Mouton M, Rootman-Le Grange I (2020) Scientific discourse: can our first-year students express themselves in science? 6th International Conference on Higher Education Advances (HEAd'20). Universitat Politècnica de València, València. https://doi.org/10.4995/HEAd20.2020.11110

Navarro-Perez M, Tidball KG (2012) Challenges of biodiversity education: a review of education strategies for biodiversity education. Int Electron J Environ Educ 2(1):13–30

O'Brien K, Reams J, Caspari A, Dugmore A, Faghihimani M, Fazey I, Hackmann H, Manuel-Navarrete D, Marks J, Miller R, Raivio K, Romero-Lankao P, Virji H, Vogel C, Winiwarter V (2013) You say you want a revolution? Transforming education and capacity building in response to global change. Environ Sci Pol 28:48–59. https://doi.org/10.1016/j.envsci.2012.11.011

Odell V, Molthan-Hill P, Martin S, Sterling S (2019) Transformative education to address all sustainable development goals. In: Filho WL, Azul A, Brandli L, Özuyar P, Wall T (eds) Quality education. Encyclopedia of the UN Sustainable Development Goals. https://doi.org/10.1007/978-3-319-69902-8_106-1

Organisation for Economic Co-operation and Development (OECD) (2018) The future of education and skills Education 2030. https://www.oecd.org/education/2030/E2030%20Position%20Paper%20(05.04.2018).pdf

Organisation for Economic Co-operation and Development (OECD) (2019) Future of education and skills 2030—OECD learning compass 2030: a series of concept notes. https://www.oecd.org/education/2030-project/teaching-and-learning/learning/learning-compass-2030/OECD_Learning_Compass_2030_Concept_Note_Series.pdf

Ortega-Torres E, Moncholí V (2021) "Expliquem l' Albufera": Transformar una salida de campo en un proyecto interdisciplinar ["Explain the Albufera" transform a field trip into an interdisciplinary project]. Enseñanza de las Ciencias 39(2):241–252. https://doi.org/10.5565/rev/ensciencias.3241

Osguthorpe TD, Jensen B (2023) Transformative practice and the moral rewards of teaching. In: Osguthorpe RD, Jensen B (eds) Teacher education quarterly. Caddo Gap Press, pp 7–29

Ottander C, Grelsson G (2006) Laboratory work: the teachers' perspective. J Biol Educ 40(3):113–118

Page MJ, McKenzie JE, Bossuyt PM, Boutron I, Hoffmann TC, Mulrow CD, Shamseer L, Tetzlaff JM, Akl EA, Brennan SE, Chou R, Glanville J, Grimshaw JM, Hróbjartsson A, Lalu MM, Li T, Loder EW, Mayo-Wilson E, McDonald S et al (2021) The PRISMA 2020 statement: an updated guideline for reporting systematic reviews. Br Med J 372:n71. https://doi.org/10.1136/bmj.n71

Palmberg I, Hermans M, Jeronen E, Kärkkäinen S, Persson C, Yli-Panula E (2018) Nordic student teachers' views on the importance of species and species identification. J Sci Teach Educ 29:397–419. https://doi.org/10.1080/1046560X.2018.1468167

Pathan A, Bröckl M, Oja L, Ahvenharju S, Raivio T (2013) Kansallisten kestävää kehitystä edistävien kasvatuksen ja koulutuksen strategioiden toimeenpanon arviointi [Strategy evaluation of sustainability education and implementation]. Final Report 11.10.2012. Gaia Consulting Oy

Paul LA (2020) Transformative education. Educ Theory 70(5):561–579

Peters E, Västfjäll D, Gärling T, Slovic P (2006) Affect and decision making: a "hot" topic. J Behav Decis Mak 19:79–85. https://doi.org/10.1002/bdm.528

Pyle R (2003) Nature matrix: reconnecting people and nature. Oryx 37(2):206–214. https://doi.org/10.1017/S0030605303000383

Randler C (2008) Teaching species identification: a prerequisite for learning biodiversity and understanding ecology. EURASIA J Math Sci Technol Educ 4:223–231

Reed MS, Evely AC, Cundill G, Fazey I, Glass J, Laing A, Newig J, Parrish B, Prell C, Raymond C, Stringer LC (2010) What is social learning? Ecol Soc 15(4):10

Renn O (2008) Risk governance: coping with uncertainty in a complex world. Earthscan

Reynolds M (2011) Reflective practice: origins and interpretations. Action Learn Res Pract 8(1):5–13

Rodríguez JG, Barth M (2020) Transformative learning in the field of sustainability: a systematic literature review (1999–2019). Int J Sustain High Educ 21(5):993–1013. https://doi.org/10.1108/IJSHE-05-2019-0168

Rüdisser J, Tasser E, Walde J, Huemer P, Lechner K, Ortner A, Tappeiner U (2017) Simplified and still meaningful: assessing butterfly habitat quality in grasslands with data collected by pupils. J Insect Conserv 21:677–688. https://doi.org/10.1007/s10841-017-0010-3

Salami MO (2015) Impact of modified laboratory learning environment on transformative biology process skills among secondary school students in Osun state. Nigeria Mediterranean J Soc Sci 6(5):84. https://doi.org/10.5901/mjss.2015.v6n5s1p84

Salas E, Cooke NJ, Rosen MA (2008) On teams, teamwork, as well as team performance: discoveries and developments. Hum Factors 50(3):540–547. https://doi.org/10.1518/001872008X288457

Sánchez-Meca J, Botella J (2010) Systematic reviews and meta-analyses: tools for professional practice. Psychologist Pap 31(1):7–17

Savery JR (2006) Overview of problem-based learning: definitions and distinctions. Interdisciplinary J Problem-Based Learn 1:8–20. https://doi.org/10.7771/1541-5015.1002

Sjöström J, Eilks I (2018) Reconsidering different visions of scientific literacy and science education based on the concept of Bildung. In: Dori YJ, Mevarech ZR, Baker DR (eds) Cognition, metacognition and culture in STEM education. Innovations in science education and technology, vol 24. Springer. https://doi.org/10.1007/978-3-319-66659-4_4

Slavich GM, Zimbardo PG (2012) Transformational teaching: theoretical underpinnings, basic principles, and core methods. Educ Psychol Rev 24:569–608. https://doi.org/10.1007/s10648-012-9199-6

Smith N, Leiserowitz A (2014) The role of emotion in global warming policy support and opposition. Risk Anal 34:937–948. https://doi.org/10.1111/risa.12140

Spangenberg JH (2011) Sustainability science: a review, an analysis and some empirical lessons. Environ Conserv 38(3):275–287. https://doi.org/10.1017/S0376892911000270

Sterling S (2001) Sustainable education: re-visioning learning and change. In: Schumacher society briefing, 6. Green Books

Sterling S (2003) Whole systems thinking as a basis for paradigm change in education: explorations in the context of sustainability. http://www.bath.ac.uk/cree/sterling/sterlingthesis.pdf

Sterling S (2004) An analysis of the development of sustainability education internationally. Evolution, interpretation and transformative potential. In Cullingford C, Blewitt J (Eds), The sustainability curriculum. The challenge for Higher Education (pp 43–62). Routledge

Sterling S (2008) Sustainable education—towards a deep learning response to unsustainability. Policy Pract 6:63–68

Sterling S (2010) Learning for resilience, or the resilient learner? Towards a necessary reconciliation in a paradigm of sustainable education. Environ Educ Res 16:511–528

Sterling S (2011) Transformative learning and sustainability: sketching the conceptual ground. J Learn Teach Higher Educ 5:17–33

Sterling S, Blake J, Goodson I (2013) Transformative learning for a sustainable future: an exploration of pedagogies for change at an alternative college. Sustainability 5:1–26. https://doi.org/10.3390/su5125347

Sterling S, Dawson J, Warwick P (2018) Transforming sustainability education at the creative edge of the mainstream: a case study of Schumacher college. J Transform Educ 16(4):323–343. https://doi.org/10.1177/1541344618784375

Stevenson RB, Wals AEJ, Dillon J, Brody M (2012) Introduction. An orientation to environmental education and the handbook. In: Stevenson RB, Brody M, Dillon J, Wals AEJ (eds) International handbook of research on environmental education. Taylor and Francis, pp 1–6

Tam K, Leung AK, Clayton S (2021) Research on climate change in social psychology publications: a systematic review. Asian J Soc Psychol 24:117–143

Taylor EW (1997) Building upon the theoretical debate: a critical review of the empirical studies of Mezirow's transformative learning theory. Adult Educ Q 48(1):34–59

Tilbury D (2007) Learning based change for sustainability: perspectives and pathways. In: Wals AEJ (ed) Social learning towards a sustainable world. Principles, perspectives, and praxis. Wageningen, Wageningen Academic Publishers, pp 117–131

Trott CD (2019) Children's constructive climate change engagement: empowering awareness, agency, and action. Environ Educ Res 26(4):532–554

Turner JC, Fulmer SM (2013) Observing interpersonal regulation of engagement during instruction in middle school classrooms. In: Volet S, Vauras M (eds) Interpersonal regulation of learning and motivation: methodological advances. Routledge, pp 147–169

UNESCO (2017) UNESCO moving forward the 2030 Agenda for Sustainable Development. https://unesdoc.unesco.org/ark:/48223/pf0000247785

United Nations (UN) (2022) Convention of Biological Biodiversity. United Nations Biodiversity Conference. COP 15. 2022, 7–19 December 2022, Montreal, Canada

Van de Walle JA, Karp KS, Lovin LH, Bay-Williams JM (2014) Teaching student-centered mathematics: developmentally appropriate instruction for grades 3–5, 2nd edn. Pearson

Van Weelie D, Wals AEJ (2002) Making biodiversity meaningful through environmental education. Int J Sci Educ 24(11):1143–1156. https://doi.org/10.1080/09500690210134839

Vasquez LG, Lopez CRH, Mujica D (2020) Experiencia de educación ambiental Para la sostenibilidad con el enfoque, ciencia, tecnología y sociedad [Experience of environmental education for sustainability with the focus, science, technology and society]. REMEA-Revista Eletrônica do Mestrado em Educação Ambiental 37(3):184–207

Volet S, Vauras M, Khosa D, Iiskala T (2013) Metacognitive regulation in collaborative learning: conceptual developments and methodological contextualizations. In: Volet S, Vauras M (eds) Interpersonal regulation of learning and motivation. Routledge, pp 67–101

Weinberg EA, Trott CD, Wakefield W, Merritt EG, Archambault L (2020) Looking inward, outward and forward: exploring the process of transformative learning in teacher education for a sustainable future. Sustain Sci 15:1767–1787. https://doi.org/10.1007/s11625-020-00831-9

Wickenberg B, Kiss B, McCormick K, Palgan YV (2022) Seeds of transformative learning: investigating past experiences from implementing nature-based solutions. Front Sustainable Cities 4:835511. https://doi.org/10.3389/frsc.2022.835511

Wolff L-A (2022) Biodiversity education. In: Idowu S, Schmidpeter R, Capaldi N, Zu L, Del Baldo M, Abreu R (eds) Encyclopedia of sustainable management. Springer. https://doi.org/10.1007/978-3-030-02006-4_1136-1

Xie B, Brewer MB, Hayes BK, McDonald RI, Newell BR (2019) Predicting climate change risk perception and willingness to act. J Environ Psychol 65:101331. https://doi.org/10.1016/j.jenvp.2019.101331

Yacek DW (2021) The transformative classroom: philosophical foundations and practical applications. Routledge

Yacek D, Rödel S, Karcher M (2020) Transformative education: philosophical, psychological, and pedagogical dimensions. Educational Theory 70:529-537. https://doi.org/10.1111/edth.12442

Zoller U (2015) Research-based transformative science/STEM/STES/STESEP education for "sustainability thinking": from teaching to "know" to learning to "think". Sustainability 7(4):4474–4491. https://doi.org/10.3390/su704447

Eija Yli-Panula, PhD, is a senior researcher (emerita) at the University of Turku, Finland and is acting as a Visiting Professor at the University of Murcia, Spain. She is an Associate Professor in sustainability education and also in aerobiology at the University of Turku. In addition to education she has a degree in biology and geography. Her research interests are in sustainability education, e.g. teacher competencies in science education, in didactics of biology and geography and in teacher education. Yli-Panula has acted as an editor for national (Publications of the Finnish Research Association of Subject Didactics) and international journals and special issues, and she has published several international scientific articles and book chapters. She has been part of international scientific research groups and networking groups.

Isabel Banos-González, PhD, is a lecturer in the Department of Didactics of Experimental Sciences at the Faculty of Education (University of Murcia, UM). She is main researcher of the 'E013-02 Research and innovation in science education' research group of the UM. She has a strong background in environmental science. Her PhD was in the field of Ecology, and she has worked as secondary school teacher. Her particular interest relies on sustainability education. Isabel has published in both natural and educational science journals, as well as chapters and books.

Eila Jeronen, PhD, is emerita in biology and geography didactics at the University of Oulu in Finland. She is also associate professor in environmental education at the University of Oulu, in biology education at the University of Lapland in Finland and in sustainable development education at the University of Helsinki in Finland. In addition, she has Ph.L. and M.Sc. degrees in biology and teacher certificate in biology and geography in the University of Oulu. Her research interest includes science education, environmental education, and sustainability education. She has supervised dissertation theses and published and worked as an editor in international journals, an editor-in-chief in Natura (the Finnish journal of Biologian ja maantieteen opettajien liitto) and in multicultural projects in several countries.

Sandra Bueno-Férez, MSc, is a part time teacher in science education at the Faculty of Education in the University of Murcia and she also does research in teacher training. She has a technical profile developed over 20 years in the field of water management, as well as formal training as a science teacher. Her research is mainly based on the competencies in primary education towards sustainability and in relation to the current socio-ecological problems.

Chapter 9
Collaboration Across Subjects as a Way to Engage Students in Wicked Sustainability Issues

Eldri Scheie

Abstract Education for sustainable development provides crucial opportunities for young people to get involved in addressing authentic socio-ecological challenges in their local and global communities. Issues of sustainable development are often wicked problems that require new didactic approaches. Teaching science from a wicked sustainability perspective may prove challenging for science teachers because such a perspective draws on knowledge structures and teaching traditions that lie outside the realm of science. A possible way to overcome these challenges is to apply a cross-curricular approach to the organization of education for sustainable development in which teachers who specialize in different subjects collaborate. The case study presented here describes the integration of wicked sustainability issues into educational practices over a 3-year period, demonstrating significant shifts towards interdisciplinary and problem-oriented teaching approaches. This study focused on teachers working in interdisciplinary teams to design curriculum units for students at a rural Norwegian upper secondary school, and it illustrates how teachers collaborated across various subjects in order to engage students in addressing a wicked sustainability issue within their community. Through multiple school subjects, the school project tackled the local and global decline in pollinators. Over the 3-year period of developing and reflecting on their project, the teachers observed significant changes in engagement levels for both themselves and their students. This improvement was driven by the transition from a subject-oriented to a problem-oriented teaching approach, with a specific focus on the decline in pollinators as a complex sustainability issue.

Keywords Education for sustainable development (ESD) · Interdisciplinarity · Professional development · Wicked problems · Problem-oriented

---

E. Scheie (✉)
Norwegian Centre for Science Education, University of Oslo, Oslo, Norway
e-mail: eldri.scheie@naturfagsenteret.no

F. Skarstein, L.-A. Wolff (eds.), *Sustainability and Science Education*, Environmental Challenges and Solutions,
https://doi.org/10.1007/978-3-032-24408-6_9

9.1 Introduction

Sustainable development (SD) is generally understood through the three dimensions of environmental, social, and economic sustainability (Giddings et al. 2002). Addressing sustainability challenges often necessitates taking an integrated approach to all of these dimensions, which can create dilemmas when prioritizing decisions in specific contexts (Öhman and Öhman 2012). For these reasons, sustainability issues are considered wicked problems, that is, they are problems marked by profound uncertainties and controversies about the knowledge required to solve them, as well as by conflicts over the values and norms that underpin various problematic definitions and proposed solutions (Block et al. 2019).

Teaching science from a sustainability perspective can be challenging for science teachers because this approach utilizes knowledge structures and teaching traditions that range beyond the typical scope of science education (Borg et al. 2012). For instance, such a teaching approach involves integrating values and emotions along with scientific concepts and adopting an anthropocentric perspective (White et al. 2023). Additionally, the complexity of sustainability issues necessitates interdisciplinary perspectives, and the wicked nature of sustainable development—as opposed to the tame nature of other kinds of problems, including those commonly taught in traditional science education—makes it difficult to apply predefined solutions (Block et al. 2019).

One way to address these challenges is through a cross-curricular approach to organizing ESD in which teachers from different subjects or disciplines collaborate (Berglund and Gericke 2016; Munkebye et al. 2020). If this approach is taken, however, it is crucial that teachers receive comprehensive training during their professional education to equip them with the skills and knowledge necessary to teach wicked problems effectively and thus to empower students (Sjöblom et al. 2024).

From 2009 to 2022, Norwegian schools had the opportunity to participate in the Sustainable Backpack Program (SBP), a professional development initiative designed to support teacher teams in implementing authentic, interdisciplinary, and community-based education for sustainable development (ESD) on a national scale (Scheie and Stromholt 2019). The school and teachers I report on here were part of the SBP for 3 years. This chapter explores the teachers' transition from a monodisciplinary to an interdisciplinary approach to ESD. Their perceptions of how their self-designed teaching units promoted problem-oriented teaching focused on wicked sustainability issues were analysed as an attempt to answer the following research questions:

RQ1: How do the teachers describe self-designed sustainability issue in relation to the characteristics of wicked problems?
RQ2: How do sustainability issues affect the choice of school subjects and competence aims in the teachers' curriculum units?

9.2 Literature Review and Background

This section begins by exploring the research on interdisciplinarity in ESD and then examines the characteristics of wicked problems and explains why sustainability issues exhibit many of these properties. I present the five key properties that define wicked problems in SD (Seager et al. 2012) and describe how these properties serve as the primary analytical tool in this study.

9.2.1 Interdisciplinarity in ESD

ESD is a policy concept, and the critical role of education in achieving sustainable development has been acknowledged in recent years, particularly during the United Nations' Decade of ESD (UNESCO 2014). ESD can be defined as "education which empowers learners to make informed decisions and responsible actions for environmental integrity, economic viability and a just society, for present and future generations, while respecting cultural diversity" (UNESCO 2014, p. 12). The objective of ESD, according to UNESCO, is to develop competencies that empower individuals to reflect on their own actions while considering their current and future social, cultural, economic, and environmental impacts, from both local and global perspectives (UNESCO 2017, p. 7). Although ESD has been gaining momentum worldwide, it can be challenging for teachers to impart such education to students (Borg et al. 2014). Schools are by default disciplinary, which can make it hard to see how the interdisciplinary nature of sustainability issues can be approached and taught, which in turn makes it difficult to authentically address real-world problems.

There is strong consensus about the importance of including a holistic approach to ESD (Walshe 2017). Researchers have long argued that issues of sustainability are too complex for any single discipline (Ling et al. 2009), and they have similarly argued that education for sustainability is too complex for any single academic discipline (Schmitz et al. 2010). However, subject teachers rarely have sufficient knowledge about how to teach ESD with an interdisciplinary approach (Hofman 2012; Pepper and Wildy 2008). Teachers are unsure of which approach they should take in teaching ESD and of which interconnections to make, and many researchers have concluded that teachers' uses of discipline-specific approaches and content constitute narrower approaches to ESD (Jonsson 2008). Many attempts have been made to overcome the traditional teaching approach, in which subjects are taught separately, by emphasizing a holistic approach (Root-Bernstein 2018).

Uitto and Saloranta (2017) argue that subject teachers would benefit from professional development in the topic of how to teach sustainability more holistically. The authors claim that it is likely that teachers are unaware of their own competencies because they do not fully understand SD, and they argue that teacher education courses can make the background and pedagogy of ESD more explicit and thus build teachers' awareness of the issue. Borg et al. (2012, 2014) recommend that teachers do their planning as part of cross-disciplinary teams in order to gain a common understanding of SD and to share knowledge and practices with each other, although the authors recognize that this approach is uncommon in many places.

In a review of the role of interdisciplinarity argumentation, Crujeiras-Pérez and Jiménez-Aleixandre highlight that "interdisciplinary thinking is understood as the capacity to integrate knowledge and modes of thinking from two or more disciplines or established areas of expertise to produce a cognitive advancement in ways that would have been impossible or unlikely through single disciplinary means" (2019, p. 33). The authors also draw attention to the distinction of "interdisciplinarity" from related concepts such as "multidisciplinarity" and "transdisciplinarity". Multidisciplinarity draws on perspectives from individual disciplines while acknowledging clear disciplinary boundaries, whereas transdisciplinarity focuses on addressing solely holistic problems or issues that involve some form of multidisciplinary knowledge. The authors caution that many problems of the twenty-first century require interdisciplinary thinking, necessitating complex reasoning that can draw on the knowledge bases of multiple disciplines.

In a study of ESD teaching programmes in Norwegian schools, Munkebye et al. (2020) looked at the subjects and competence aims that teachers included in their ESD teaching programmes, and they concluded that the schools' approach to SD was to some extent holistic. However, the authors suggest that teachers should pay more attention to the untapped potential of various subjects in terms of exploration, argumentation, critical thinking, and ethical issues when teaching ESD (Munkebye et al. 2020).

One approach for improving the teaching of ESD is to encourage a more pluralistic approach to teaching and to emphasize students' argumentation and critical thinking skills by introducing the idea of wicked problems (Scheie and Stromholt 2019).

9.2.2 Wicked Problems and Sustainable Development

Sustainability issues such as climate change, biodiversity loss, resource scarcity, and global health challenges are profoundly complex, multidimensional, and often intractable. Sustainability issues are therefore often referred to as unstructured problems (Hisschmöller and Hoppe 2001) or wicked problems (Rittel and Webber 1973). The term "wicked problem" has been widely used in connection with sustainability issues (Lönngren and Van Poeck 2021), and it is a useful concept in ESD because it can be used to highlight approaches for addressing complex societal and environmental problems in teaching and learning (Block et al. 2019; Lönngren et al. 2017).

Rittel and Webber (1973) argue that at least 10 distinguishing properties characterize wicked problems generally (although they did not assert that all 10 need to be present for a problem to be considered wicked). Subsequently, Pryshlakivsky and Searcy (2013) suggested that sustainability problems in particular do indeed involve all 10 characteristics. At around the same time, Seager et al. (2012) reduced the list of wicked problem characteristics (at least as they pertain to SD) by proposing five key properties that define wicked problems in SD, namely difficulties in problem formulation, multiple but incompatible solutions, open-ended time frames, uniqueness or novelty, and competing value systems.

The complexity of SD arises from the multidimensional nature of sustainability issues, which span multiple levels across both time and space (local to global) and necessitate the integrated consideration of environmental, social, and economic dimensions (Öhman and Öhman 2012). Wicked sustainability problems are especially challenging, if not impossible, to solve in a satisfactory manner because they involve intricate interdependencies among environmental, social, and economic interests. There is often no clear right or wrong answer, and decisions must be made that prioritize certain groups and goals over others based on norms, values, and available knowledge (Block et al. 2019). To give a tangible example, establishing a wind farm might negatively affect local biodiversity, but it might also boost local employment and provide valuable renewable energy, which would help reduce the greenhouse gas emissions driving climate change. In such scenarios, simple solutions are elusive, requiring the prioritization of certain values over others when making decisions. Many sustainability issues include such wicked complexity, and, as underscored in the discourse on ESD, these issues must be taught through interdisciplinary teaching and learning approaches.

Lönngren and Van Poeck (2021) discuss the use of the concept of wicked problems in the sustainability literature and suggest that it can be a critical tool, but they note that the term has often been given different meanings by different authors and that they have not always made the epistemological assumptions behind it explicit. In contrast, several other scholars in the field of ESD advocate for using the concept of wicked problems as a means to address the highly complex issues of sustainability that face society (Block et al. 2019; Gericke 2022; Lozano and Barreiro-Gen 2022).

According to Pryshlakivsky and Searcy (2013), students must be engaged in determining a pluralistic strategy in which they critically approach different perspectives rather than reflecting solely on predefined solutions to problems. Erduran (2023) calls for including more to science education than the "objectives about scientific knowledge and investigations" that have dominated (and still dominate) school science curricula in many parts of the world. This recommendation is in line with the PISA Science Framework 2025, which calls for improving students' abilities to interpret scientific data and evidence critically so that they can take informed action regarding sustainability issues (OECD 2023).

Interdisciplinary teaching to address wicked problems benefits from a community-based approach, which is when a larger problem is placed in and restricted to the context of a local (and therefore more tangible) community. Nikitina (2006, p. 266) describes the community-based approach as problem-oriented and proposes the approach as a strategy for interdisciplinary teaching that focuses on examining "messy real-life problems" and that "easily brings together a wide range of disciplines in the curriculum. Learning becomes personally meaningful and highly motivated by a desire to resolve an important social concern".

9.2.3 Framework for Analysing Wicked Problems in Teaching and Learning

Tame problems are those that are well defined and can be resolved through traditional linear processes such as data collection, analysis, explanations, and the implementation of a solution (Block et al. 2019; Hisschmöller and Hoppe 2001; Rittel and Webber 1973). In contrast, wicked problems are complex issues that are difficult to define, have open-ended timeframes, involve multiple and often conflicting solutions, and exhibit high levels of uncertainty and complexity. These problems engage multiple stakeholders and decision-makers, each of which might hold values that conflict with the values of other stakeholders/decision-makers, and each of which might propose solutions that conflict with other proposed solutions (Block et al. 2019; Hisschmöller and Hoppe 2001; Rittel and Webber 1973). As mentioned previously, Seager et al. (2012) outline five characteristics that define wicked problems within the context of SD. I will use these to delve deeper into the issue of how teachers can approach wicked problems in their sustainability teaching:

1. *Difficulties in Problem Formulations*: Wicked problems are difficult to define clearly because they often involve numerous stakeholders with various perspectives and priorities. The complexity and interdependency of the issues at hand make it nearly impossible to craft a universally accepted problem statement.
2. *Multiple but Incompatible Solutions*: Given the complexity of wicked problems, there is typically no single, universally accepted way to address them. Instead, there are multiple possible solutions, often the result of conflicting approaches to solving the problem. These different solutions may cater to different stakeholders, making it hard to implement a solution that satisfies everyone.
3. *Open-Ended Time Frames*: The time frames are open-ended, meaning there is no single point in time to definitively evaluate the adequacy of a proposed solution because the circumstances of both the problem and the solution are subject to continuous change and evolution.
4. *Uniqueness or Novelty*: Each wicked problem is distinct and exists within its specific context, which means that standard, pre-developed solutions are usually insufficient. Innovative and context-sensitive approaches are required because solutions need to be tailored specifically to the nuances of the individual problem.
5. *Competing Value Systems*: Wicked problems are often deeply entangled with societal values and objectives that can be in direct opposition to each other. What constitutes a desirable outcome can vary drastically among stakeholders, making it hard or impossible to find an unequivocally good single solution to a particular problem.

9.2.4 Context and the Sustainable Backpack Programme

The SBP was initiated by the Norwegian Ministries of Education and Research and the Ministry of Climate and Environment (2009–2022). Targeting Norwegian primary, secondary, and upper secondary schools, the SBP's aim was to increase sustainability awareness in students based on the understandings and competencies of the teachers and students (Scheie and Stromholt 2019). Teachers designed curriculum units in interdisciplinary teams in the subject areas of science, social studies, art, language, and so on. Each year, approximately 140 schools were selected to participate, with at least two teachers per school supported by the school leader. Teachers committed to a 3-year programme to enhance their teaching plans.

The schools participating in the SBP represented a great variety in terms of student population, teaching traditions, and local context. The school in focus here, Borg School, participated for a total of 3 years and focused on improving ESD practices for eleventh-grade students (ages 16–17). Teachers attended regional and national SBP meetings, becoming part of a professional learning community facilitated by teacher educators and researchers who acted as mentors. These meetings served as a collaborative network for professional development, where teachers exchanged feedback and guidance. Twice a year, teachers submitted reflection reports on their curriculum units, which were subsequently reviewed by mentors.

9.3 Methods

This section documents the data collection and analysis of teachers' reports after the first and third years of the SBP. Following that, the limitations and ethical considerations of the study are described.

9.3.1 Data Collection and Analysis

The data were collected from teachers' final reports after year one (two teachers actively involved in the writing) and year three (four teachers actively involved in the writing) of SBP participation, which described the curriculum units and included teachers' reflections about the SBP. Each report was 2500–3000 words. The final reports followed a template (Table 9.1) that requested specific information about the sustainability theme, the subjects involved, and student outcomes. The teachers were also asked about how they collaborated across subjects and for their reflections related to teaching and collaboration in ESD at their school. Borg School serves students in a small rural community in Norway. The selection of Borg School, one of many that participated in the SBP, can be categorized as convenience sampling, with access to supplementary information being crucial (Robson 2002). Borg

Table 9.1 Excerpts from the template for the final report written by teachers

The theme and the sustainability issue that the students will be involved in
Interdisciplinary work: Describe the subjects that are part of the project
Competence aims from the curriculum
Action competence: What knowledge, skills, attitudes, and actions will the students practise
ESD: Describe the ways in which the teaching programme helps to promote students' understanding of SD
The curriculum unit: Describe the methods, teaching, and student activities
Assessments of student outcomes
Teachers' reflection on their ESD and collaboration between colleagues in different subjects

School was selected from a group of approximately 30 upper secondary schools that participated in the SBP over a 3-year period. The criteria for selecting the school included the submission of comprehensive descriptions and reflections on the work they developed and implemented with their own students. Additionally, the teachers gave two oral presentations at the SBP meetings at which the author of this study was present. These presentations and the teachers' reflections were used in this study as support material to describe and reflect on the school's ESD development. The same teachers participated all 3 years, although in the final year they included additional teachers in the design process. During the analysis, teachers were contacted once by e-mail for clarifications about their interdisciplinary collaborations.

This study is not intended to evaluate the outcomes achieved by the school after participating in the SBP; instead, it aims to provide an analysis of how teachers describe their self-designed sustainability issues in relation to the characteristics of wicked problems, and how the characteristics of the sustainability issue can influence the choice of school subjects and competency goals within the teachers' ESD. The written reports were analysed by content analysis (Castleberry and Nolen 2018), focusing on the differences between the first and third years of teachers' writing and reflections about their ESD programme (Tables 9.2). To analyse the sustainability issues in the curriculum units, we used the framework of Seager et al. (2012), which outlines five characteristics that define wicked problems within the context of SD: difficulties in problem formulations, multiple but incompatible solutions, open-ended time frames, uniqueness or novelty, and competing value systems. The unit of analysis included teacher teams' written reflections on developing and implementing their self-designed ESD.

9.3.2 Limitations and Ethical Considerations

The analysis was based on teachers' written reports, statements, and reflections on their teaching and the SBP rather than on direct observations. Consequently, there may be inherent uncertainty regarding whether the teachers' perceptions accurately reflect what occurred. By taking on the roles of researcher and mentor in the SBP, I had multiple opportunities over 3 years to both listen to and converse with the

Table 9.2 Short descriptions from the curriculum units after the first and third years

Issues in the reports	First year of the curriculum unit	Third year of the curriculum unit
Theme	Pollinators	Pollinators
Sustainability issue	How important are willows (willow bushes, *Salix* spp.) for the population size of pollinators?	Should we do more to increase the population of pollinators in our local surroundings?
Learning arena	At school and in the school's local area	At school (inside and outside) and in the school's local area, at the Botanical Garden, and in the municipality
Collaborators	Researchers from the university	Researchers from the university, staff at the Botanical Garden, local politicians, a newspaper published in the summer, and an author (Maja Lunde)
Subjects (and teachers) involved	Science, mathematics, physical education	Science, mathematics, social studies, Norwegian language
Student outcome	"Students have planned and carried out fieldwork. They have had research meetings and discussions. The results were presented to other students, teachers, and the researcher leading this project"	"The students have gained a thorough understanding of pollinators. The students have expertise in how to influence their society by using various channels. They have also planned and optimized the plant selection and planted them in the school's entrance area"

The text in both columns consists of excerpts and some quotes from the teachers' descriptions of their teaching in the first and third years

participating teachers. Additionally, I attended two presentations in which they specifically discussed their ESD and the outcomes for their students. This interaction enhances the credibility of the analysed data, as there was no significant difference between what the teachers communicated to me and what they had written.

The study was conducted in accordance with ethical guidelines for research, including obtaining informed consent from all participants, ensuring their confidentiality and anonymity, and securing approval from the Norwegian National Research Ethics Committee. Borg School is a pseudonym.

9.4 Results

This study investigated the educational practices to approaching ESD that teachers developed for their students during a 3-year programme in SBP based on curriculum units designed for upper secondary students. The teachers themselves selected the sustainability issue, school subjects, and curriculum units and set objectives for student learning outcomes.

9.4.1 Selection of a Sustainability Issue

This section addresses the first research question, "How do the teachers describe their self-designed sustainability issue in relation to characteristics of wicked problems?"

The ESD project focused on the issues of local and global declines in pollinators. The description in their first-year report (first year of the curriculum unit, Table 9.2) mostly focused on science and the issues of pollinators and willows. In the first year, teachers included science, mathematics, and physical education in their teaching about sustainability. In the third year, the teachers included science, mathematics, social studies, and Norwegian language (Table 9.2).

In the first year, science teachers decided to focus on the question "How important are willows (willow bushes, *Salix* spp.) for the population size of pollinators?" and designed a project around it. During meetings in the SBP, teachers participated in the discussions and in deep reflections about complex, controversial, and wicked problems and interdisciplinary teaching in ESD.

In the third year, the science, social studies, and Norwegian language teachers collaborated, and through discussion at their school, they changed the question to: "Should we do more to increase the population of pollinators in our local surroundings?" The teachers' writings about the new question were analysed through the lens of each of the five characteristics of a wicked problem as outlined by Seager et al. (2012). Here are excerpts from their analysis:

1. *Difficulties in Problem Formulations*: The problem statement itself is not straightforward. The phrase "do more" is subjective and can mean different things to different people. Determining which specific pollinators to focus on, understanding the factors contributing to their decline, and defining the geographical scope of local surroundings add to the ambiguity.
2. *Multiple but Incompatible Solutions*: There are several ways to address the issue, such as planting pollinator-friendly plants, reducing local pesticide use, creating habitat corridors, or engaging in public education. These solutions can sometimes conflict with one another or with other local needs and priorities. For example, reducing local use of pesticides may conflict with agricultural practices that rely on these chemicals for crop production.
3. *Open-Ended Time Frames*: The effectiveness of initiatives to increase pollinator populations can take years to assess, and conditions will continue to evolve due to climate change, urban development, and other environmental factors. Continuous monitoring and adaptation are needed, making it hard to determine an end solution or to measure success definitively.
4. *Uniqueness or Novelty*: The specific context of local surroundings makes each instance of the problem unique. What works in one area may not be effective in another due to differences in local ecosystems, climate, community values, and existing biodiversity. This fact requires tailored solutions rather than a one-size-fits-all solution.

5. *Competing Value Systems*: Different stakeholders may have conflicting interests and priorities. For example, some farmers might prioritize crop yields over pollinator protection due to economic pressures (Even different types of farmers will have competing interests as some crops are dependent on pollinators), while environmentalists and local residents may prioritize biodiversity and ecosystem health. Balancing these competing values and objectives makes it challenging to find a universally accepted solution.

To sum up, the sustainability issue "Should we do more to increase the population of pollinators in our local surroundings?" involves complex and interdependent factors, makes unambiguous problem definition difficult, requires solutions that may not be universally applicable or acceptable, and demands flexible, adaptive management over time in the face of competing values and interests.

9.4.2 The Choice of School Subjects

This section addresses the second research question "How do sustainability issues affect the choice of school subjects and competence aims in the teachers' curriculum units?"

The school subjects involved in the sustainability project in the first year of the project were science, mathematics, and physical education; while in year three, the latter was omitted in favour of social studies and Norwegian language (Table 9.3). The competence aims in science were the same in the first and third years. In mathematics, the competence aims were reduced from two to one, omitting teaching about solving and rearranging various types of equations and inequalities using algebra and digital tools. In physical education, the competence aim in the first year was about outdoor life. In addition to their own subject area, the school subjects that the science teachers included in the first year were mathematics and physical education. In year three, the science teachers decided to discuss the issue with other colleagues at their school, and together they decided to replace physical education with social studies and Norwegian language. The collaborations were driven by the science teachers, especially in the first year, but by the third year, all subjects and teachers had a clearer sense of ownership in the planning and teaching. A shift to problem-oriented teaching based on the question "Should we do more to increase the population of pollinators in our local surroundings?" resulted in the work becoming more complex, and teachers from both social studies and Norwegian language became more centrally involved, which resulted in competence aims from those areas being added to the curriculum unit in year three.

One teacher noted: *[p]articipation in the SBP means that we are forced to think more interdisciplinary, something we unfortunately rarely normally do* (Teacher 1). Another teacher said: *[s]itting down and working [on] interdisciplinarity with other teachers about the same project has been very educational for us as teachers* (Teacher 2). After discussion with researchers in the SBP and with colleagues

Table 9.3 Teachers' choice of competence aims from subject curricula after eleventh grade used in the first year and third year of the curriculum unit (Norwegian Directorate for Education and Training 2020)

Subject curriculum	Competence aims The aims of the training are to enable the apprentice to:	
	First year	Third year
Science	S1: Plan and carry out various types of investigations by identifying variables, gathering and processing data, and writing a report that includes discussions about estimating uncertainties of measurements and assessing possible sources of errors S3: Discuss and elaborate on authentic problems related to science based on practical investigations or systematized information from different sources S5: Provide an account of the concept of sustainable development S6: Elaborate on factors that influence the size of a population S18: Explain the greenhouse effect and elaborate on and analyse how human activity is altering the energy balance of the atmosphere	
Mathematics	M2: Evaluate, select and use mathematical methods and tools to solve problems from various subjects and social areas and reflect on, evaluate, and present solutions in a purposeful manner M4: Reformulate expressions and solve equations, inequalities, and systems of equations of the first and second order and simple equations with exponential and logarithmic functions using algebra and digital aids	M2: Evaluate, select, and use mathematical methods and tools to solve problems from various subjects and social areas and to reflect on, evaluate, and present solutions in a purposeful manner
Physical education	PE5: Practise outdoor life in different natural environments based on local traditions	
Social studies		SS2: Explore local, national, or global problems of current interest and discuss and elaborate on the different recommendations for solutions, orally and in writing, with precise use of social studies terminology SS3: Use a variety of digital search strategies to find and compare information that describes problems from different points of view and evaluate the objectives and relevance of one's sources SS4: Use concurrent and contradictory information from statistics to discuss and elaborate on a social studies issue SS21: Explore and discuss how one can participate in and influence the political system through different means and channels

(continued)

Table 9.3 (continued)

Subject curriculum	Competence aims The aims of the training are to enable the apprentice to:	
	First year	Third year
Norwegian language		NL3: Apply specialist knowledge and terminology from one's own study programme in conversations, discussions, and presentations about school, society, and work NL6: Master different roles in conversations, discussions, dramatizations, and presentations NL18: Discuss cross-cultural encounters and cultural conflicts on the basis of selected contemporary texts NL20: Interpret and evaluate the relationships among content, form, and purpose in composite texts

Competence aim labels refer to their subject (science/S, social studies/SS, Norwegian language/NL, mathematics/M) and are numbered in order of their appearance in the curricula for eleventh grade (Norwegian Directorate for Education and Training 2020)

teaching various subjects at their own school, Teacher 1 reflected on the experience in their third year: *The development of the project into a complex problem has led the problem to become more socially oriented and contemporary*. The teachers also reported that students had a better idea about what they were going to work on because the teachers understood what the contributions of each discipline should be. In their third-year report, the teachers wrote:

> We hope to create engagement with the students in terms of giving them ownership of local environmental issues through a school project. We would like to improve awareness and attitudes among students about a sustainable future for their local community. We will use the local community to provide students with an opportunity for data/information gathering and for processing and presenting this information.

The teachers recognized that by adopting a broad interdisciplinary approach to ESD, they also received valuable support from the school leader. Teacher 1 said during their presentation: *The project was coordinated across subjects at the school. [As a result of our participation in the SBP], school leaders have focused more on interdisciplinary work as a means of reaching the school's strategic goals, and as a goal in itself to develop deep learning and to see the links among [the various subjects]*.

In their third-year report, the teachers wrote about the collaboration as beneficial because *teachers have varying approaches and, as a result, diverse ideas on how to address the focal question*. This example shows that teachers recognize that they are learning from each other when they collaborate in this way.

Finally, in their report, the teachers also added some reflections and quotes from their students' written works about their experiences with the pollinators project. In the magazine publication for the public audience, the students wrote: *Our gardens can be designed into oases for the insect pollinators by growing pollinator-friendly plants. You can contribute with own garden!*

An example from one student's reflection on the outcome is (teacher's report, third year):

> Through the project I have learned a lot about the significance of pollinators in the world, and with this connection I have had new hooks to hang my knowledge of sustainability and the anatomy of pollen-bearing plants [on]. Through reading the book "The History of Bees" I have also made a mental picture of what can happen if we do not take care of our pollinating friends.

Two teachers said that through these activities, students had the opportunity to delve deeply into a wicked sustainability issue about insect pollinators and to explore the issue in various ways.

9.5 Discussion

The results of this study, which examined the educational practices developed by teachers over 3 years of the SBP, provide insights into how teachers approached and integrated wicked sustainability issues into their teaching across various subjects. The discussion presented here delves into how teachers adjusted their instructional focus in accordance with wicked sustainability issues and how these issues affected the design of their educational curricula.

9.5.1 A Shift to Problem-Oriented Teaching

Using the five characteristics identified by Seager et al. (2012) that underscore the complex nature of problems in SD (that is, the factors that make problems in SD wicked) helps us understand and discuss the intricacies of sustainability issues both in general and in the specific sustainability topics that teachers choose to focus on in their teaching. This shift in focus eventually resulted in teachers transitioning their teaching from dealing with straightforward, structured (tame) problems to addressing more complex and challenging (wicked) issues in their lessons about sustainability.

In the teachers at Borg School's first-year writings about ESD, the guiding question of their curriculum design was "How important are willows (willow bushes, *Salix* spp.) for the population size of pollinators?" This specific and researchable question aimed to understand an ecological relationship through empirical study. The problem was well structured and well understood, and it involved traditional linear processes such as data collection and analysis, explanation, and the implementation of solutions (Block et al. 2019; Hisschmöller and Hoppe 2001; Rittel and

Webber 1973). While the findings might eventually inform broader discussions and decisions involving complex, wicked problems, the question itself is tame; it is relatively straightforward and does not embody the multifaceted nature of wicked problems.

A challenge related to ESD is that this type of question does not cover the complexity of the sustainability concept because the focus remains on the ecological dimension of development and does not really include the social and economic dimensions. Focusing on relatively straightforward questions like this one has also been pointed out by other researchers as a potential weakness in schools' implementations of ESD (Munkebye et al. 2020).

In contemporary education, addressing complex and multifaceted challenges is crucial to preparing students for the real world. Wicked problems—those issues that are difficult to define, have open-ended timeframes, involve multiple, often conflicting solutions, and exhibit high levels of uncertainty and complexity—are emblematic of the kinds of challenges students are likely to face in their professional and personal lives. Therefore, using wicked problems as a means to address the highly complex sustainability issues facing society is recommended in teaching and learning (Block et al. 2019; Gericke 2022; Lozano and Barreiro-Gen 2022).

Addressing complex and value-laden societal challenges often involves navigating a myriad of conflicting interests and areas. When planning teaching strategies, it can be beneficial to explore the characteristics of wicked sustainability issues. By doing so, educators can identify and develop pertinent issues suitable for educational use, as demonstrated by the teachers in this study.

The more complicated question "Should we do more to increase the population of pollinators in our local surroundings?" was chosen by the science, social studies, and Norwegian language teachers at Borg School in their third year of designing ESD curricula. According to the five characteristics outlined by Seager et al. (2012), this question fits the profile of a wicked problem because it involves complex, interdependent factors that make it difficult to define the problem unambiguously. The solutions required may not be universally applicable or acceptable, necessitating flexible and adaptive management over time amidst competing values and interests.

Reflecting on the experience of teaching the third-year question, teachers noted that developing the project into a complex problem made it more socially relevant and contemporary. The teachers reported that students gained a clearer understanding of their tasks because the educators had a shared understanding of the complex problem that served as the foundation for their teaching. This approach shifted the focus of their instruction from being subject-oriented to being problem-oriented. The teachers reported significant developments over time in how students were engaged in a problem-oriented and interdisciplinary project focused on pollinators.

9.5.2 Interdisciplinarity and Student Outcomes

In discussing the complex and multifaceted nature of wicked problems within the context of SD, the five characteristics outlined by Seager et al. (2012) provide valuable insights. These characteristics emphasize the necessity of adaptive, interdisciplinary approaches and continuous dialogue among stakeholders in order to effectively tackle wicked problems in SD.

This perspective is echoed in UNESCO's goal for ESD, which advocates for the need to understand the complex world in which we live. UNESCO emphasizes that teachers must be prepared to facilitate ESD through a range of innovative teaching and learning practices, including taking interdisciplinary perspectives on global change issues and their local manifestations (UNESCO 2017). This viewpoint is supported by Jones et al. (2010, p. 19), who argue that "sustainability presents an overarching and complex socio-economic-ecological context. In this context, interdisciplinarity—viewed as a holistic model for understanding, organizing knowledge, and conducting inquiries—appears particularly appropriate".

Upon analysing the first-year ESD focus at Borg School, in which teachers planned sustainability teaching based on the question "How important are willows (willow bushes, *Salix* spp.) for the population size of pollinators?", it is not clear that an interdisciplinary approach was taken. Although several subjects were reportedly involved, we interpret the question and the tasks the students engaged in as having a primary focus on the knowledge, methods, and perspectives of science. The teachers' goals for their students were subject-specific, with a monodisciplinary focus aimed at achieving competence goals in science.

A pivotal moment in the project was the decision to change the first-year (tame) question to the third-year (wicked) question "Should we do more to increase the population of pollinators in our local surroundings?" Addressing this question necessitated broad interdisciplinary involvement, adding several competence aims from both social studies and Norwegian language to the curriculum. This approach enriched the teaching and learning experience by drawing on diverse perspectives, knowledge, and skills. It seems that integrating interdisciplinary teaching methods can play a crucial role in addressing wicked sustainability issues.

The teachers' reflections on the project highlighted the educational benefits of interdisciplinary collaboration. Teachers noted that participation in the SBP and collaborations with colleagues necessitated a more interdisciplinary approach, a practice often neglected in traditional settings. The success of this interdisciplinary approach was aided by the school leadership, which viewed interdisciplinarity as a means to achieve the school's strategic goals and foster deep learning. Interdisciplinary teaching and learning in ESD requires teachers to not simply cooperate on teaching goals in parallel, but to collaborate on shared goals. According to Klein (2010), there are some common characteristics in interdisciplinarity; for example, interdisciplinarity requires a common problem, that connections are made among disciplines, and that the disciplines are integrated. Addressing the complex or wicked real-life problems central to ESD requires accessing the knowledge, practices, and ways of thinking of several disciplines.

However, teachers are rarely fully supported in teaching ESD in interdisciplinary ways (Munkebye et al. 2020). Feng (2012, p. 35) argues that "[i]t is important to recognize that whilst interdisciplinarity seems desirable in principle, achieving it on a practical level is far from straightforward". Feng later (2012) points to the difference between multidisciplinary and interdisciplinary approaches, arguing that accomplishing an interdisciplinary approach takes more than teachers with different disciplinary backgrounds working together; teachers also have to reflect, discuss, and find ways to integrate and explore different perspectives concerning the sustainability challenge into their teaching. In this study, we showed that teachers moved from identifying and solving individual instructional problems to identifying and solving shared instructional problems. It appears that work in the SBP supports teachers in examining the interconnections between subjects, practices, and ways of thinking that are common and/or unique to various disciplines, helping them to choose and use the appropriate tools for addressing wicked problems in sustainability (Scheie and Stromholt 2019).

The SBP also supports teachers in choosing and collaboratively developing useful "artifacts" that can be shared both within and outside the teaching team, which Meirink et al. (2010) argue increases the level of interdependency that teachers experience. Collaborations that centre on joint work or instructional problem-solving and planning have the highest potential for teachers' professional development because "teachers feel a collective responsibility for the work of teaching. They may either agree to act in a similar way in their own practice or agree on general principles that guide their individual actions in teaching practice" (Meirink et al. 2010, p. 164). Teacher teams with higher levels of interdependency not only exchange ideas for alternative teaching methods but also discuss their experiences with experimentation. Additionally, these teams identify and solve shared instructional problems.

The teachers in this study recognized the usefulness of collaboration for their own professional growth, discovering that working with colleagues from different disciplines on the same project was highly educational. Collaborative effort resulted in a more socially relevant and contemporary project, with students gaining a clearer understanding of their tasks due to cohesive interdisciplinary planning. By exposing learners to a variety of perspectives, interdisciplinary teaching also allows students to develop their own viewpoints on sustainability (Jickling 2003). Moreover, such projects can significantly increase student engagement by giving them a sense of ownership over local environmental issues. By focusing on local communities, students were able to gather, process, and present relevant data.

Students had the opportunity to read about, reflect on, and discuss SD issues through the novel *The History of Bees* (Maja Lunde 2015), which (among other things) hypothesizes a future world without bees and imagines what agricultural practices might be like in such a world. This experience of exploring future scenarios and envisioning outcomes beyond the students' immediate perceptions can enhance their understanding of wicked problems. Engaging with literature in this way helps students grasp the complexities and far-reaching implications of SD issues. Managing wicked sustainability problems requires imaginative thinking that extends beyond the confines of any single discipline (e.g. Brown et al. 2010; Cantor et al. 2015). By integrating discussions based on literary works, students learn to approach these problems with creativity and critical thinking, fostering a deeper, more comprehensive understanding of the interconnected challenges they face.

The students also engaged in discussions about the issue with botanists and local politicians and shared their findings with the entire community through a local newspaper. This multifaceted approach and hands-on experience with real-world problems appeared to foster a deeper connection to their learning and its impact on their community.

Although student interviews were not conducted as part of this study, teachers reported engaging in numerous observations of student activities. Future research should examine student outcomes associated with these interdisciplinary teaching methods, which will provide a more comprehensive understanding of the effectiveness and benefits of interdisciplinary and practical learning approaches in enhancing students' comprehension of complex sustainability issues.

9.5.3 Conclusion and Implications

This study explored the integration of wicked sustainability issues into educational practices over a 3-year period, highlighting significant shifts towards interdisciplinary and problem-oriented teaching. Initially focused on a well-defined, monodisciplinary question, the project evolved to embrace a broader, more complex sustainability challenge that required the collaborative efforts of multiple disciplines. This transformation not only enriched the educational experience, but also fostered a deeper engagement and understanding among students and teachers alike.

Key findings underscore the importance of adaptive, interdisciplinary approaches to tackle wicked sustainability issues. By addressing more intricate and multifaceted questions, teachers at Borg School developed curricula that incorporated diverse perspectives and knowledge areas, thereby creating a more socially relevant and contemporary learning environment.

Interdisciplinary collaboration proved highly beneficial, enhancing professional growth for teachers and improving student outcomes. Engaging students in real-world, community-based projects encouraged a sense of ownership and responsibility towards local sustainability issues, while hands-on experiences and literature discussions deepened their understanding of sustainability's complexity.

This study demonstrates the potential of interdisciplinary teaching methods in fostering comprehensive education on sustainability and underscores the necessity for continuous support and professional development if teachers are to effectively implement these approaches. Future research should focus on assessing student outcomes related to interdisciplinary teaching in order to provide further insights into its effectiveness and benefits. Emphasizing collaborative problem-solving and innovative instructional strategies is essential for preparing students to navigate and address the sustainability challenges of tomorrow.

References

Berglund T, Gericke N (2016) Separated and integrated perspectives on environmental, economic, and social dimensions: an investigation of student views on sustainable development. Environ Educ Res 22(8):1115–1138. https://doi.org/10.1080/13504622.2015.1063589

Block T, Van Poeck K, Østman L (2019) Tackling wicked problems in teaching and learning. Sustainability issues as knowledge, ethical and political challenges. In: Van Poeck K, Östman L, Öhman J (eds) Sustainable development teaching: ethical and political challenges. Routledge, pp 28–39

Borg C, Gericke N, Höglund H-O, Bergman E (2012) The barriers encountered by teachers implementing education for sustainable development: discipline bound differences and teaching traditions. Res Sci Technol Educ 30(2):185–207

Borg C, Gericke N, Höglund HO, Bergman E (2014) Subject- and experience-bound differences in teachers' conceptual understanding of sustainable development. Environ Educ Res 20(4):526–551. https://doi.org/10.1080/13504622.2013.833584

Brown VA, Deane PM, Harris JA, Russell JY (2010) Towards a just and sustainable future. In: Harris JA, Brown VA, Russel JY (eds) Tackling wicked problems. Routledge, pp 3–15

Cantor A, DeLauer V, Martin D, Rogan J (2015) Training interdisciplinary "wicked problem" solvers: applying lessons from HERO in community-based research experiences for undergraduates. J Geogr High Educ 39(3):407–419. https://doi.org/10.1080/03098265.2015.1048508

Castleberry A, Nolen A (2018) Thematic analysis of qualitative research data: is it as easy as it sounds? Curr Pharm Teach Learn 10(6):807–815. https://doi.org/10.1016/j.cptl.2018.03.019

Crujeiras PB, Jimenez-Aleixandre MP (2019) Interdisciplinarity and argumentation in chemistry education. In: Erduran S (ed) Argumentation in chemistry education: research, policy and practice. Royal Society of Chemistry, pp 32–61

Erduran S (2023) Social and institutional dimensions of science: the forgotten components of the science curriculum? Science 381(6659). https://doi.org/10.1126/science.adk1509

Feng L (2012) Teacher and student responses to interdisciplinary aspects of sustainability education: what do we really know? Environ Educ Res 18(1):31–43

Gericke N (2022) Implementation of education for sustainable development through a whole school approach. In: Karaarslan-Semiz G (ed) Education for sustainable in primary and secondary schools. Springer

Giddings B, Hopwood B, O'Brien G (2002) Environment, economy and society: fitting them together into sustainable development. Sustain Dev 10(4):187–196. https://doi.org/10.1002/sd.199

Hisschmöller M, Hoppe R (2001) Coping with intractable controversies: the case for problem structuring in policy design and analysis. In: Hoppe R, Hisschmöller M, Dunn WN, Ravetz JR (eds) Knowledge, power and participating in environmental policy analysis. Routledge, pp 47–72

Hofman M (2012) Sustainable development in the Finnish teacher education–political rhetoric or reality. Nord Stud Sci Educ 8(3):303–303

Jickling B (2003) Environmental education and environmental advocacy: revisited. J Environ Educ 34(2):20–27. https://doi.org/10.1080/00958960309603496

Jones P, Selby D, Sterling S (2010) Sustainability education (s. 17–37). Routledge. https://doi.org/10.4324/9781849776516

Jonsson G (2008) An approach full of nuances–on student teachers' understanding of and teaching for sustainable development. In: Öhman (ed) Values and democracy in education for sustainable development: contributions from Swedish research. Liber, pp 93–108

Klein JT (2010) A taxonomy of interdisciplinarity. In: Frodeman R, Klein JT, Pacheco RCDS (eds) The Oxford handbook of interdisciplinarity. Oxford University Press, pp 15–30

Ling C, Hanna K, Dale A (2009) A template for integrated community sustainability planning. Environ Manag 44(2):228–242

Lönngren J, Van Poeck K (2021) Wicked problems: a mapping review of the literature. Int J Sustain Dev World Ecol 28(6):481–502. https://doi.org/10.1080/13504509.2020.1859415

Lönngren J, Ingerman Å, Svanström M (2017) Avoid, control, succumb, or balance: engineering students' approaches to a wicked sustainability problem. Res Sci Educ 47:805–831
Lozano R, Barreiro-Gen M (2022) Connections between sustainable development competences and pedagogical approaches. In: Vare PE, Lausselet NE, Rieckmann ME (eds) Competences in education for sustainable development. Springer International Publishing, pp 139–144
Maja L (2015) Bienes historie (The history of bees). Aschehoug
Meirink JA, Imants J, Meijer PC, Verloop N (2010) Teacher learning and collaboration in innovative teams. Camb J Educ 40(2):161–181. https://doi.org/10.1080/0305764X.2010.481256
Munkebye E, Scheie E, Gabrielsen A, Jordet A, Misund S, Nergård T, Bergljot Øyehaug A (2020) Interdisciplinary primary school curriculum units for sustainable development. Environ Educ Res. https://doi.org/10.1080/13504622.2020.1750568
Nikitina S (2006) Three strategies for interdisciplinary teaching: contextualizing, conceptualizing, and problem-centering. J Curric Stud 38(3):251–271
Norwegian directorate for education and training (2020) The National Curriculum. https://www.udir.no/in-english/curricula-in-english/
OECD (2023) PISA 2025 science framework: PISA 2015 assessment and analytical framework. https://pisa-framework.oecd.org/science-2025/
Öhman M, Öhman J (2012) Harmoni eller konflikt?–en fallstudie av meningsinnehållet i utbildning för hållbar utveckling. [Harmony or conflict?–a case study of the conceptual meaning of education for sustainable development.]. Nord Stud Sci Educ 8(1):59–72
Pepper C, Wildy H (2008) Leading for sustainability: is surface understanding enough? J Educ Adm 46(5):613–629
Pryshlakivsky J, Searcy C (2013) Sustainable development as a wicked problem. In: Kovacic SF, Sousa-Posa A (eds) Managing and engineering in complex situations, Topics in safety, risk, reliability and quality, vol 21. Springer Science+Business Media, pp 109–128. https://doi.org/10.1007/978-94-007-5515-4_6
Rittel HWJ, Webber MM (1973) Dilemma in a general theory of planning. Policy Sci 2:155–169
Robson C (2002) Real world research: a resource for social scientists and practitioner-researchers. Blackwell
Root-Bernstein R (2018) STEMM education should get "HACD". Science 361(6397):22–23. https://doi.org/10.1126/science.aat8566
Scheie E, Stromholt S (2019) "The sustainable backpack": exploring possibilities in education for sustainable development through a nationwide professional development program. Acta Didact Nor 13(2). https://doi.org/10.5617/adno.6473
Schmitz CL, Stinson C, James C (2010) Community and environmental sustainability: collaboration and interdisciplinary education. Crit Soc Work 11(3):83–100. https://doi.org/10.22329/csw.v11i3.5834
Seager T, Selinger E, Wiek A (2012) Sustainable engineering science for resolving wicked problems. J Agric Environ Ethics 25:467–484. https://doi.org/10.1007/s10806-011-9342-2
Sjöblom P, Wolf LA, Sundman J (2024) Climate change as a socio-scientific issue in upper secondary education. Addressing wicked problems through crosscurricular approaches. In: Klausen SH, Mård N (eds) Developing a didactic framework across and beyond school subjects. Routledge, pp 182–196
Uitto A, Saloranta S (2017) Subject teachers as educators for sustainability: a survey study. Educ Sci 7(1):8. https://doi.org/10.3390/educsci7010008
UNESCO (2014) Roadmap for implementing the Global Action Programme on Education for Sustainable Development. https://sustainabledevelopment.un.org/content/documents/1674unescoroadmap.pdf

UNESCO (2017) Education for sustainable development goals. Learning objectives. http://unesdoc.unesco.org/images/0024/002474/247444e.pdf
Walshe N (2017) An interdisciplinary approach to environmental and sustainability education: developing geography students' understandings of sustainable development using poetry. Environ Educ Res 23(8):1130–1149. https://doi.org/10.1080/13504622.2016.1221887
White PJ, Ardoin NA, Eames C, Monroe MC (2023) Agency in the Anthropocene: supporting document to the PISA 2025 science framework. OECD education working papers, no. 297. OECD Publishing. https://doi.org/10.1787/8d3b6cfa-en

Eldri Scheie, PhD, is Associate Professor in Science Education at the Norwegian Centre for Science Education. Her research interests include dialogue, critical thinking, and interdisciplinarity within the context of science, environmental, and sustainability education. Additionally, she is involved in professional development for teachers and explores implementation strategies in these fields.

Part III
Students, Nature, and Worldviews

Chapter 10
Children's Agency and Sense-Making in Minecraft: Virtual Practices for Learning Science and Sustainable Development

Maria Impedovo, Anne-Marie Cederqvist, and Slaven Gasparovic

Abstract In this study, we explore the link between children's agentic virtual making in the digital game Minecraft, and science education and children's views of nature concerning sustainable development. The research questions are as follows: Which agency do children express in their virtual making in relation to science and sustainable development? The study adopts a qualitative analytical framework based on agency and activity. Three dimensions are discussed: agentic virtual making focusing on body movement (avatars versus real body); context manipulation; and content mixing (knowledge, notions, formal learning, and media content). The findings indicate that Minecraft offers a unique platform for children to connect with nature and develop environmental awareness. Overall, the study underscores the potential of informal virtual making in Minecraft to activate interest and learning in science education, offering educators new avenues for engaging students and addressing misconceptions.

Keywords Science education • Sustainable development • Plant blindness • Virtual worlds • Children's agency • Minecraft

M. Impedovo (✉)
Adef, Aix Marseille University, Marseille, France
e-mail: maria-antonietta.IMPEDOVO@univ-amu.fr

A.-M. Cederqvist
Halmstad University, Halmstad, Sweden

S. Gasparovic
Department of Geography, Faculty of Science, University of Zagreb, Zagreb, Croatia

F. Skarstein, L.-A. Wolff (eds.), *Sustainability and Science Education*, Environmental Challenges and Solutions,
https://doi.org/10.1007/978-3-032-24408-6_10

10.1 Introduction

Education for sustainability raises an urgent call to act and become active in defending the natural environment. This mobilizes the notion of agency and that education itself needs to be transformed, adopting new approaches that focus on human well-being and global development and educating individuals as agents of change (Leicht et al. 2018).

Increasingly, children spend less time in nature than before (Sandberg 2009). Instead, they spend more time indoors and their social worlds are saturated with digital technologies (Danby et al. 2018a, 2018b). In their leisure time, they socialize with friends online, play digital games, and watch digital media more than before. This means children's everyday lives incorporate virtual worlds, which influence how they experience the physical world. The combination of spending less time outdoors and more time in virtual worlds is also likely to affect how children experience natural phenomena related to nature concerning plants and animals. A primary concern regarding this is that reduced direct contact with the natural environment could lead to plant blindness, which is described as the inability to notice plants in the environment and understanding their importance to life on our planet (Wandersee and Schussler 2001).

Accordingly, digital technologies have changed the nature of childhood, and it is necessary to recognize children's experiences based on the use of digital technologies (Bond 2014). Children's prior understanding of science phenomena is an important prerequisite when organizing teaching in science education. Children make sense of scientific phenomena based on their experiences in everyday life, which they later bring into class (Driver et al. 1985). That is, children form their understanding of scientific phenomena from activities in everyday life, both from physical and digital activities. However, children's increased use of digital media and technologies such as digital games provide them with virtual worlds that might affect their understanding of phenomena in the natural world.

In this chapter, we explore the link between children's agentic virtual making in the digital game Minecraft, science education, and children's views of nature concerning sustainable development. The research question is: *Which agency do children express in their virtual making in relation to science and sustainable development?* Virtual making in Minecraft refers to the creative process of constructing digital structures, objects, or landscapes within the game's virtual environment. This involves using the game's various building blocks, tools, and mechanics to design and build anything from simple houses to complex machinery, sculptures, or entire cities. The act of creating is the main activity in Minecraft. The digital object is shaped by the conditions present in both the digital environments and children's creativity, opening implications for learning (Culpepper and Gauntlett 2020).

The study adopts a qualitative analytical framework to analyse the emerging phenomena of learning through informal play to draw out some insights that would otherwise be difficult to grasp. Some potentially helpful recommendations for teaching are also discussed.

We first propose a theoretical framework on children's agency, science education, and education for sustainable development (ESD). We then focus on agentic making and the game Minecraft. We take departure from two previous studies: "Minecraft as a Hybrid Boundary Object: Exploring Nature in Squares" (Cederqvist and Impedovo 2023); "Socio-(im) material-making activities in Minecraft: retracing digital literacy applied to ESD" (Impedovo and Cederqvist 2023). In the first study (Cederqvist and Impedovo 2023), we investigated children's ways of experiencing virtual worlds and how this may affect their understanding of nature, scientific phenomena, and sustainable development; that is to say, Minecraft was explored as a boundary object in children's making activities with Minecraft. In the second study (Impedovo and Cederqvist 2023), starting from analysing some episodes recorded with a child aged seven, this paper aims to better understand the maker process in a virtual setting with implications for digital literacy and sustainable development education. These studies are part of a broader project about digital literacy and ESD inside the European project, COST Action, titled "Protection, Resilience, Rehabilitation of Damaged Environment" (PHOENIX CA19123) Finally, we question and extend the reflection on how children's making process in formal and informal contexts, also with virtual resources, could affect agency concerning naturalistic and digital environments.

10.2 Children's Agency and Sense-Making of Nature

Specific reflections have emerged on children's agency, which is what some describe as the "ownership" that children experience of their learning process and their involvement in human–material–spatial–temporal assemblages (Kumpulainen and Kajamaa 2020). Children's agency is an interactive process between a child, other people, the context, and the environment (Vygotsky 1978; Sairanen et al. 2022). In a study by Sairanen et al. (2022), eight 5-year-old children were followed with two Early Childhood Education (ECE) practitioners in a Finnish ECE centre. The results show that children's agency is an interactive process between the child, the provided materials, other people, and the socio-cultural context. When the child is allowed to take the initiative, the practitioner responds to it with acceptance. This suggests that a dialogical process develops in the interaction between the child and the environment, and this interaction constitutes the children's agency. Emirbayer and Mische (1998) suggest that children's agency is shaped over time when they encounter various situations and experiences. In the interactive process, children react upon question, and elaborate on how other people, the context, and the environment respond (Sairanen and Kumpulainen 2014; Sairanen et al. 2022). This is not a new phenomenon; however, increasingly, children experience this interactive process in situations and environments that involve digital technologies and virtual worlds such as video games, which shape children's agency in the natural world.

Some studies have explored how children carry out activities in virtual worlds and related this to their orientation towards nature. Kajamaa and Kumpulainen

(2019) investigated the development of children's agency at the individual level in a digital maker space. The results indicate that the socio-material context and children's activities in the digital maker space shaped and developed their agency. In another study, Kumpulainen et al. (2020) explored children's augmented storying in nature using the augmented reality application MyAR Julle. The results indicate a relationship between how children position themselves in relation to nature and the children's augmented storying about nature, which triggers children's agency towards sustainable action.

Considering these results, there is an urgent need to investigate further how digital-making activities in virtual environments can promote children's agency and impact children's sense-making of nature and sustainable development in the real world.

10.3 Science Education and ESD

Some of the most pressing SD issues are how to mitigate climate change and how to preserve biodiversity. Yet, many people find it challenging to understand the importance of these issues. Magntorn and Helldén (2006) outline the concept of "reading nature" as an essential aspect of understanding nature. Magntorn and Helldén carried out a study in a grade 3–4 class (years 10–11) with 23 students. They describe how students' ability to "read nature" means understanding nature outdoors, understanding ecosystems by recognizing their organisms and relationships within an ecosystem, and understanding how ecosystems are influenced by abiotic factors such as sunlight and water. A major concern is linked to the students' ability to "read nature" and its opposite, plant blindness, which is described as the inability to notice plants in one's own environment and understand their importance to life on planet Earth (Wandersee and Schussler 2001). This means that many people do not appreciate the diversity of plants and do not understand the importance of their existence to humanity. However, Wandersee and Schussler (2001) suggest that early educational experiences that include plant exposure could be a key to overcoming plant blindness.

Primary school curricula in most countries include teaching children about common animals and plants in their immediate environment and about various ecosystems. However, there is a growing concern that children spend less time in nature and that their experience of familiar animals and plants in their immediate environment is limited (Bebbington 2005; Sandberg 2009). The fact is that children are spending more time indoors, and their social lives are saturated with the use of digital technologies (Danby et al. 2018a, 2018b). Children play digital games, watch digital media, and socialize with friends online more than before. This means they gain many of their everyday experiences in the virtual world rather than the natural one. An important question is how the experience of virtual worlds influences how students experience the natural world, and how this will affect how children experience phenomena related to nature.

For decades, science education research has highlighted that children make sense of their everyday experiences by building mental representations that they can later use or discard to make sense of scientific phenomena in the classroom context (von Glaserfeld 1989; Driver et al. 1985). Knowledge of the prior understanding of science phenomena that children bring to the classroom is seen as necessary in organizing instruction to support children's learning of science and to help them overcome misconceptions (Driver et al. 1985).

Children increasingly navigate their daily lives within virtual realms, and it is therefore important to consider the part of their experiences that are based on technology (Bond 2014). This implies that children understand the natural world and its phenomena both based on experiences in virtual worlds and experiences in the physical world (Cederqvist and Impedovo 2023). Moreover, the experiences may include content related to the school curricula, such as science knowledge and knowledge of SD (Impedovo and Cederqvist 2023). Therefore, it is important to investigate children's experiences in the virtual world and how these experiences are connected to the physical world when it comes to understanding nature. The next section presents the use of Minecraft and its capacity to bridge the virtual and physical world.

10.4 Agentic Virtual Making in Minecraft

Minecraft is a popular sandbox video game where players explore a block-based, procedurally generated 3D world. Players can build structures, collect resources, craft items, and interact with various animals and environments. The game offers several modes such as Creative, Adventure, and Survival. The game presents a green space open to possibilities, and the environments are composed of cubes representing various elements, such as water, earth, sand, rock, wood, animals, and plants. Nebel et al. (2016) present some interesting features of Minecraft, such as the simple ecological system with plants and animals that the player can elaborate to create farms, produce food, clothes, et cetera. In this way, the player can interact with and influence the environment's ecological representations (Ekaputra et al. 2013).

As mentioned earlier, virtual worlds such as Minecraft are part of many children's everyday lives. They spend a lot of time in these virtual worlds with friends online. They solve problems alone or with others and try to make sense of the challenges and environments they encounter. Making sense of problem-solving situations in various biomes requires children to apply scientific knowledge when manipulating digital material (Dezuanni 2020). These experiences can then be linked to school science curricula. In this sense, a game like Minecraft has the potential to bridge children's experiences in the physical and virtual worlds (Cederqvist and Impedovo 2023).

When children play Minecraft, they develop and use the skills needed to manage nature resources such as resources and food to survive and solve problems when facing challenges in the biomes. In this sense, Minecraft can be seen as a community where children can act according to the expectations that a position in the community entails (Gillen and Hall 2012). Furthermore, children constantly merge their sense-making in the virtual world with experiences in the physical world, thus bridging the virtual and physical worlds, and simultaneously developing their understanding of phenomena (Deuze 2012). Therefore, virtual making in Minecraft could be considered an important activity that can promote children's agency and sense-making of nature and sustainable development in the physical world. This study aims to further explore the relationship between children's agentic virtual making in the digital game Minecraft, science education, and children's views of nature concerning sustainable development.

10.5 Methods

Inside the project COST project CA19123 (PHOENIX (https://www.cost-phoenix.eu/) Protection, Resilience, Rehabilitation of damaged environment), a qualitative data collection protocol was developed for children who play or have played Minecraft.

This study revisits the data, which partly has been presented in two previous studies (Cederqvist and Impedovo 2023; Impedovo and Cederqvist 2023).

In conducting interviews with children about their experiences with Minecraft, the study adheres strictly to ethical standards to ensure the protection of participants. Anonymity is meticulously maintained by assigning pseudonyms and removing any identifying information from the data. Privacy is safeguarded through secure data storage and restricted access to sensitive information. The study also carefully considers the ethical implications of involving children, obtaining informed consent from both the participants and their guardians, and ensuring that the interview process is non-intrusive and respectful. By prioritizing these ethical principles, the research upholds the integrity of its findings while protecting the well-being and rights of the child participants. All the participants in the study were boys due to the specific sampling criteria used. The selection process might have targeted a particular demographic where boys were more likely to engage with Minecraft, reflecting the game's popularity among this group. Additionally, it could be an unintended outcome of recruitment methods, such as reaching out to communities or networks predominantly comprising boys. This gender distribution does not imply a lack of interest or participation from girls but highlights the need for further research to ensure a more balanced representation.

To understand children's use of Minecraft in various contexts, the data was collected in several locations (Italy, Croatia, and Sweden) by educators and researchers involved in the main project. Six semi-structured interviews were conducted with young people (all boys) aged between 7 and 18 years (Table 10.1), in which a prepared interview protocol was used. This was supplemented by an in-depth observation of one of the participants after the interviews using a think-aloud method while the child played Minecraft for about 30 min (Atkinson and Hammersley 1994).

Table 10.1 The participant data on the play of Minecraft at the moment of the interview

	Age	Country	Duration in years	Time/ day	Where/when	Tool
Mark	7	Italy	2	>1 h	Living room, afternoon	iPad
Rem	11	Croatia	4	30 min	Bedroom, evening	iPad, mobile phone, PlayStation
Vito	11	Croatia	4	1–2 h	Parents' room, weekends and evenings	Computer
Andrea	15	Croatia	7	1–2 h	Bedroom, afternoon	Computer
Theo	15	Croatia	4	>1 h	Bedroom, later at night	Computer
Oleg	18	Sweden	10	2 h	Bedroom, evenings	Computer

Adapted from Cederqvist and Impedovo (2023)

While the current sample may not be ideal for making sweeping generalizations, it provides valuable insights into the topic under study in this first phase of the project.

All participants used Minecraft regularly, with various levels of expertise. The data was collected based on volunteer participation.

The interviews were conducted in the children's mother tongue and were held in their home environment. The interviews were based on questions from the interview protocol to explore the agentic virtual making in Minecraft (M):

- Time: How long have you been playing M.? How many hours? Where do you play, and when?
- Like: What do you like about M? What do you like to do in M? What can you do in M that you cannot do in a real place?
- Nature: What about nature in M? Do you like to stay in nature of M? Nature changes in M, or is it always the same?
- Minerals: Which kind of minerals are there in M? Did you discover something about the minerals? What can you do with them in M?
- Vegetation: Are there trees in M? What types of trees do you know in M? What can you do with the trees in M?
- Animals: Are there any animals in Minecraft? If so, which ones? What can you do with the animals in M? Which animal do you like best in M.? Have you ever interacted with that animal in real life?
- Formal learning: What new do you think you have learned in M? And about nature? Do you like the idea to use M in the classroom?
- Informal learning: Did you explore something in nature outside thanks to M? What can you do in M that you cannot do in reality? Have you ever used something discovered in M in other contexts? (talking with friends, in family, to do some activities…).

The interviews were audio-recorded, then transcribed verbatim and translated into English. The three researchers then re-read the transcripts to get an overall picture of the content. In supplementary, we also chose to make an in-depth analysis of one of the interviews collected in Italy that of a 7-year-old child, here called Mark.

10.5.1 Data Analysis

In the study, we conducted qualitative content analysis on the interview transcripts following Mayring (2014). Employing deductive content analysis, aligning our study's conceptual framework with the data. The main goal was to identify and explore agentic virtual making in Minecraft. This involved identifying active verbs (e.g. I do, I will do, I did) when children described their actions. We initially categorized the main aspects and proceeded to read and code the data accordingly. Through multiple readings, we identified meaningful segments of text and assigned them labels corresponding to their respective categories. Additionally, we noted keywords characterizing responses within each category, merging similar categories after completing 50% of the coding process. The inter-rater agreement between researchers stood at 72%, ranging from 50% to 80% in individual instances. In cases of disagreement, we engaged in discussions to achieve consensus, resulting in a 95% agreement rate thereafter. NVIVO facilitated this process. Upon completing the deductive data analysis, we defined three primary dimensions of agentic virtual making related to the body, context, and content. Below, we present the three dimensions with the definitions and examples (Table 10.2):

10.6 Results from the Interviews

In this section, we present the results together with examples from interviews concerning children's experiences of agentic virtual making in terms of acting with the virtual body, manipulation of the virtual context, and of mixing virtual content.

10.6.1 Agentic Virtual Making: Acting with the Virtual Body

The children experienced that Minecraft gave them the opportunity to do all sorts of activities such as spending time in nature, building various artefacts, and collecting materials. They also suggested using animals to achieve certain goals such as getting food. They expressed these activities as using their virtual body. Below are examples from the data that show the agentic virtual making where the children are acting with their virtual body:

- I like to be in nature and build something with a natural theme (like houses or castles). For example, if I build a castle, I build it overgrown to merge with nature. (Theo, 15 years)
- You can build a house with rocks, although blocks are not so nice, so I didn't like them so much. I made shields and armor and weapons using iron and diamonds. (Theo, 15 years)

Table 10.2 The three dimensions of agentic virtual making

Dimension	Definitions	Examples
Body	Agentic virtual making with a focus on body movement (avatars, real body, etc.)	I made shields and Armour and weapons using iron and diamonds
Context	Agentic virtual making with a focus on context manipulation (settings, environment)	You can excavate the rock, and then you get a stone, and then you can build the house, some tool, or weapon
Content	Agentic virtual making with a focus on content mixing (knowledge, notions, formal learning, and media content)	Later, when we learned about it in school, I was able to easily relate it to Minecraft

- I don't build so much, but more I collect materials and play with them to survive. […] You can build and make tools. (Rem, 11 years)
- I used parrots as a pet and using cows and sheep to get food. (Andrea, 15 years)

Further, the children experienced that the greater the number of activities they engaged in, the more options to act there were. This characteristic distinguishes Minecraft from traditional video games: while in most games, once you've exhausted all possible paths and completed the game, your playtime comes to an end. In the examples below, one child expressed the endless possibilities of building and creatively expressing himself while playing Minecraft.

- […] it has endless possibilities of building, and you can express yourself in a very creative way in the game." (Andrea, 15 years)

Additionally, the agentic virtual making is appreciated by the children in terms of that they can do other kinds of activities than in traditional video games. In this study, mining seems to be the preferred activity, granting Minecraft a unique position in the gaming world.

- I like it because it is not a classic shooting game, but you can build something […] (Theo, 15 years)
- I like to build houses and look and excavate the ore and diamonds […] (Theo, 15 years)

Despite the many features of the game, the children are aware that there is a distinction of what they can or cannot do in the real world in compared to the virtual world, which is expressed by the child in the excerpt below:

- Definitely, you cannot fly. Also, you cannot make a stone block, you cannot excavate stones and make blocks, or you can not break the trees. (Vito, 11 years)

The playful aspect remains predominant in terms of moving, exploring, and journeying. Children experienced that the adventurous side of the game gave them the opportunity to act and solve problems by using the environmental features in the virtual context of the game.

10.6.2 Agentic Virtual Making: Manipulation of the Virtual Context

The participants recognized that the game takes place in nature, with animals, woods, and savannah. They recognize different types of trees, like birch, oak and Pine, which is expressed like this: *In Minecraft there are many trees but the only one I can recognize is the fir… the pine* (Mark, 7 years) or *[…] I think almost the whole world in Minecraft is nature. Especially when there is a village, then around is nature, like the sea, beaches, woods, there are different types of woods, then there are swamps,* etc. (Vito, 11 years).

In Minecraft, the children also have the opportunity to experience different animals such as expressed in the following excerpt: *Sheep, cows, dolphins, turtles, tropical fishes, polar bears, chickens* (Remy, 11 years), and different biomes such as savannah and deserts: *I have learned how,* e.g. *what a desert looks like* (Andrea, 15 years).

In the game, the children recognized the different contextual aspects and learnt to use this knowledge in their virtual activities for manipulation of the virtual context. Further, they recognized the difference between the real nature and the virtual in their relationship with the animals and environments: *What I can't do in reality, what I can do in Minecraft, is that I can see everything in squares and go into environments that I've never actually seen before* (Mark, 7 years). Another child explained the difference between what he had been experienced in nature in Minecraft and experienced in real life: *You can get food, ride a horse, use the wolf as a protection if you tame it* (Vito, 15 years). On the other hand, the child explained that he has never experienced this in real life. The use of nature and the contextual aspect of the game opens for agentic virtual making where the children can solve problems by manipulating the virtual context, as expressed in the examples below.

- *You can destroy* trees, you can excavate something, and you can make flat terrain. (Rem, 11 years)
- *You can excavate* the rock, and then you get a stone, and then you can build the house, some tool, or weapon. (Rem, 11 years)

Children are also active in the manipulation of the structure, like seasons

- *You can install* different modules (e.g. summer, autumn, winter, spring), and then nature is changing. For example, leaves fall, or animals hibernate during the winter or so. (Theo, 15 years)

10.6.3 Agentic Virtual Making: Mixing Virtual Content

Many of the contexts that the children meet in Minecraft include content that can be linked to real life science. That is, children experience facts such as nature changes in various biomes.

- Nature changes [...], nature is different (like woods, grass, animals, etc.). (Rem, 11 years)

A closer look into the relationship with Minerals shows that all children spend a lot of time understanding minerals such as diamonds, copper, ore, and obsidian which also occur in the real world.

- [...] I learned that the diamond is the hardest mineral in the world. (Rem, 11 years)
- [...] copper conducts current. (Oleg, 18 years)
- [...] ores are important for making machines. (Theo, 15 years)
- By sudden cooling of lava gives you obsidian. [...] in school, when we learned about the obsidian, I already knew about the term because I saw it in Minecraft. (Rem, 15 years)

In Minecraft, the children face problems which they solve by using contextual aspects such as the minerals, animals, plants, and other materials they encounter. In their making activities they become aware of the properties of the materials since they need this knowledge to be able to solve the problems at hand. The virtual content, as presented in Minecraft, is in many ways similar to how it is represented in real life. This implies the possibility that the children's experiences of the materials are transferred into the real-world context in school, as expressed in one of the excerpts above (Andrea, 15 years).

10.7 Results from Mark's Virtual Agentic Making

The virtual realm of Minecraft offers ample chances for interaction and exploration amidst its digital landscape. Its simulated nature presents a plethora of opportunities to engage with and explore, akin to navigating through a vast garden teeming with possibilities. In the following, we present the results of the in-depth analysis of one of the interviews with a 7-year-old child, here called Mark.

10.7.1 Agentic Virtual Making: Acting with the Virtual Body

The interview with Mark indicates that he was aware that the animals in Minecraft need food such as plants, and that he needed to provide them with that otherwise the animals would have died. He also discussed that Minecraft nature makes him aware of specific materials such as rubber, diamonds, wood, and other materials which he could use for building things.

> When I play Minecraft, I like to dig, build, and cut wood in nature. [...] I've never cut wood in the garden. [...]. When I build in Minecraft nature, I like trees, animals, sheep, horses, flowers ... and monsters. [...] I like both worlds, the real one and the virtual one. When I'm

> in Minecraft nature, the great thing is that you can do many things, like go underground, dig, and build many things. On the other hand, I don't know everything I can do in Minecraft when I'm in this garden, so I can only do a few things. […] If I dig the underground, I think I'll find something… I think sometimes something can be found.

The excerpt indicates that the virtual making with the virtual body activated Marks interest and engagement of acting in nature. These are activities that he may not yet have experienced in real life, but now he had the opportunity to experience in terms of his virtual body.

10.7.2 Agentic Virtual Making: Manipulation of the Virtual Context

In the context of Minecraft, Mark recreated the jumping box in his neighbours' garden. To do this, he had to explore the materials presented in Minecraft. He described concluding that the only material that could be used was the rubber available in the Minecraft context. This suggests that his previous knowledge of materials helped him to come up with a solution for how to build the jumping box. However, it can also be seen because of exploring materials in the virtual context function in a similar way in the real context. That is, the manipulation of materials in the virtual context may have provided new insights that could be used in the real world. Agentic virtual making is prominent, Mark has successfully created a digital rendition of the jumping box, a creation that is both innovative in the virtual world and relevant to his real-world environment.

10.7.3 Agentic Virtual Making: Mixing Content

Mark described his interaction with animals, plants, and various materials within the game, which illustrates the requisite understanding of scientific concepts needed to make digital objects. The results indicate that Mark understood the necessity of feeding the animals in Minecraft with plants to ensure their survival, recognizing that without proper nourishment, the animals would perish. This suggests that he wanted to care for the animals and that he had a basic understanding of food chains and the interdependence between animals and plants, which is applicable knowledge also in the real-world context.

Further, Mark presented his work of constructing digital creations inspired by the "Poppy Playtime" game, which is developed by the American indie studio Mob Entertainment. In this game, players assume the role of a former employee of Playtime Co., tasked with exploring the abandoned toy factory years after the disappearance of its staff. As Mark set out to recreate the factory and its characters, he encountered challenges concerning the proportions and dimensions required for authenticity within Minecraft. These obstacles prompted him to ingeniously utilize his understanding of the game's materials. Within Minecraft, he found a solution through the manipulation of digital objects, which he repurposed to generate new digital elements, effectively integrating and expanding upon the Poppy Playtime universe within the Minecraft environment.

10.8 Discussion

The results of this study show that virtual making in Minecraft could be considered as a relevant activity that can promote children's agency and sense-making of nature and learning of science concepts in the physical world. In the following section, the three dimensions are discussed, focusing on the body, context, and content by the double set of data: from the group interviews and the deeper focus on one child, Mark.

Regarding the agentic virtual making focusing on body movement, the acting is oriented to the real body and the avatar. In the interview, the children expressed how they liked to position themselves in real nature. They also liked to stay in a virtual nature setting to do things. The virtual natural setting probably gave them more resources and activities than is possible in a real natural context. Mark's virtual agentic making is oriented to explore and discover new resources. Surviving is linked to curiosity and the endless opportunity to move or travel inside virtual settings or to build and destroy things. Mark expressed his freedom in performing activities that are impossible for him in real life, for instead of cutting wood in his garden outside of his house.

The data shows the conditions of "moving a lot" and at the same time staying "still" on the chair in the bedroom or living room while playing the game. The discovery of the external environment is mediated by the touch on the screen. In philosophy, touch is a concept developed by authors like Merleau-Ponty (1962) or Nancy (2008). For the last one, touch is a sensation and a feeling. This contact reduces the distances of "between": information is transmitted through tactile interaction, translating sensation into meaning. In Minecraft, the act of mining serves as a primary activity. Here, we observed a dynamic interaction between the practical manipulation required for survival or further exploration by touching the screen and the individual preferences inherent in the act of movement. However, it's important to remember that although children may experience simulated movement in virtual environments, real-life physical activity is still essential for their development (Coetzee 2018). It supports motor skills, social interaction, and cognitive growth (Pellegrini and Smith 1998) that digital experiences cannot fully replicate.

Games such as Minecraft hold appeal as they afford students a sense of creative agency, allowing them to harness limitless resources to craft and reshape digital objects according to their imagination (Kumpulainen et al. 2020). A new time and space condition is open, thanks to the interactions with Minecraft concerning the agentic virtual making with a focus on context manipulation. Children can explore new environments and engage in more complex or age-appropriate experiences—such as playing with or taming exotic animals—while staying in their home. For example, we recognized how children could grow, create, select animals, or interact with them in Minecraft before they had similar experiences in real life. Moreover, when they discovered that familiar animals, like cows, are friendly in the virtual environment, they discovered new facts, for instead about their food, their reproduction cycle, and implications of the food chain. The agentic action is retraceable inside the willingness to explore new environments and settings.

In Minecraft, children adapt to a virtual environment, in which the realm of possible actions diverge. Within this digital space, opportunities abound, readily available for immediate execution, unlike the real world where opportunities exist but necessitate specific skills and contextual adjustments. As illustrated by Mark's examples, real-world objects like the jumping box are creatively repurposed as resources tailored to fit the virtual context. Similarly, the real world is approached from a fresh perspective, with materials serving as the building blocks for replication, dimensions offering opportunities for continual structural reconstruction, and perspectives demanding adherence. This blurs the boundary between digital and non-digital realms (Deuze 2012).

Through play, children can become familiar with science content like changes in nature, species, and awareness of various biomes. As seen from the interviews, this also involves knowledge of materials and how they can use them in their agentic virtual making through acting with their avatar body. During the interview with Mark, we delved into the intricate blend of real and virtual content. Furthermore, in navigating these blurred boundaries between reality and the digital realm, there was an implicit requirement for spatial awareness to manipulate and envision three-dimensional spaces. This necessitated problem-solving abilities and critical thinking skills, particularly when constructing and adapting across dimensions. The practical applications of this content (Gillen and Hall 2012) serve as catalysts for problem-solving and for thriving in the digital environment.

The results indicate that children's engagements with digital worlds like Minecraft can significantly enhance their learning of science. Their regular interactions within Minecraft may facilitate the construction of mental models, which they can subsequently apply to comprehend scientific concepts within classroom settings (see von Glaserfeld 1989; Driver et al. 1985). There is a rising concern regarding children's diminishing exposure to nature, as highlighted by Bebbington (2005) and Sandberg (2009). However, their experiences with familiar animals and plants within digital worlds could, to some degree, compensate for this deficit by fostering a sense of connection with nature.

Knowledge of the prior understanding of science phenomena that children bring to the classroom is seen as necessary in organizing instruction to support children's learning of science and to help them overcome misconceptions (Driver et al. 1985). The informal virtual making presented in this study could be a ground for activating interest and learning in science education. For instance, educators might frame science content within the context of problem solving in Minecraft, tasks such as growing crops for in-game animals' sustenance can serve as focal points. Further, initiating discussions with students can unearth their informal gameplay preconceptions, particularly regarding ecological concepts. Encouraging students to translate academic content into Minecraft scenarios fosters critical analysis of the game's capabilities and limitations. Additionally, educators can guide children in exploring Minecraft's depiction of nature, prompting reflection on science concepts and sustainability issues, and thereby nurturing a deeper environmental consciousness.

10.9 Conclusion

A virtual environment like Minecraft can play a role in defending the natural environment. It makes possible an integration between the positionality of the body, the aesthetic value of moving and exploring, and the creative pleasure of creating digital objects. The three activities could be interpreted as "post-digital play" (Pettersen et al. 2022), in which children stay in contact with the game when concretely not in the *game*. This study highlights the significance of children's engagement with virtual making in Minecraft, indicating its potential to foster agency and understanding of nature, as well as science concepts in the physical world. Through an exploration of three dimensions—body, context, and content—observed through group interviews and individual focus on one of the interviewed, Mark, the study unveils examples of how children can navigate between the virtual and the real worlds. Their interactions with Minecraft not only enable them to explore new environments and complex experiences but also deepen their understanding of science phenomena. The immersive nature of Minecraft blurs the boundary between digital and non-digital realms, allowing children to creatively repurpose real-world objects within the virtual context. Moreover, Minecraft offers a unique platform for children to connect with nature and develop a deeper environmental consciousness. Overall, the study underscores the potential of informal virtual making in Minecraft to activate interest and learning in science education, offering educators new avenues for engaging students and addressing misconceptions. However, while the study on children's experiences with Minecraft provides valuable insights, it is crucial to acknowledge the darker aspects of gamification. Relying solely on virtual movement can pose significant risks to children's development. Without real-life physical activity, children may experience delays in motor skill development, increased risk of weight problems, and poor physical health. Further, commercial aims often drive game design, prioritizing profit over the well-being of players. This focus can lead to the inclusion of addictive elements designed to maximize user engagement and spending. Additionally, excessive gaming poses risks, including addiction, which can negatively impact children's physical and mental health. It is also important to recognize that virtual experiences cannot replace the essential benefits of real-world nature interactions. Engaging with the natural environment is vital for the holistic development of children, offering sensory experiences and physical activities that digital games cannot replicate. Thus, while Minecraft and similar games can be beneficial educational tools, they should be balanced with outdoor activities and monitored to mitigate potential negative effects.

While the main focus of this chapter is on the potential benefits of integrating students' experiences from virtual worlds like Minecraft into science education, it is equally important to address the challenges and limitations inherent in these digital environments. One significant challenge is the inability of virtual worlds to accurately convey complex real-world issues such as pollution and social dynamics. For instance, Minecraft's simplified ecosystems do not adequately represent the multifaceted and interconnected nature of environmental pollution or its impact on living organisms and their habitats. Furthermore, the game's anthropocentric perspective may lead students to develop a skewed understanding of ecological relationships, as interactions with animals and plants in Minecraft often lack the depth and nuance found in natural ecosystems. This can foster misconceptions about the roles and importance of species and ecological processes. Moreover, the representations in Minecraft, while educationally valuable, can sometimes be at odds with reality, leading to further misconceptions. For example, the simplistic mechanics of farming, animal husbandry, and resource management in the game might give students an oversimplified view of these complex activities. To counteract these challenges, educators must supplement Minecraft-based learning with real-world examples, discussions on environmental and social issues, and critical analysis of the differences between virtual and real-life experiences. By doing so, they can ensure that students gain a more holistic and accurate understanding of scientific concepts and their real-world applications.

We acknowledge the limits of the study. The small sample size, variation in participant demographics, and lack of consistent parameters for participant selection could potentially affect the generalizability of the findings. The study's scope has been constrained by practical considerations such as access to participants and resources. While the sample size and diversity may not be ideal for drawing broad conclusions, the study provides insights into the potential educational benefits of integrating Minecraft into learning environments. Ultimately, while the current study's findings should be interpreted with caution due to sample limitations, they can still serve as a starting point for future research and discussions on the educational implications of incorporating digital gaming platforms like Minecraft into pedagogical practices.

References

Atkinson P, Hammersley M (1994) Ethnography and participant observation. In: Denzin NK, Lincoln YS (eds) Handbook of qualitative research. Sage, pp 248–261

Bebbington A (2005) The ability of A-level students to name plants. J Biol Educ 39(2):63–67. https://doi.org/10.1080/00219266.2005.9655963

Bond E (2014) Understanding childhood. In: Childhood, mobile technologies and everyday experiences. Studies in childhood and youth. Palgrave Macmillan. https://doi.org/10.1057/9781137292537_2
Cederqvist AM, Impedovo M (2023) Minecraft as a hybrid boundary object: exploring nature in squares. Educ Sci 13(9):952
Coetzee MH (2018) Embodied knowledge (s), embodied pedagogies and performance. S Afr Theatre J 31(1):1–4
Culpepper MK, Gauntlett D (2020) Making and learning together: where the makerspace mindset meets platforms for creativity. Glob Stud Child 10(3):264–274. https://doi.org/10.1177/2043610620941868
Danby SJ, Fleer M, Davidson C, Hatzigianni M (2018a) Digital childhoods across contexts and countries. In: Danby S, Fleer M, Davidson C, Hatzigianni M (eds) Digital childhoods, vol 22. Springer, pp 1–14. https://doi.org/10.1007/978-981-10-6484-5_1
Danby S, Fleer M, Davidson C, Hatzigianni M (2018b) Digital childhood. Springer
Deuze M (2012) Media life. Polity, London
Dezuanni M (2020) Minecraft 'worldness' in family life. In: The Routledge handbook of digital literacies in early childhood. Routledge, pp 366–376
Driver R, Guesne E, Tiberghien A (1985) Children's ideas in science. Open University Press
Ekaputra G, Lim C, Eng KI (2013) Minecraft: a game as an education and scientific learning tool. ISICO, pp 237–242
Emirbayer M, Mische A (1998) What is agency? Am J Sociol 103(4):962–1023. https://doi.org/10.1086/231294
Gillen J, Hall N (2012) The emergence of early childhood literacy. In: Larson J, Marsh J (eds) The SAGE handbook of early childhood literacy. Sage, pp 3–17
Impedovo M, Cederqvist AM (2023) Socio-(im)material-making activities in Minecraft: retracing digital literacy applied to ESD. Res Sci Technol Educ:1–21. https://doi.org/10.1080/02635143.2023.2254460
Kajamaa A, Kumpulainen K (2019) Agency in the making: analyzing students' transformative agency in a school-based makerspace. Mind Cult Act 26(3):266–281. https://doi.org/10.1080/10749039.2019.1647547
Kumpulainen K, Kajamaa A (2020) Sociomaterial movements of students' engagement in a school's makerspace. Br J Educ Technol 51(4):1292–1307. https://doi.org/10.1111/bjet.12932
Kumpulainen K, Byman J, Renlund J, Wong CC (2020) Children's augmented storying in, with and for nature. Educ Sci 10(6):Article 149. https://doi.org/10.3390/educsci10060149
Leicht A, Heiss J, Byun WJ (2018) Issues and trends in education for sustainable development, vol 5. UNESCO Publishing
Magntorn O, Helldén G (2006) Reading nature—experienced teachers' reflections on a teaching sequence in ecology: implications for future teacher training. NorDiNa 5:67–81. https://doi.org/10.5617/nordina.415
Mayring P (2014) Qualitative content analysis: theoretical foundation, basic procedures and software solution. http://nbn-resolving.de/urn:nbn:de:0168-ssoar-395173
Merleau-Ponty M (1962) Un inédit de Maurice Merleau-Ponty. Rev Metaphys Morale 67(4):401–409
Nancy JL (2008) *Corpus* (no. 67). Fordham University Press
Nebel S, Schneider S, Rey GD (2016) Mining learning and crafting scientific experiments: a literature review on the use of Minecraft in education and research. J Educ Technol Soc 19(2):355–366. https://www.jstor.org/stable/jeductechsoci.19.2.355
Pellegrini AD, Smith PK (1998) Physical activity play: the nature and function of a neglected aspect of play. Child Dev 69(3):577–598. https://doi.org/10.2307/1132187
Pettersen K, Arnseth HC, Silseth K (2022) Playing Minecraft: young children's postdigital play. J Early Child Lit 25:133–157. https://doi.org/10.1177/14687984221118977

Sairanen H, Kumpulainen K (2014) A visual narrative inquiry into children's sense of agency in preschool and first grade. Int J Educ Psychol 3(2):141–174. https://doi.org/10.4471/ijep.2014.09

Sairanen H, Kumpulainen K, Kajamaa A (2022) An investigation into children's agency: children's initiatives and practitioners' responses in Finnish early childhood education. Early Child Dev Care 192(1):112–123. https://doi.org/10.1080/03004430.2020.1739030

Sandberg M (2009) Barn och natur i storstaden: En studie av barns förhållande till naturområden i hemmets närhet—med exempel från Stockholm och Göteborg. Doctoral dissertation, University of Gothenburg, Choros. http://hdl.handle.net/2077/20095

von Glaserfeld E (1989) Cognition, construction of knowledge, and teaching. Synthese 80(1):121–140

Vygotsky L (1978) Mind in society: the development of higher mental processes. Harvard University Press

Wandersee JH, Schussler EE (2001) Toward a theory of plant blindness. Plant Sci Bull 47(1):2K9

Maria Impedovo, PhD, is an Associate Professor in psychology of education at Aix-Marseille University, France. Her academic background combines educational sciences and psychology, with a particular focus on teacher professional development, learning in higher education, and digital innovation in teaching and learning contexts. Her research interests lie at the intersection of professional learning, sociocultural and dialogical perspectives, and educational technology. Impedovo has extensive experience in coordinating international projects and blended learning programs, and she actively contributes to European initiatives on teacher education and virtual exchange. She has published widely in international journals and edited volumes and collaborates regularly with interdisciplinary research networks across Europe.

Anne-Marie Cederqvist She is a Doctor of Philosophy in Subject Matter Didactics with a focus on technology. Her research interests encompass three main areas: subject matter didactics in science and technology, digital learning, and sustainable development. She serves as the Associate Head of the Department of Learning at the School of Education, Humanities and Social Sciences and teaches in the Master's Programme in Digital Learning. Additionally, she teaches technology and science in the Primary Teacher Education program, for both years F–3 and years 4–6.

Slaven Gasparovic, PhD, is an Associate Professor of geography at the University of Zagreb, Faculty of Science, Department of Geography, Croatia. In addition to transport geography, his main research interests include urban planning, with a primary focus on complex problems of human societies related to sustainable development in urban areas, with particular emphasis on the possibilities for adopting environmental citizenship concept. His research also addresses the geographical significance of sustainable development in the educational process, particularly from the perspective of digital solutions. He has published more than thirty articles and chapters in geographical, social, and educational science journals and books.

Chapter 11
Primary School Students' Knowledge About Water, and Their Environmental Worldviews and Action

Gregor Torkar, Iris Zajc, and Bruce Johnson

Abstract Children enrolled in primary schools today will have to act as informed citizens to protect water ecosystems and water quality in the future. Children themselves must make personal choices on a daily basis, and they can be agents of change and active participants in the present and in the future. This study investigates Slovenian primary school students' system knowledge of the water cycle, their environmental worldviews, and self-reported environmental actions. The students showed poor system knowledge of the water cycle, but positive environmental worldviews and actions. The results show no statistically significant correlation between primary school students' system knowledge of the water cycle and their environmental worldviews or actions. Structural equation modeling showed strong correlations between environmental worldviews and actions. Furthermore, the results showed that there is no clear correlation between grade level and students' system knowledge of the water cycle. Younger students had a more positive environmental worldview than older students. In favor of older students, there were statistically significant differences in grade level for environmental actions, in which students help or encourage others to conduct environmental actions. Gender differences were found only for environmental actions, in favor of girls. These findings call for an in-depth examination of the learning and teaching strategies that promote sustainable development in Slovenian schools, as well as further studies of the role of specific water-related attitudes and environmental actions in relation to knowledge about the water cycle.

Keywords Environmental worldviews · Environmental actions · Knowledge of the water cycle · Primary school students · Grade level · Gender · Slovenia

---

G. Torkar (✉) · I. Zajc
University of Ljubljana Faculty of Education, Ljubljana, Slovenia
e-mail: gregor.torkar@pef.uni-lj.si

B. Johnson
University of Arizona, College of Education, Tucson, AZ, USA
e-mail: brucej@arizona.edu

F. Skarstein, L.-A. Wolff (eds.), *Sustainability and Science Education*, Environmental Challenges and Solutions,
https://doi.org/10.1007/978-3-032-24408-6_11

11.1 Introduction

Human actions increasingly affect the natural systems that all life depends on, including the Earth's hydrologic systems (Peters and Meybeck 2000). The consequences of the human impact on the Earth's biophysical boundaries are alarming in the present era, which some have termed the Anthropocene. Wolff et al. (2020) argue that the Anthropocene demands education in a more authentic environment and a change toward a more holistic, transformative, and sustainability-oriented approach. Children enrolled in primary schools today will have to act as environmentally informed citizens in the near future (Palmer and Neal 1994, pp. 3–18; Torkar 2014). According to UN policy, actions are required to protect water ecosystems and water quality, and to ensure the availability and sustainable management of water and sanitation for all (Sustainable Development Goal 6; UN 2015). They will need to protect water resources in their public roles (e.g., as advocates and voters) and private roles (e.g., as learners, consumers, and workers; Gunckel et al. 2012). The global response of science educators to the increasing impact of humans is their call for greater emphasis on environmental science literacy in the curriculum (Bybee 2008). That is why it is important to assess students' knowledge and views, and to build a student-focused participatory culture that also includes their ideas in present decisions.

Slovenia is one of the smallest countries in Europe, but it is also one of the richest in water resources. Even though water quantity and quality in Slovenia are at the very top compared to other European countries, there are still problems with water quality (Ambrožič et al. 2008). Some sections of watercourses are significantly or even excessively polluted with industrial waste and sewage. The quality of Slovenian watercourses in 2019 showed a good chemical status for 73 measuring points and a poor chemical status for 19 measuring points, which represents 20.7% of all watercourses in Slovenia (Cvitanič et al. 2020). Pollution was also detected in groundwater, which is a source of drinking water for approximately 97% of Slovenia's population. This indicates the exceptional importance of groundwater for Slovenia (Ambrožič et al. 2008). Public survey data show that Slovenians are concerned about the environment and are aware of the pollution of water resources (European Commission 2012). This raises the question of why there is a gap between environmental knowledge, attitudes, and the evident water pollution.

The primary and lower secondary school levels are critical stages in receiving water education because that is when growing teens learn information that will shape their thinking about relevant water issues in their local environment (Schaap and van Steenbergen 2002). Students' knowledge of the water cycle has already been studied in various countries (e.g., Amahmid et al. 2019; Ben-zvi-Assarf and Orion 2005a; Barrutia et al. 2021; Cardak 2009; Koomson and Owusu-Fordjour 2018; Malleus and Kikas 2019). Many studies have also dealt with students' environmental attitudes and actions (e.g., Baierl et al. 2021; Johnson and Činčera 2015; Maurer et al. 2020; Roczen et al. 2014; Torkar et al. 2020). However, there is a lack of studies that connect knowledge of the water cycle, environmental attitudes, and environmental actions, which was the main aim of this study.

In Slovenia, learning modules dealing with the human impact on the environment have been implemented in science subjects in primary and secondary schools. In the Slovenian primary school curriculum, the water cycle is extensively discussed in various grades, mainly in science subjects (Kolar et al. 2011; Skvarč et al. 2011; Vodopivec et al. 2011; Vilhar et al. 2011). The curriculum deals not only with the water cycle, but also with various phenomena and processes in the living and non--living natural systems (e.g., transpiration, photosynthesis, evaporation, and groundwater), and human influences on the water cycle (e.g., water pollution, field irrigation, saving water, and renewable energy). Science subjects also emphasize the importance of the nature of science, which gradually develops students' scientific and critical thinking. Because the water cycle and human impacts on the environment are extensively covered in the Slovenian primary school curriculum, they are an appropriate learning topic for environmental education. Gunckel et al. (2012) argue that providing model-based accounts (explanations and predictions) of water and substances in water moving through environmental systems is an important practice in environmental education and an instructive model for raising awareness about the global and local water quantity and quality issues.

This study investigates whether a more in-depth understanding of the water cycle, which is extensively discussed in the Slovenian primary school curriculum, can be transferred to general environmental attitudes and actions.

The research questions (RQs) that guided this study are as follows:

RQ1. What are primary school students' knowledge of the water cycle, environmental worldviews, and environmental actions?
RQ2. How are primary school students' knowledge of the water cycle, environmental attitudes, and environmental actions interrelated?
RQ3. How are the aforementioned variables influenced by students' grade?
RQ4. How are the aforementioned variables influenced by students' gender?

11.2 Literature Review

This section presents a short review of the literature to introduce the three main areas that were of interest in this study: knowledge about the water cycle, environmental worldviews, and environmental actions.

11.2.1 Students' Knowledge About the Water Cycle

Studies report that primary and secondary school students understand various hydrological, biological, and geological processes, but they often have difficulty understanding their dynamic, cyclical, and systemic nature at the system level (e.g.,

Amahmid et al. 2019; Barrutia et al. 2021; Ben-zvi-Assarf and Orion 2005a). For example, Amahmid et al. (2019) report that Moroccan primary and lower secondary students' knowledge regarding water, such as the water cycle, water chemistry, and water treatment, is limited. Malleus and Kikas (2019) argue that the answer to the question "What is the water cycle?" is very complex, and few students provide sufficient scientific explanations. Students often describe only the atmospheric component of the water cycle and ignore the groundwater component (Ben-zvi-Assarf and Orion 2005a; Barrutia et al. 2021). Studies also show that students omit the biosphere components (humans, animals, and plants) and do not take into account the sun, which drives the water cycle (e.g., Barrutia et al. 2021; Koomson and Owusu--Fordjour 2018). Barrutia et al. (2021) suggest that most young students do not contemplate the role of water in biotic systems. Jacobson and Wilensky (2006) argue that students must have enough opportunities to learn about complex systems because this will help them understand these systems. Ecological phenomena and processes are better taught and understood by students when their systemic nature becomes explicit (Korfiatis 2018). Frick et al. (2004) define the understanding of the natural states of ecosystems and the processes within them as system knowledge. Ben-zvi-Assarf and Orion (2005b) emphasize that the first steps of system thinking should be introduced at the primary school level. System thinking is a set of skills that can be taught separately and a learning strategy that explicitly considers system characteristics when attempting to understand and predict natural phenomena or complex man-made systems (Verhoeff et al. 2008). Ben-zvi-Assarf and Orion (2005b) summarize eight key characteristics of system thinking that are hierarchically arranged and explained in the context of the water cycle system, emphasizing (1) the ability to identify the components and processes in the system: in the context of the water cycle, this means recognizing components such as oceans, rivers, and clouds, and processes such as evaporation, sublimation, and runoff; (2) the ability to identify the relationships between the system's components: this means understanding that polluted rivers could directly affect water quality; (3) the ability to organize the system's components and processes within the framework of relationships: this means being able to show the components, processes, and relationships with a drawing; (4) the ability to make generalizations: this can be achieved by showing an understanding of the dynamic and cyclical nature of the water cycle; (5) the ability to recognize dynamic relationships within the system: for example, recognizing how people affect groundwater through the use of fertilizers and pesticides; (6) understanding the system's hidden dimensions: the ability to recognize and understand patterns or connections that are not initially visible, such as the concept of the hydrosphere; (7) the ability to understand the cyclical nature of the system: this means understanding that the water cycle has a cyclical nature and contains several subsystems, which are also cyclical; and (8) retrospection and prediction: understanding that present events are the results of the past, whereas the future depends on present actions (e.g., the current water quality is the result of past events).

11.2.2 Environmental Attitudes and Actions, and the Relevance of Knowledge

An environmental attitude is a collection of beliefs, affect, and intentions related to the environment (Schultz et al. 2004). Milfont and Duckitt (2004) describe environmental attitudes as a psychological tendency expressed as an inclination toward concepts related to the environment. The importance of measuring environmental attitudes lies in the expectation of obtaining a valuable predictor of environmental actions (Torkar et al. 2020). Roczen et al. (2014) emphasize that a person's environmental attitude is the force that drives his or her environmental actions. Empirical researchers regard environmental attitudes as unidimensional or multidimensional constructs (e.g., Manoli et al. 2019; Munoz et al. 2009; Wiseman and Bogner 2003), and a series of closely related attitudes defines a value (Rokeach 1973). They argue that individuals who feel more connected to nature and believe they are part of the natural environment tend to have a more comprehensive view of ecological problems and are more committed to the environment.

Kollmuss and Agyeman (2002) define environmental actions as behavior change that strives to reduce negative impacts on the environment. After comparing various frameworks for analyzing environmental actions, they defined three groups of influential factors: demographic, external, and internal factors. They found that the most influential demographic factors were gender and years of education. Among the most influential external factors were institutional, economic, social, and cultural factors, such as educational institutions, socioeconomic status, and family. The third group of factors were internal factors, which included motivation, environmental knowledge, values, attitudes, environmental awareness, emotional involvement, locus of control, and responsibility and priorities (Kollmuss and Agyeman 2002).

In contrast to the above assumption of a causal relationship between attitude and action in the "behavioral explanation paradigm," an evidence-based competence model for environmental education has been proposed (Kaiser et al. 2008) and empirically tested (Baierl et al. 2021; Roczen et al. 2014). In this model, environmental knowledge forms the intellectual basis, and environmental attitudes are the motivational source for a person's environmental actions. According to Roczen et al. (2014), environmental knowledge has a modest effect on environmental actions, whereas environmental attitudes are a stronger determinant of environmental actions. They argue that an individual who values nature and the environment has more motivation to seek information related to the environment, environmental problems, and solutions. Frick et al. (2004) believe that different types of knowledge are required to act in an environmentally friendly manner: system knowledge (the understanding of the natural states of ecosystems and the processes within them), action-related knowledge (how to solve environmental problems), and effectiveness knowledge (knowledge about the benefit of environmentally responsible actions). Environmental system knowledge does not trigger environmental actions directly (Baierl et al. 2021; Roczen et al. 2014), but it can motivate the acquisition of action-related knowledge and effectiveness knowledge, and these two significantly promote a person's environmental actions. Environmental attitudes have mediating effects on the influence of knowledge on actions (Roczen et al. 2014).

11.3 Material and Methods

The following section provides an overview of the steps we took to collect empirical data and the measures taken to ensure their quality. The methodological approach and research instruments are presented below.

11.3.1 Procedure and Sample

The administrations of seven primary schools in central and southeastern Slovenia were contacted over the phone to confirm or reject participation in the survey. Approval from the schools' head offices was acquired before the survey, and written parental consent was obtained. The schools and parents were informed that the study would focus on students' environmental awareness. All the schools agreed to participate in the survey. Teachers were given written guidelines to collect the data from the study coordinator. They distributed the questionnaires to their students during regular classes, collected the questionnaires, and mailed them to us. The study involved a convenience sample of 259 students from fourth (n = 93), fifth (n = 73), and sixth (n = 93) grade. There were more girls (n = 133) than boys (n = 126). The respondents were 9–12 years old, with the exception of one 14-year-old student. Data for the research were collected through an anonymous paper-and-pencil questionnaire from October 2019 to April 2022, largely after the first COVID-19 wave in the spring of 2020. The long period of data collection was a consequence of the COVID-19 pandemic.

11.3.2 Instruments

Water cycle knowledge test: students' knowledge of the water cycle was tested using an adapted version of the test developed by Ben-zvi-Assarf and Orion (2005a). The test items were originally designed for slightly older students. Students were instructed to indicate their agreement with 12 items (1 = disagree, 2 = not sure, 3 = agree), which are presented in Table 11.1.

Environmental worldviews: students' environmental worldviews were explored using the Slovenian version of the New Ecological Paradigm (NEP) Scale for Children (Manoli et al. 2007; Torkar et al. 2021). Torkar et al. (2021) suggest that a unidimensional and a three-dimensional (Rights of Nature, Eco-Crisis, and Human

Exemptionalism) modified NEP Scale for Children, with 10 instead of 15 items and revised wording, are suitable for use with students 10 to 12 years old. Torkar et al. (2021) report that a three-dimensional model is an appropriate structure for use in the Slovenian context, where the NEP scale was used with students 9–13 years old. The 10-item NEP Scale for Children on a five-point Likert scale (1 = strongly disagree, 2 = disagree, 3 = not sure, 4 = agree, 5 = strongly agree) allows for investigations of how children's environmental worldviews develop, how they change as a result of new experiences or educational programs, and how the environmental worldviews of children from different socioeconomic or cultural backgrounds compare (Manoli et al. 2007).

Environmental actions: 10 items from the General Ecological Behavior (GEB) scale (Kaiser and Wilson 2000; Kaiser et al. 2007; Roczen et al. 2014; Torkar et al. 2020) were used. We used items relevant for primary school students (e.g., "I collect and recycle used paper" and "I have pointed out unecological behavior to someone"). The items were ranked on a five-point Likert scale (1 = never, 2 = rarely, 3 = occasionally, 4 = often, 5 = always).

11.3.3 Statistical Analysis

Data entry and analysis were conducted using the Statistical Package for the Social Sciences (SPSS). Descriptive and inferential statistics were computed. At the descriptive level, frequency distribution and basic descriptive statistics were used. At the inferential level, structural equation modeling (SEM) using AMOS version 28 was used.

11.4 Results

The primary school students in this study had difficulties understanding the water cycle systems (Table 11.1), especially in statements concerning the cycling of matter (Items 1, 2, and 4). Half of the students did not understand the role of the sun in the water cycle (Item 10), and many students struggled to link the water cycle with biological processes (Items 8 and 11). Students also had difficulties recognizing the importance of the groundwater system (Items 5, 6, 7, 9, and 12).

Student responses to the 10 NEP items are presented in Table 11.2. There was strong overall agreement on several of the environmental items, particularly Item 1 *Plants and animals have as much right as people to live* (90.0%), Item 4 *People must still obey the laws of nature* (97.3%), Item 8 *People are treating nature badly.* (81.1%), and Item 10 *If things do not change, we will soon have a big environmental disaster* (85.3%). There was less agreement on other items, except that most of the students (66.4%) disagreed with Item 7 *People are supposed to rule over the rest of nature.*

Table 11.1 Frequency distribution of responses to questions about the water cycle

Items	*f* (%)		
	Disagree	Not sure	Agree
1. In the water cycle, the beginning point is the ocean, and the ending point is underground water.	34.0	27.0	38.6
2. *The amount of water in the ocean grows from day to day because rivers are continually flowing into the ocean.*	42.9	19.7	37.5
3. If the population on earth continues to grow, water consumption will increase, and so the amount of drinking water will decrease.	71.0	14.3	14.7
4. *The amount of water that evaporates from the surface of the Earth into the atmosphere is not equal to the amount of rain that falls on the Earth's surface.*	28.2	32.0	39.4
5. Groundwater is formed by the infiltration of rainwater into the soil.	11.2	15.4	73.0
6. *Groundwater can only be found in rainy climate areas.*	47.5	24.7	27.0
7. Agriculture causes groundwater pollution.	21.6	26.6	51.0
8. *Trees usually store water pumped by their roots for many years.*	32.8	34.0	32.0
9. *Water can penetrate rocks only through cracks.*	33.2	23.2	41.3
10. Without the sun, there would be no precipitation (rain and snow) on Earth.	24.3	21.6	54.1
11. When air is exhaled from the lungs, the body loses water.	24.7	21.2	53.7
12. Rivers form karst caves.	7.7	11.6	79.9

Note. Items in italics are incorrect statements and were reversed to calculate the total score of correct answers

Table 11.3 shows students' environmental actions that were measured with 10 items from the GEB. From the responses, it can be seen that students' engagement in these actions varies widely. The least common among all the actions is that they try to persuade their parents to buy energy-efficient appliances. However, they are more likely to implement actions that they can control themselves; for example, 58.7% of the respondents always turn off the light if they are the last to leave the room, 69.9% often or always collect and recycle paper, 74.1% often or always take empty bottles to a recycling bin, and 71.0% often or always walk or ride a bike to travel short distances.

Table 11.2 Frequency distribution of responses to the NEP items

Items	Strongly disagree $f(\%)$	Disagree $f(\%)$	Not sure $f(\%)$	Agree $f(\%)$	Strongly agree $f(\%)$	M	SD
1. Plants and animals have as much right as people to live.	2.7	2.7	4.2	12.0	78.0	4.60	0.90
2. There are too many people (or almost too many people) on earth.	6.2	10.8	30.5	30.1	21.6	3.51	1.13
3. People are clever enough to keep from ruining the Earth.	14.7	20.8	22.0	22.8	18.1	3.09	1.33
4. People must still obey the laws of nature.	1.2	0.8	0.4	11.2	86.1	4.81	0.59
5. When people interfere with nature, the results are bad.	3.1	3.5	22.8	27.8	40.9	4.02	1.03
6. Nature is strong enough to handle the bad effects of our modern lifestyle.	21.2	22.8	31.7	13.5	10.0	2.68	1.24
7. People are supposed to rule over the rest of nature[a]	47.1	19.3	18.9	6.6	6.6	2.05	1.24
8. People are treating nature badly.	2.7	5.4	9.7	26.3	54.8	4.27	1.02
9. People will someday know enough about how nature works to be able to control it.	18.5	16.6	35.9	17.0	10.0	2.83	1.22
10. If things do not change, we will soon have a big environmental disaster.	1.9	3.9	8.5	23.9	61.4	4.39	0.94

[a]All items are displayed as students responded. For calculating mean scores, Item 7 was reverse--scored because it is negatively phrased in relation to the other items in its factor

11.4.1 Correlations Between Knowledge, Worldview, and Actions

A structural equation model of correlations between water cycle knowledge (WC score), environmental worldviews, and environmental actions was tested. In the first model, the 10 NEP items were grouped into three subfactors applied in a previous study (Torkar et al. 2021): Items 1, 4, and 7 in Rights of Nature; Items 2, 5, 8, and 10 in Eco-Crisis, and Items 3, 6, and 9 in Human Exceptionalism. The 10 GEB items were grouped under one factor. The fit indices varied, with some reaching recommended levels (χ^2/df ratio = 1.253, RMSEA = 0.031) and others not (CFI = 0.880, TLI = 0.846). Elimination of three NEP items (2, 5, and 7) and three GEB items (1, 7, and 10) with a poor fit improved the model's fit, but not to a sufficient level (χ^2/df ratio = 1.387, CFI = 0.890, TLI = 0.840, RMSEA = 0.039). In the next model, the seven remaining GEB items were grouped into two subfactors. Five items (2, 3, 4, 5, and 8) dealt with actions that students could take, and so they were grouped together under the Personal Action factor. The other two items (6 and 9)

Table 11.3 Frequency distribution of responses to the GEB items

Items	Never f (%)	Rarely f (%)	Occasionally f (%)	Often f (%)	Always f (%)	*M*	*SD*
1. I ride a bike, take the bus, or walk to school.	11.2	12.7	11.2	21.6	42.1	3.71	1.41
2. I collect and recycle paper.	5.0	8.5	16.2	30.1	39.8	3.91	1.16
3. If I am the last to leave the room, I turn off the lights.	1.9	4.2	5.8	26.6	58.7	4.40	0.92
4. I drink water or other beverages in reusable packaging.	4.6	7.3	22.8	32.4	30.1	3.78	1.11
5. I take empty bottles to a recycling bin.	7.7	6.2	9.7	16.6	57.5	4.13	1.28
6. I try to convince parents to buy energy-efficient appliances.	41.3	21.2	16.6	11.6	6.9	2.20	1.29
7. I walk or ride a bicycle for shorter distances (up to 15 minutes).	5.0	8.5	13.5	30.5	40.5	3.95	1.67
8. I use reusable bags when shopping.	2.7	4.6	10.0	25.5	55.6	4.29	1.01
9. I have pointed out unecological behavior to someone.	6.6	15.1	24.7	27.0	23.9	3.48	1.21
10. I learn about environmental content from the media (internet, TV, newspapers, etc.).	14.3	22.8	32.4	20.5	7.7	2.84	1.15

Note. Cronbach's alpha = 0.59

dealt with talking to others about their actions and were grouped together into the Helping Others factor. In addition, Human Exceptionalism was changed from a NEP factor to a factor negatively correlated to the NEP subfactor Eco-Crisis. These changes resulted in the model shown in Fig. 11.1.

The fit indices for this revised model were improved (χ^2/df ratio = 1.316, CFI = 0.913, TLI = 0.869, RMSEA = 0.035). All correlations shown in Fig. 11.1 are statistically significant except for the correlations (1) between water cycle knowledge (WC) and environmental worldviews (NEP), and (2) between water knowledge and actions (GEB); water cycle knowledge in this model had no influence on either environmental worldviews or actions. Human Exceptionalism is negatively correlated with Eco-Crisis, as expected. People who view humans as being exempt from natural systems and limitations are unlikely to be concerned about ecological crises. Environmental worldviews (NEP) are strongly correlated with environmental actions (GEB) with its subfactors Personal Actions and Helping Others.

To further increase the model's fit to the data, water cycle knowledge (WC) was eliminated because it had no statistically significant correlation with any other factors in the model. The results are presented in Fig. 11.2.

As expected, fit indices increased in this model (χ^2/df ratio = 1.114, CFI = 0.972, TLI = 0.956, RMSEA = 0.021), resulting in an excellent fit of the data to the model. Removing the water cycle knowledge score because it had no correlation with other factors resulted in an improved model. Of course, this model only includes the correlations between environmental worldviews and actions.

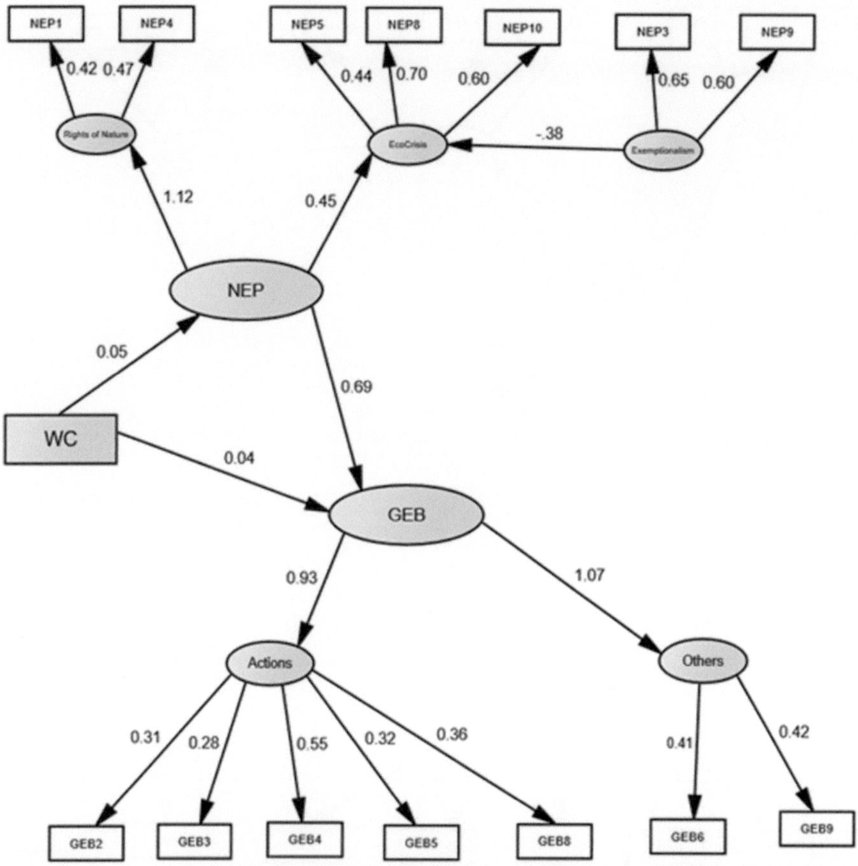

Fig. 11.1 Model of correlations between water cycle knowledge (WC), environmental worldviews (NEP), and environmental actions (GEB). *Note*. Statistically significant correlations are in bold

11.4.2 Grade and Gender Differences

The mean score for water cycle knowledge was 3.70 of the total of 12 points ($SD = 2.01$), an average of 30.8% correct. There were differences by grade level. ANOVA revealed a statistically significant difference between the mean scores for the three grades studied ($F = 4.12$, $p < 0.05$). Fifth-grade students' scores ($M = 3.18$, $SD = 1.82$) were significantly lower than those for both fourth grade ($M = 4.06$, $SD = 1.74$) and sixth grade ($M = 3.74$, $SD = 2.31$). There was no statistically significant difference between the scores of girls and boys.

For environmental worldviews, mean scores were calculated for the three NEP factors as applied in a previous study using the NEP Scale in Slovenia (Torkar et al. 2021): Rights of Nature, Eco-Crisis, and Human Exceptionalism. Students' views

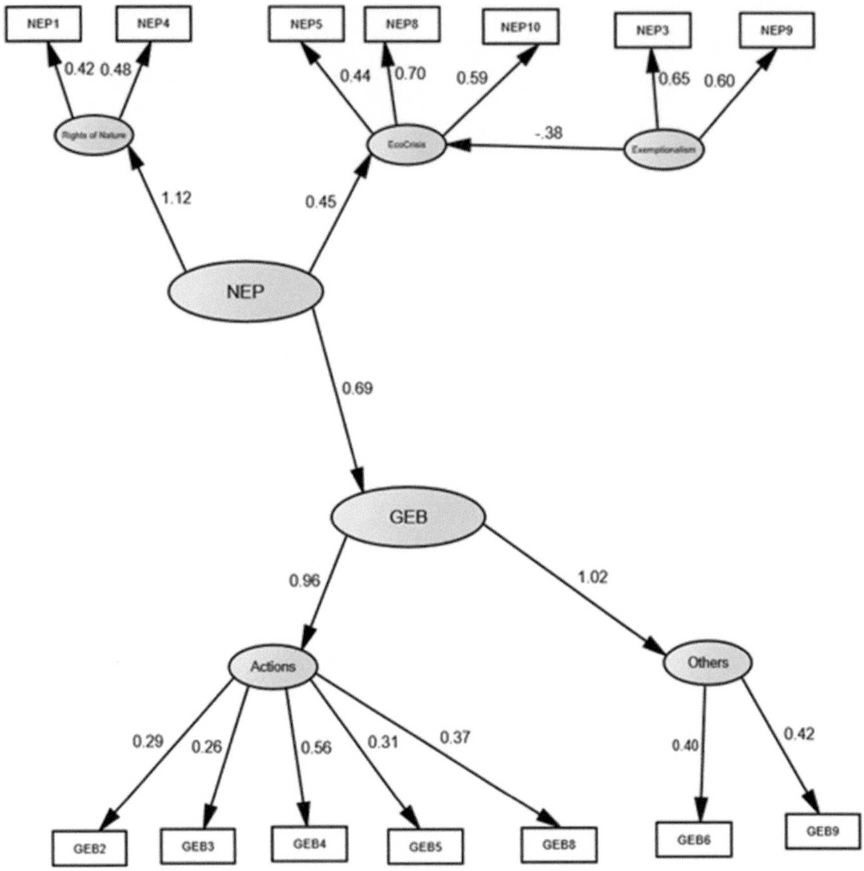

Fig. 11.2 Model of correlations between environmental worldviews and actions. *Note*. Statistically significant correlations are in bold

on the Rights of Nature ($M = 4.71$, $SD = 0.59$) were stronger than their views on the extent to which we have an Eco-Crisis ($M = 4.23$, $SD = 0.75$), but both were environmentally positive. Student views on Human Exceptionalism ($M = 2.95$, $SD = 1.07$) were slightly below 3.0, indicating that many students viewed humans as separate from nature. However, the higher standard deviation shows that there was more variation in student views than there was for Rights of Nature or Eco-Crisis, and, as can be seen below, the results for Human Exceptionalism varied by grade level. The only statistically significant difference in worldviews by grade level was in Human Exceptionalism ($F = 6.250$, $p = 0.002$). Students in fourth grade had higher scores ($M = 3.25$, $SD = 1.11$) than those in fifth grade ($M = 2.86$, $SD = 1.01$) and sixth grade ($M = 2.71$, $SD = 1.00$). The scores of students in fifth and sixth grades were not statistically different from one other. There were no statistically significant differences in the mean scores for girls and boys for any of the three NEP factors.

For environmental actions, items were grouped into the factor Personal Actions or the factor Helping Others, based on the results of model testing. The means for these two factors were calculated. Students reported substantially higher rates for taking Personal Actions (M = 4.11, SD = 0.62) than Helping Others (M = 2.85, SD = 0.96). There was no statistically significant difference by grade level for Personal Actions, but there was for Helping Others (F = 3.475, p = 0.023). Sixth-grade students had a higher Helping Others score (M = 3.07, SD = 0.97) than either fourth graders (M = 2.75, SD = 0.91) or fifth graders (M = 2.70, SD = 0.98). Comparing gender, girls had higher scores than boys for both factors. For Personal Action (F = 13.12, $p < 0.001$), girls' scores (M = 4.24, SD = 0.59) were higher than the boys' (M = 3.96, SD = 0.63). For Helping Others, girls' scores (M = 3.05, SD = 0.97) were also higher than the boys' (M = 2.65, SD = 0.92).

11.5 Discussion

The findings indicate that most of the students have very limited knowledge and understanding of the water cycle with several misconceptions, which are evident from their responses to the knowledge statements. Students showed a lack of understanding of the cyclical nature of water. More than half of them did not realize that in a cyclical process the overall amount of water is conserved. They struggled to link the water cycle with biological processes in plants and animals. Respondents were mostly aware that an increase in population would affect the amount of drinking water. Although the statements in Table 11.1 do not cover all the components of the water cycle, it is evident that students have problems with system thinking—that is, with organizing components and processes of the system within the framework of relationships (Ben-zvi-Assarf and Orion 2005a). Many studies (e.g., Cardak 2009; Koomson and Owusu-Fordjour 2018) reported that students have difficulties understanding the dynamic and cyclical nature of the water cycle and often do not include the biosphere component in descriptions and drawings.

Gunckel et al. (2012) argue that in making decisions about environmental issues, such as protecting fresh water supplies, students should be able to identify driving forces and consider the effect of constraining factors on the pathways for water and substances in water, and to provide explanations and predictions at the atomic-molecular scale. For understanding the water system, they need to trace water and substances in water along multiple pathways across visible and invisible boundaries, including pathways into and through biotic systems, human engineered systems, and connections between natural and human-engineered systems. Many Slovenian primary school students were not aware of the importance of the sun's energy for the water cycle. The same findings are reported by other researchers (e.g., Barrutia et al. 2021; Koomson and Owusu-Fordjour 2018), even though the sun's energy is essential for water circulation. We assume that a lack of understanding may be related to the treatment of the topic in primary school textbooks. A review of textbooks has shown that representations of the water cycle are highly simplified, limited only to abiotic factors, and there is a very limited or no explanation of the importance of solar energy in the cycle. This could also explain why students have misconceptions regarding the groundwater system because it is often not represented in images. Other studies also concluded that students have an incomplete understanding of the groundwater system (e.g., Barrutia et al. 2021; Ben-zvi-Assarf and Orion 2005a; Cardak 2009).

Students showed the highest agreement with environmental worldview items in the factor Rights of Nature, followed by their views of an Eco-Crisis (both views were very environmentally positive), and the lowest with Human Exceptionalism, which is in line with previous studies (e.g., Boeve-de Pauw and Van Petegem 2012; Manoli et al. 2007; Torkar et al. 2021). The results for general environmental behaviors (GEB) are interesting because not many studies have focused on children's environmental actions. On average, students engaged occasionally to often in the listed actions. These actions were carefully selected because many of the actions predicted by the GEB instrument (Kaiser and Wilson 2000; Kaiser et al. 2007) cannot be performed by children themselves or are in the domain of adults. It is not surprising that in this study, students were more likely to carry out environmental actions that they can control themselves than those for which they must rely on others. In the future, it would be meaningful to further explore why some students preferred some actions over others by using focus interviews or open-ended questions for which students could elaborate their views on specific environmental actions.

The main objective of the study was to explore the correlations between students' water cycle knowledge, environmental worldviews, and environmental actions. An evidence-based competence model for environmental education (Kaiser et al. 2008; Roczen et al. 2014; Baierl et al. 2021) assumes that environmental knowledge catalyzes the development of environmental attitudes and actions. The results indicate that there is no statistically significant correlation between students' knowledge of the water cycle and environmental attitudes or environmental actions. Other researchers report that environmental knowledge is typically not directly related to environmental actions and that the relationship is weak or nonexistent (Kollmuss and Agyeman 2002; Roczen et al. 2014; Suárez-Perales et al. 2021). A possible reason for the absence of correlation is also the low level of knowledge about the water cycle. Johnson and Činčera (2015) emphasized that, despite the knowledge students acquire, additional encouragement and help are needed to maintain environmental actions in the future. However, some studies have shown that knowledge about environmental issues may be associated with more positive environmental actions and greater concern for nature (Theodori and Luloff 2002). In this study, a system knowledge scale has been used without action-related and effectiveness knowledge items, which may be more strongly related to environmental attitudes and actions, as reported by Roczen et al. (2014) and Baierl et al. (2021). An important issue worthy of discussion is the transferability of specific environmental system knowledge to general environmental attitudes and actions. This study was based on a curriculum in which specific learning objectives include understanding the water cycle, water pollution, and water conservation, while at the same time, the formation of positive and responsible attitudes toward nature and respect for all forms of life are expected to be general goals of science education. Therefore, we were interested in whether such a transfer from specific knowledge to general environmental awareness and actions can be expected. Our research does not support this conclusion because water cycle knowledge did not predict environmental worldviews or actions.

As mentioned in the introduction, the water cycle is extensively covered in science subjects in Slovenia's primary school curricula (Kolar et al. 2011; Skvarč et al. 2011; Vodopivec et al. 2011; Vilhar et al. 2011). For this reason, we also paid attention to the differences between grade levels. The water cycle is included in the curriculum from first grade: starting with weather phenomena, states of water, water in living beings, and water pollutants are covered in the first 3 years of primary school. In the fourth-grade curriculum, knowledge of the states of water and water in the human body is expanded. In the fifth-grade curriculum, students' knowledge is expanded and consolidated the most: they learn about the water cycle, surface and underground waters, the states of water, water in living systems, and water pollution and purification, and so it was expected that their knowledge would significantly improve from fifth grade onward. In sixth grade, content related to the water cycle is also covered, such as hydropower, water in plants, adaptations of plants to water ecosystems, use of water resources, and sustainable use of water. Contrary to our expectations, knowledge of the water cycle varied to some extent among students in the three grade levels studied (fourth, fifth, and sixth grades). Fifth-grade students' scores were significantly lower than those for fourth and sixth grades. In contrast, Barrutia et al. (2021) reported that older students had more knowledge about the water cycle than younger students. The results indicate that knowledge does not increase progressively from fourth to sixth grade, which could have been affected to some extent by the COVID-19 pandemic and the quality of science education during that period when students remained home for several months.

The results show statistically significant differences in environmental worldviews by grade level only in Human Exceptionalism, for which students in fourth grade had higher scores than those in fifth and sixth grades. Students in sixth grade were more likely to implement environmental actions in which they have to rely on others than fourth- and fifth-grade students, but no significant differences in grade level were found for personal environmental actions. Previous studies carried out among Slovenian primary school students found that older students were characterized by more negative environmental attitudes and actions (Torkar and Krašovec 2019; Torkar et al. 2020). All of this testifies to the complexity of researching environmental attitudes and actions in children and adolescents, whose environmental worldviews are emerging and are subject to various demographic, external, and internal factors (Kollmuss and Agyeman 2002). Two demographic factors that have been found to influence environmental attitudes and environmental actions are gender and years of education (Kollmuss and Agyeman 2002).

Some studies report no correlation between environmental worldviews and gender (e.g., Evans et al. 2007; Torkar and Bogner 2019; Torkar et al. 2021), and others report that female students have stronger environmental views and concerns than male students (Torkar 2016; Zelezny et al. 2000). We found no gender differences in environmental knowledge and worldviews. Our results confirm that girls are more likely to take environmental action than boys, a finding that aligns with Vicente-Molina et al. (2013) but contradicts with Hines et al. (1987) and Anderson and Krettenauer (2021). These differences have traditionally been explained in terms of the sociocultural emphasis on stereotypical female traits, such as empathy and caring, versus the emphasis on stereotypical male traits, such as independence and dominance. Chawla (1988) bases this theory on the developmental timing at which these differences occur, such as in sixth grade, which is known to be the period when gender roles are most pronounced in childhood (Chawla 1988). Sociological theories of gender emphasize gender differences in the socialization process and/or social roles and status in society (e.g., Davidson and Freudenburg 1996; Xiao and McCright 2015).

11.6 Conclusion

The water cycle is covered extensively in the Slovenian primary school curriculum: knowledge is strengthened in science subjects, enabling students to follow the path of water and substances contained in water along various pathways, including the pathways into and through biotic systems and human-engineered systems. Furthermore, the water cycle is closely related to human activities, such as water pollution, field irrigation, water conservation, drinking water, or renewable energy. Therefore, it is a suitable learning topic for environmental education and allows students to develop their system thinking about the water cycle, which hopefully will then be transferable to other complex natural and social science problems and to desired environmental attitudes and actions. By working across disciplines, we are more likely to achieve the social transformation (Braus et al. 2022) required for a sustainable future. However, Slovenian primary school students do not demonstrate sufficient system thinking. Importantly, specific system knowledge of the water cycle did not translate into environmental worldviews and environmental actions. In other words, the results show that there was no transfer of water knowledge into their environmental attitudes and actions. Older students should know more about the water cycle because this is required in the primary school curriculum. However, the results show no clear correlation between the students' grade level and their system knowledge about the water cycle. This calls for a thorough examination of the learning and teaching strategies applied in environmental education in Slovenian schools.

Research on environmental attitudes and actions has repeatedly shown that it can be subject to strong biases (Bratt et al. 2015; Hadler et al. 2022), and it is problematic that some scales, although only one person is interviewed, target the household level (Hadler et al. 2022). For example, individual environmental actions often depend on context and socio-demographic characteristics, which are rarely explored in surveys (Hadler et al. 2022). In our case, students' response to the item "I ride a bike, take the bus, or walk to school" could depend on how far from school they live, how busy the roads are, their parents' occupation, and so on. We were also aware of this when selecting the items. For this particular item, given that the majority (98%) of students in Slovenia attend the public primary school system, in which students are required to attend a local primary school, the distances to school are relatively short for most students.

Most studies suggest that attitudes do have some impact on actions (e.g., Kollmuss and Agyeman 2002); therefore, it is important to explore this when pursuing behavior change. Maybe the weak alignment of the knowledge–attitude–action items found in research is a consequence of the fact that the questions were not more specific to individual water consumption (e.g., the amount of saved drinkable water). Furthermore, Heimlich and Ardoin (2008) explain that environmental actions that researchers see as logically fitting together in constellations of environmental behaviors or attitudes may not mesh in someone else's worldview. Concerning future research, we want to investigate the relationship between knowledge of the water cycle, general environmental attitudes and actions, and environmental attitudes and behaviors specifically related to water resource conservation. This will hopefully give us a comprehensive picture of how school curricula can be developed to ensure the development of students' environmental knowledge, attitudes, and actions.

References

Amahmid O, El Guamri Y, Yazidi M, Razoki B, Kaid Rassou K, Rakibi Y, Knini G, El Ouardi T (2019) Water education in school curricula: impact on children knowledge, attitudes and behaviours towards water use. Int Res Geogr Environ Educ 28(3):178–193. https://doi.org/10.1080/10382046.2018.1513446

Ambrožič Š, Cvitanič I, Dobnikar Tehovnik M, Gacin M, Grbović J, Jesenovec B, Kozak Legiša Š, Krajnc M, Mihorko P, Poje M, Remec Rekar Š, Rotar B, Sodja E (2008) Kakovost voda v Sloveniji. Agencija RS za okolje. https://www.arso.gov.si/vode/poro%C4%8Dila%20in%20publikacije/kakovost%20voda/Kakovost%20voda-SLO.pdf

Anderson DJ, Krettenauer T (2021) Connectedness to nature and pro-environmental behaviour from early adolescence to adulthood: a comparison of urban and rural Canada. Sustainability 13(7):3655. https://doi.org/10.3390/su13073655

Baierl TM, Johnson B, Bogner FX (2021) Assessing environmental attitudes and cognitive achievement within 9 years of informal earth education. Sustainability 13(7):3622. https://doi.org/10.3390/su13073622

Barrutia O, Ruíz-González A, Villarroel JD, Díez JR (2021) Primary and secondary students' understanding of the rainfall phenomenon and related water systems: a comparative study of two methodological approaches. Res Sci Educ 51(2):823–844. https://doi.org/10.1007/s11165-019-9831-2

Ben-zvi-Assarf O, Orion N (2005a) A study of junior high students' perceptions of the water cycle. J Geosci Educ 53(4):366–373. https://doi.org/10.5408/1089-9995-53.4.366

Ben-zvi-Assarf O, Orion N (2005b) Development of system thinking skills in the context of earth system education. J Res Sci Teach 42(5):518–560. https://doi.org/10.1002/tea.20061

Boeve-de Pauw J, Van Petegem P (2012) Cultural differences in the environmental worldview of children. Int Electron J Environ Educ 2(1):1–11

Bratt C, Stern PC, Matthies E, Nenseth V (2015) Home, car use, and vacation: the structure of environmentally significant individual behavior. Environ Behav 47(4):436–473. https://doi.org/10.1177/0013916514525038

Braus JA, Heimlich JE, Ardoin NM, Clark CR (2022) Building bridges, not walls: exploring the environmental education ecosystem. Appl Environ Educ Commun 21(4):319–330. https://doi.org/10.1080/1533015X.2022.2115226

Bybee RW (2008) Scientific literacy, environmental issues, and PISA 2006: the 2008 Paul F-Brandwein lecture. J Sci Educ Technol 17(6):566–585. https://doi.org/10.1007/s10956-008-9124-4

Cardak O (2009) Science students' misconceptions of the water cycle according to their drawings. J Appl Sci 9(5):865–873

Chawla L (1988) Children's concern for the natural environment. Children's Environments Quarterly 5(3):13–20. https://www.jstor.org/stable/41514681

Cvitanič I, Dobnikar Tehovnik M, Gacin M, Jesenovec B, Plestenjak E, Poje M, Sodja E, Velikonja M (2020) Kemijsko stanje površinskih voda v Sloveniji. [Chemical status of surface waters in Slovenia]. Agencija RS za okolje. http://www.arso.gov.si/vode/poro%c4%8dila%20in%20publikacije/publikacije%20in%20poro%c4%8dila/Kemijsko%20Poro%c4%8dilo%20povr%c5%a1inske%20vode%202019.pdf

Davidson DJ, Freudenburg WR (1996) Gender and environmental risk concerns: a review and analysis of available research. Environ Behav 28(3):302–339. https://doi.org/10.1177/0013916596283003

European Commission (2012) Attitudes of Europeans towards water. https://europa.eu/eurobarometer/surveys/detail/1047

Evans GW, Brauchle G, Haq A, Stecker R, Wong K, Shapiro E (2007) Young children's environmental attitudes and behaviors. Environ Behav 39(5):635–658. https://doi.org/10.1177/0013916506294252

Frick J, Kaiser FG, Wilson M (2004) Environmental knowledge and conservation behavior: exploring prevalence and structure in a representative sample. Personal Individ Differ 37(8):1597–1613. https://doi.org/10.1016/j.paid.2004.02.015

Gunckel KL, Covitt BA, Salinas I, Anderson CW (2012) A learning progression for water in socio-ecological systems. J Res Sci Teach 49(7):843–868. https://doi.org/10.1002/tea.21024

Hadler, M., Klösch, B., Schwarzinger, S., Schweighart, M., Wardana, R., & Bird, D. N. (2022). Surveying climate-relevant behavior: measurements, obstacles, and implications. Palgrave Macmillan https://doi.org/10.1007/978-3-030-85796-7_2

Heimlich JE, Ardoin NM (2008) Understanding behavior to understand behavior change: a literature review. Environ Educ Res 14(3):215–237. https://doi.org/10.1080/13504620802148881

Hines JM, Hungerford HR, Tomera AN (1987) Analysis and synthesis of research on responsible environmental behavior: a meta-analysis. J Environ Educ 18(2):1–8. https://doi.org/10.1080/00958964.1987.9943482

Jacobson MJ, Wilensky U (2006) Complex systems in education: scientific and educational importance and implications for the learning sciences. J Learn Sci 15(1):11–34. https://doi.org/10.1207/s15327809jls1501_4

Johnson B, Činčera J (2015) Examining the relationship between environmental attitudes and behaviour in education programmes. Sociální Studia / Social Studies 12(3):97–111. https://doi.org/10.5817/SOC2015-3-97

Kaiser FG, Wilson M (2000) Assessing people's general ecological behaviour: a crosscultural measure. J Appl Soc Psychol 30:952–978. https://doi.org/10.1111/j.1559-1816.2000.tb02505.x

Kaiser FG, Oerke B, Bogner FX (2007) Behaviour-based environmental attitude: development of an instrument for adolescents. J Environ Psychol 27:242–251. https://doi.org/10.1016/j.jenvp.2007.06.004

Kaiser FG, Roczen N, Bogner FX (2008) Competence formation in environmental education: advancing ecology-specific rather than general abilities. Umweltpsychologie 12(2):56–70. https://doi.org/10.5167/uzh-9249

Kolar M, Krnel D, Velkavrh A (2011) Program osnovna šola. Spoznavanje okolja. Učni načrt. [Elementary school program. Getting to know the environment. Syllabus]. Zavod RS za šolstvo. https://www.gov.si/assets/ministrstva/MIZS/Dokumenti/Osnovna-sola/Ucni-nacrti/obvezni/UN_spoznavanje_okolja_pop.pdf

Kollmuss A, Agyeman J (2002) Mind the gap: why do people act environmentally and what are the barriers to pro-environmental behavior? Environ Educ Res 8(3):239–260. https://doi.org/10.1080/13504620220145401

Koomson CK, Owusu-Fordjour C (2018) Misconceptions of senior high school science students on evaporation and water cycle. Europ J Res Reflect Educ Sci 6(5):13–28

Korfiatis K (2018) Ecology. In: Kampourakis K, Reiss MJ (eds) Teaching biology in schools: global research, issues, and trends. Routledge, pp 153–163

Malleus E, Kikas E (2019) Student knowledge about the water cycle and its change in different interest groups. J Educ Develop 3(1):37–51

Manoli CC, Johnson B, Dunlap RE (2007) Assessing children's environmental worldviews: modifying and validating the new ecological paradigm scale for use with children. J Environ Educ 38(4):3–13. https://doi.org/10.3200/JOEE.38.4.3-13

Manoli CC, Johnson B, Buxner S, Bogner F (2019) Measuring environmental perceptions grounded on different theoretical models: the 2-major environmental values (2-MEV) model in comparison with the new ecological paradigm (NEP) scale. Sustainability 11(5):1286. https://doi.org/10.3390/su11051286

Maurer M, Koulouris P, Bogner FX (2020) Green awareness in action; how energy conservation action forces on environmental knowledge, values and behaviour in adolescents' school life. Sustainability 12(3):955. https://doi.org/10.3390/su12030955

Milfont TL, Duckitt J (2004) The structure of environmental attitudes: a first-and second-order confirmatory factor analysis. J Environ Psychol 24(3):289–303. https://doi.org/10.1016/j.jenvp.2004.09.001

Munoz F, Bogner F, Clement P, Carvalho GS (2009) Teachers' conceptions of nature and environment in 16 countries. J Environ Psychol 29(4):407–413. https://doi.org/10.1016/j.jenvp.2009.05.007

Palmer J, Neal P (1994) The handbook of environmental education. Routledge

Peters NE, Meybeck M (2000) Water quality degradation effects on freshwater availability: impacts of human activities. Water Int 25(2):185–193. https://doi.org/10.1080/02508060008686817

Roczen N, Kaiser FG, Bogner FX, Wilson M (2014) A competence model for environmental education. Environ Behav 46(8):972–992

Rokeach M (1973) The nature of human values. Free Press. https://doi.org/10.1177/0013916513492416

Schaap W, van Steenbergen F (2002) Ideas for water awareness campaigns. Global Water Partnership

Schultz PW, Shriver C, Tabanico JJ, Khazian AM (2004) Implicit connections with nature. J Environ Psychol 24(1):31–42. https://doi.org/10.1016/S0272-4944(03)00022-7

Skvarč M, Glažar SA, Marhl M, Skribe Dimec D, Zupan A, Cvahte M (2011) Program osnovna šola. Naravoslovje Učni načrt [Elementary school program Science Syllabus]. Zavod RS za šolstvo. http://www.mizs.gov.si/fileadmin/mizs.gov.si/pageuploads/podrocje/os/prenovljeni_UN/UN_naravoslovje.pdf

Suárez-Perales I, Valero-Gil J, Leyva-de la Hiz DI, Rivera-Torres P, Garces-Ayerbe C (2021) Educating for the future: How higher education in environmental management affects pro-environmental behaviour. J Cleaner Prod 321:128972

Theodori GL, Luloff AE (2002) Position on environmental issues and engagement in proenvironmental behaviors. Soc Nat Resour 15(6):471–482. https://doi.org/10.1080/08941920290069128

Torkar G (2014) Learning experiences that produce environmentally active and informed minds. NJAS-Wageningen J Life Sci 69:49–55. https://doi.org/10.1016/j.njas.2014.03.002

Torkar G (2016) Secondary school students' environmental concerns and attitudes toward forest ecosystem services: implications for biodiversity education. Int J Environ Sci Educ 11(18):11019–11031

Torkar G, Bogner FX (2019) Environmental values and environmental concern. Environ Educ Res 25(10):1570–1581. https://doi.org/10.1080/13504622.2019.1649367

Torkar G, Krašovec U (2019) Students' attitudes toward forest ecosystem services, knowledge about ecology, and direct experience with forests. Ecosyst Serv 37:100916. https://doi.org/10.1016/j.ecoser.2019.100916

Torkar G, Fabijan T, Bogner FX (2020) Students' care for dogs, environmental attitudes, and behaviour. Sustainability 12(4):1317. https://doi.org/10.3390/su12041317

Torkar G, Debevec V, Johnson B, Manoli CC (2021) Assessing children's environmental worldviews and concerns. Center Educ Policy Stud J 11(1):49–65. https://doi.org/10.26529/cepsj.793

UN (2015) Transforming our world: the 2030 agenda for sustainable development. United Nations. https://sustainabledevelopment.un.org/post2015/transformingourworld

Verhoeff RP, Waarlo AJ, Boersma KT (2008) Systems modelling and the development of coherent understanding of cell biology. Int J Sci Educ 30:543–568. https://doi.org/10.1080/09500690701237780

Vicente-Molina MA, Fernández-Sáinz A, Izagirre-Olaizola J (2013) Environmental knowledge and other variables affecting pro-environmental behaviour: comparison of university students from emerging and advanced countries. J Clean Prod 61:130–138. https://doi.org/10.1016/j.jclepro.2013.05.015

Vilhar B, Zupančič G, Gilčvert Berdnik D, Vičar M, Zupan A, Sobočan V (2011) Program osnovna šola. Biologija, Učni načrt. [Elementary school program. Biology. Syllabus]. Zavod RS za šolstvo. https://www.gov.si/assets/ministrstva/MIZS/Dokumenti/Osnovna-sola/Ucni-nacrti/obvezni/UN_Biologija.pdf

Vodopivec I, Papotnik A, Gostinčar Blagotinšek A, Skribe Dimec D, Balon A (2011) Program osnovna šola. Naravoslovje in tehnika. Učni načrt [Elementary school program Science and Technology Syllabus]. Zavod RS za šolstvo. https://www.gov.si/assets/ministrstva/MIZS/Dokumenti/Osnovna-sola/Ucni-nacrti/obvezni/UN_naravoslovje_in_tehnika.pdf

Wiseman M, Bogner FX (2003) A higher-order model of ecological values and its relationship to personality. Personal Individ Differ 34(5):783–794. https://doi.org/10.1016/S0191-8869(02)00071-5

Wolff LA, Skarstein TH, Skarstein F (2020) The mission of early childhood education in the Anthropocene. Educ Sci 10(2):27. https://doi.org/10.3390/educsci10020027

Xiao C, McCright AM (2015) Gender differences in environmental concern: revisiting the institutional trust hypothesis in the USA. Environ Behav 47(1):17–37. https://doi.org/10.1177/0013916513491571

Zelezny LC, Chua PP, Aldrich C (2000) New ways of thinking about environmentalism: elaborating on gender differences in environmentalism. J Soc Issues 56(3):443–457. https://doi.org/10.1111/0022-4537.00177

Gregor Torkar is a Full Professor of Biology Education at the Faculty of Education, University of Ljubljana, Slovenia. His research interests include sustainability, environmental, outdoor, and science education. He is an active editor and author with numerous international publications and has extensive experience in both national and international collaboration. He is a chairholder of the UNESCO Chair in Teacher Education for Sustainable Development at the University of Ljubljana. He is also a member of the academic committee of the European Researchers in Didactics of Biology (ERIDOB) and the program coordinator for Outdoor Education Centers in Slovenia.

Iris Zajc is a first-grade teacher at Primary School Gradec Litija, Slovenia, dedicated to creating a supportive and engaging learning environment. Her teaching approach integrates nature, movement, and sustainability to promote child development and environmental awareness. She incorporates outdoor and experiential learning to enhance curiosity, responsibility, and respect for nature and the community. By connecting classroom learning with real-world experiences, she encourages students to think critically, act compassionately, and become environmentally conscious individuals prepared for a sustainable future.

Bruce Johnson is Professor Emeritus and former Dean Emeritus of the College of Education at the University of Arizona in Tucson, Arizona, USA. His research includes the teaching and learning of ecological concepts, development of environmental values/attitudes and actions, and curriculum development. His Earth Education Research and Evaluation (EERE) Team has conducted research and provided evaluation services to centers and schools in Asia, Australia, Europe, North America, and South America. As International Program Coordinator for the Institute for Earth Education (IEE), he oversees the development and implementation of earth education programs and leads sessions on earth education around the world.

Part IV
Pedagogical Practices and Challenges

Chapter 12
Sustainability and Action Competence in Science Education: A Review of Educational Research

Anna Uitto

Abstract The goals of science education are scientific literacy and competence to use scientific knowledge and skills in a changing world with many sustainability challenges. However, while the number of studies focusing on school sustainability education is increasing, there is not much research on the competence to act for sustainability in school science education. The aim of this study is to clarify how the competence is studied in the research focusing on sustainability and science education in schools. In the literature review, 21 studies using the terms related to competence to act for sustainability were found. In most cases, the concept *action competence* (AC) was used, while only two studies used the concept *sustainability competence* (SC). Eight of the studies were theoretical and had varying definitions for context-dependent AC. Most empirical studies addressed AC in the context of school science education or different-aged citizens and focused on participants' perceptions of or emotions related to science and sustainability issues and AC. Few empirical studies addressed AC in biology education, focusing on human biology, health, sexuality and equity issues in school textbooks. The results indicate that regardless of the increasing emphasis on sustainability issues and competence in educational policy documents and science curricula, there is a need of more theoretical research on competence definition, as well as empirical research on the implementation of AC for sustainability in science education.

Keywords Science education · Scientific literacy · Action competence · Sustainability competence · Education for sustainable development · Secondary school

A. Uitto (✉)
Department of Education, University of Helsinki, Helsinki, Finland
e-mail: anna.uitto@helsinki.fi

F. Skarstein, L.-A. Wolff (eds.), *Sustainability and Science Education*, Environmental Challenges and Solutions,
https://doi.org/10.1007/978-3-032-24408-6_12

12.1 Introduction

To promote sustainability transformation through education, empowerment through knowledge, skills and values is needed. The most important goal for sustainability education is to promote learners' competence to act for sustainable development (e.g. Redman and Wiek 2021; Wiek et al. 2011). According to the European Sustainability Competence Framework (GreenComp, Bianchi et al. 2022) published by the European Commission in 2022, the aim of sustainability education is to encourage learners to embody sustainability values, support fairness and promote nature. Embracing complex systems requires competencies in systems thinking, critical thinking and problem framing to take or request action through political agency, collective action or individual initiatives. The competences of future literacy, exploratory thinking and adaptability to changes are important in envisioning a sustainable future. The GreenComp goals insist that several *sustainability competences* are considered in the curricula of all disciplines and in multi- and transdisciplinary education. To this end, there is not much research on the implementation of the GreenComp framework in school education (e.g. Vesterinen and Ratinen 2023), but according to Laherto et al. (2023), issues described in GreenComp competence areas have already been addressed in science education research and practice. However, as stated by Shephard et al. (2019), it is still challenging to define the concept of sustainability competence (SC) in education. For instance, it is not clear what the competence is as an outcome and what are the pedagogical approaches to the teaching and assessment of the competence.

In many European countries, the school curricula are based on didactical thinking related to the North-European *Bildung* tradition. According to the German educational theorist Wolfgang Klafki (2000), *Bildung* draws from humanistic and philosophical traditions and values for holistic development, self-cultivation and individualisation, critical thinking and life-long learning to actively shape one's own education. Using the ideas of the German *Bildung* tradition, Sjöström et al. (2017) and Sjöström and Eilks (2018, 2020) have defined the goals for science education, not only from the viewpoints of learning on science and technology, but also from the viewpoint of competence to face socio-scientific and sustainability issues. For their theoretical work, Sjöström and Eilks have used the concept of *scientific literacy* as defined by Roberts (2007, 2011) and Roberts and Bybee (2014), who have defined this literacy with different visions. Vision I focuses on knowledge and skills as well as on understanding of the nature of science—referring to scientifically valid knowledge and a basic understanding of validity requirements for the methods by which knowledge is produced—and on understanding that scientific knowledge is tentative in nature (Schwartz et al. 2004). Scientific competence can be practised in school, ranging from traditional teacher-led instruction to learner-centred approaches such as problem-based learning and inquiry-based science education (Pedaste et al. 2015). However, school science education should not look only inwards to science by teaching "pipeline science for future scientists", which according to Vision I, is essential mostly for those students who are heading into science-related careers.

For non-scientist citizens, it is important to have the competence to identify questions relating to science and understand their local or personal implications (Feinstein et al. 2013). In contrast to Vision I, Vision II is attuned towards *Scientific literacy* (Vision II, "science for all", see Roberts 2011; Roberts and Bybee 2014), which emphasises competence for applying scientific knowledge and skills in everyday life situations. Numerous issues are related to Vision II, for instance, human biology and health issues, such as the role of the health benefits of food quality, exercise, sleep and vaccinations, or concerns related to biodiversity loss at the personal, local and societal level. Vision II includes the consideration of *socio-scientific issues* (SSI), which mean ill-defined issues and dilemmas linked to science and technology (e.g. Sadler and Zeidler 2005). Thus, the goals of science education are to increase students' competence in argumentation, evaluation of information, problem-solving and decision-making (Roberts and Bybee 2014).

Sustainability decision-making and actions include commitment to ethical questions of human dignity, the intrinsic and extrinsic value of nature and human responsibility for taking care of nature and human well-being, requiring new visions for all education (Wiek et al. 2011). A literature review by Redman and Wiek (2021) defines five separate key sustainability competencies; systems thinking, anticipatory, normative, strategic and interpersonal competency. However, according to Chen and Hsu (2020), futures, values and strategies thinking competencies are key competencies for sustainability transformation.

Sjöström and Eilks (2018, 2020) have defined Vision III for science education: "science of transformation", which emphasises transformative education and learning (Mezirow 1997; Sterling 2011). According to the *Bildung*-oriented view, the goal of science education is dialogical emancipation, critical citizenship and socio-ecojustice (Sjöström and Eilks 2018). In Sjöström and Eilks' Vision III, science education should enhance learners' *action competence* (c.f. Jensen and Schnack (1997) as active citizens to act towards sustainable transformation through personal, communal and societal change. According to Mogensen and Schnack (2010), action competence is an *educational ideal*, referring to the German concept of *Bildung* with democracy, a participatory and holistic approach, intersubjectivity and dialogue in decision-making as some of the key ingredients:

> Strictly speaking, you cannot speak of an action competence or action competencies. Action competence is an educational philosophical ideal, an overall perspective. From this viewpoint things will look different than from a behavioural change viewpoint, for example. Through the spectacles of action competence, you may look for and ask for and measure different (key) competencies, but action competence will not be one of them. Action competence will be the lens that makes some types of knowledge, skills, qualifications, competencies, abilities, and action readiness more educationally important and valuable than others (Mogensen and Schnack 2010, p. 67)
>
> Thus, it will make no sense to evaluate action competence per se and especially not when evaluation, in a summative perspective, asks 'has this and that been learned?' However, evaluation can also have a formative character in that it promotes further learning. Evaluation will, in this perspective, particularly focus on the *learning conditions* that, at the moment, are expected to develop and qualify action competence (Mogensen and Schnack 2010, p. 68)

The excerpts above describe the main difference between the concept of action competence and multiple sustainability competencies. However, in the sustainability research literature, a plural form of "competences" in addition to "competencies" are used interchangeably (c.f. Bianchi et al. 2022; Wiek et al. 2011). For clarity reasons, in this study, only the terms "action competence" (AC) and "sustainability competence" (SC) are studied, meaning a more general competence to act for sustainable development.

Moreover, according to Jensen (2002), the general aim of *action-oriented education* is to enhance students' competence to act for sustainable development in a real-world situation. Jensen (2002) defines four dimensions for action-oriented knowledge that helps teachers and learners to ask questions and to move towards action. For instance, when students want to know more about an environmental issue, the case can be studied democratically with *why, what, where* and *how* questions to clarify the situation and an action that could be taken.

The present study addresses the concepts of action-oriented teaching and learning, sustainability competence (SC), action competence (AC) and scientific literacy in science education. Figure 12.1 presents the relationships between the Visions I-III. The SC concept is included in the model (Fig. 12.1), because it is widely used in sustainability education research (Redman and Wiek 2021) and educational frameworks (Bianchi et al. 2022).

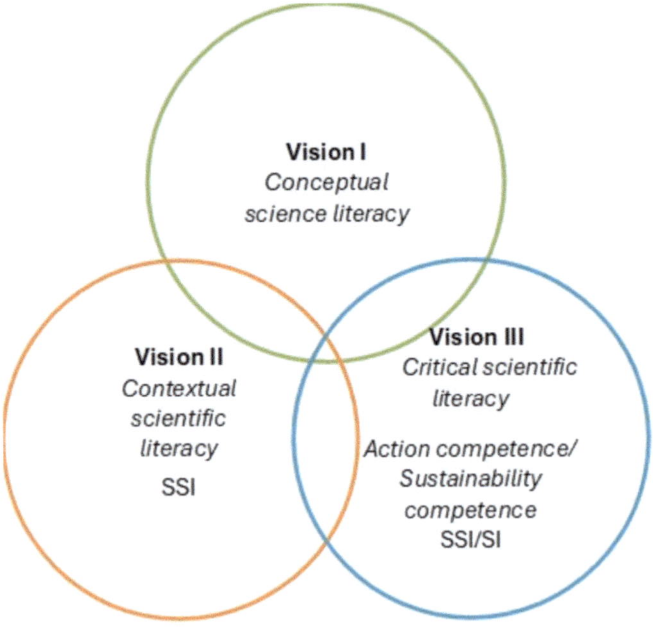

Fig. 12.1 Conceptual model of the relationship between the Visions I–III, AC and SC in this study. (Modified from Sjöström and Eilks 2018; Mogensen and Schnack 2010; Redman and Wiek 2021, Bianchi et al. 2022). SSI = socio-scientific issues, SI = sustainability issues

12.1.1 *Subject-Specific Viewpoint in Sustainability Science Education: The Case of Biology*

In general, school science education is made up of the subjects of biology, chemistry, physics and earth sciences/geology. Together with Science, Technology, Engineering, the Arts and Mathematics education (STEAM), science subjects emphasise, for instance, creativity and problem-solving skills (Perignat and Katz-Buonincontro 2019). In addition, science subjects can be combined with other school subjects in the teaching of sustainability issues. In their literature review on ESD (education for sustainable development) research, Laurie et al. (2016) found that subject-specific, multidisciplinary and interdisciplinary teaching are recommended teaching approaches. However, Tonnetti and Lentillon-Kaestner (2023) show that few interdisciplinary practices are used in secondary schools and even fewer achieve a real integration of subjects.

Because most of the teaching and learning take place during classroom subject lessons, it is important to know how the learning goals addressing sustainability issues and competence can be achieved in subject-based education (Uitto and Saloranta 2017). For instance, biology education includes teaching about dynamic living systems and offers key concepts of importance within sustainability. The biological concepts of ecology, biodiversity, species identification, ecosystems, habitats, wildlife, ecosystem services, agriculture, forestry, conservation and restoration ecology, as well as the threats posed to biodiversity such as deforestation, fragmentation, overexploitation and invasive species, are mentioned in the Learning Objectives of the Sustainability Goal "Life on Land" (SDG 15, United Nations Educational, Scientific and Cultural Organization [UNESCO] 2017). These are basic or applied concepts in biology education, especially in ecology education (c.f. Campbell et al. 2020). More specifically, as a discipline, biology has five unifying themes or "big ideas" (e.g. Dempster 2023), namely (1) biological organisation from molecules to ecosystems, (2) expression and transmission of genetic information from generation to generation, (3) transfer and transformation of energy and matter, (4) interactions between organisational systems and (5) evolution, which accounts for the unity and diversity of life (Campbell et al. 2020), which forms the main content of the biology curriculum. Thus, in biology education, sustainability is linked to scientific literacy Visions II and III, for instance in value-laden questions about the exploitation of living nature; the human-nature relationship; health issues, genetically modified food and other matters (Aivelo and Uitto 2021; Aivelo et al. 2022). The recent European Commission/EACEA/Eurydice (2024) report shows that sustainability competencies as described in the GreenComp framework are considered mostly as cross-curricular themes in science subjects and other subjects. In general, the role of subject education seems to be important for sustainability education in schools. For instance, the competence of "promoting nature" is considered particularly in the biology curriculum. There are only few cases in which sustainability education is a separate subject in European school curricula.

12.1.2 Aim of the Study and Research Questions

In general, science curricula emphasise the Vision I type of scientific literacy to explain natural phenomena, understand scientific inquiry and interpret scientific evidence (Roberts and Bybee 2014). However, socio-scientific and sustainability issues, as well as competence to act for sustainable development, are increasingly taken into account in school education, for instance, in Finland (FNAE 2016, 2020). Despite the growing number of studies on socio-scientific and sustainability issues in science education (Lederman et al. 2023), there is not much research focusing on sustainability-related competence with a subject-specific viewpoint. Thus, to find out how the competence is considered in secondary school science education, a systematic literature review was carried out in the present study to find answers to the questions:

1. What are the main trends in the research literature focusing on SC and AC in science education?
2. How are the concepts of SC and AC understood in the same research literature?

As a discipline biology provides ecological and human-related concepts for science and sustainability education. Therefore, this study pays attention to research on biology education as an example of a subject-specific approach in competence research.

12.1.3 Material and Methods

In this study, a systematic literature review was carried out using the PRISMA protocol for data acquisition (Moher et al. 2015). The Scopus, Web of Science and ERIC databases were used. The search took place in August and September 2023. The search was limited to article titles, abstracts and keywords in peer-reviewed articles and book chapters written in English. The concepts of “sustainability competence” and “action competence” were searched for, because they have been defined as core goals of ESD (education for sustainable development) or EE (environmental education) (Bianchi et al. 2022; Sjöström and Eilks 2018; Mogensen and Schnack 2010). The first search round was carried out using the terms “action competence” OR “sustainability competence” AND “science education”. In the second round, the term “biology” OR “life sciences” AND “education” was used, but it produced only four additional articles. Only articles that included the concept “school education” were studied.

The advanced search phrases were:

(TITLE-ABS-KEY (“action competence” OR “sustainability competence”) AND TITLE-ABS-KEY (science AND education)) AND (LIMIT-TO (DOCTYPE, “ar”) OR LIMIT-TO (DOCTYPE, “re”) OR LIMIT-TO (DOCTYPE, “ch”)) AND (LIMIT-TO (LANGUAGE, “English”))

Table 12.1 Coding system for data categorisation

Research type	Definition of competence	Focus of the study	Connection to competence
Theoretical Empirical Qualitative approach Quantitative approach Research methods	Sustainability competence (SC) Action competence (AC) Specified action competence (ACs)	Environmental education (EE) Education for sustainable development (ESD), or other definitions Science education (SEd) Competence Action-oriented (Ao)	Explaining the focus with keywords

(TITLE-ABS-KEY ("action competence" OR "sustainability competence") AND TITLE-ABS-KEY (biology OR "life sciences" AND education) AND (LIMIT-TO (DOCTYPE, "ar") OR LIMIT-TO (DOCTYPE, "re") OR LIMIT-TO (DOCTYPE, "ch")) AND (LIMIT-TO (LANGUAGE, "English"))

The total number of articles found was 41, and after excluding five duplicates, the number dropped to 36. However, 15 articles did not study the focus areas of the present study: school science education at grades 1 to 12, sustainability education/education for sustainable development and SC or AC. Thus, the final tally of the search was 21 research articles. The studies were categorised using content analysis, which started with a deductive analysis of the research articles: theory, methodology, definitions, focus of the study and connections to the competence (Table 12.1). The framework was used to categorise the results presented in Appendix 1. Another coder (Ph.D. university lecturer) evaluated and agreed to the logic of the coding and categorisation in addition to the author of the study.

12.2 Results

Most studies were carried out in Europe and two in Bhutan, in 1997 to 2023. More than half (62%) of the studies were published in the 2020s. In general, most studies focused on secondary school education; only the studies of Vesterinen and Ratinen (2023) and Solís-Espallargas and Morón-Monge (2020) concern primary education in Finnish and Spanish schools, respectively. The results of the categorisation of the studies are summarised in Appendix 1.

12.2.1 Theoretical Framing of AC

The concepts of environmental education (EE) and education for sustainable development (ESD) are the most used frames for the studies. In a few studies, the concepts of environmental and sustainability education (ESE) and sustainability education (SE) are mentioned. Eight of the articles are theoretical, concerning EE or ESD and their relation to science education (Jensen and Schnack 1997; Breiting and Wickenberg 2010; Bishop and Scott; Rauch 2002; Ashley; Sass et al. 2020; Gericke 2022) (Appendix 1). The articles by Jensen and Schnack (1997) and Breiting and Wickenberg (2010) describe the central idea of AC in EE as well as the development of EE and ESD in the era of rising environmentalism in Europe. These studies emphasise large differences between EE and science education. In the 1990s, school biology education was criticised because its teaching concerned mostly environmental problems and lacked the action-oriented viewpoint (Jensen and Schnack 1997; Breiting and Wickenberg 2010). The position articles by Bishop and Scott and Ashley state different views on the relationship between EE and science education, by emphasising the importance of *scientific literacy* in understanding the nature of science education in EE. For instance, Bishop and Scott argue that science has a significant role in effective EE, mostly because scientific literacy is essential in understanding environmental issues. Ashley emphasises the importance of "scientific action competence" in understanding the limits of science and that scientists are moral agents who face ethical dilemmas in their work. He also points out the role of scientific literacy by presenting the concept of *Scientific action competence for environmental education*. Rauch (2002) presents three views on AC in EE or ESD for curriculum reform.

As described in the introduction, Jensen and Schnack (1997) and Breiting and Wickenberg (2010) define the concept of action competence as an educational ideal (Appendix 1). In a more recent study, Sass et al. (2020) call the educational approach ESD, including holism, pluralism and action-orientedness in teaching and learning, and define the concept of *Action Competence in Sustainable Development* (ACiSD) to mean a generic competence in people. In psychological terms, ACiSD means "knowledgeability, willingness, capacity expectations and outcome expectancy regarding contributing to sustainability action" (Sass et al. 2020, p. 1). The study by Lampert et al. (2023) relies on the definition of Sass et al. (2020) in their theory-based conceptual framework tool that can be used in various educational institutions as an AC framework for improving education and citizen action for insect biodiversity conservation. Gericke (2022) defines *Action competence for sustainability* in the framework of subject-specific education, emphasising the European discipline-oriented educational theory of *Bildung* (cf. Klafki 2000) and the role of "powerful knowledge" (Young 2015) in teaching and learning sustainability issues.

12.2.2 Defining AC and SC in Empirical Studies

Thirteen empirical studies use the concept of AC in relation to science education (Appendix 1). Only one study uses the concept of SC (Solís-Espallargas and Morón-Monge 2020). That study focuses on primary school teachers' SC to teach climate change, and SC is defined through sustainability knowledge and skills (Knows, Knows how, Shows how and Does) at three levels of learning, considering the contextualisation of sustainability issues, sustainable use of resources and application of ethical principles.

Specific concepts for AC are reflected in the contexts of several studies (Appendix 1). For instance, Biström (2022) uses the concept *Action competence for sustainable sexuality* when studying sexuality education in secondary school biology and religion textbooks. Instead of AC, Ratinen and Linnanen (2022) use the future-oriented concept of *Anticipatory competence* and Ottander and Simon (2021) the concept *Competence for scientific literacy.* Only Ottander and Simon (2021) and Vesterinen et al. (2016) frame their studies to consider SSI. In their empirical studies, Sass et al. (2023a, b) use the definition *Action Competence in Sustainable Development (ACiSD)* defined earlier by Sass et al. (2020).

The concept *action-oriented* is mentioned in 10 studies. In the theoretical studies, the concept is defined as the philosophical educational ideal behind action-orientedness, for instance in teaching, learning, knowledge to promote AC (e.g. Bishop and Scott; Sass et al. 2020; Lampert et al. 2023, see Appendix 1). In the empirical studies, the concept is mentioned in the context of school science and EE (Mongar et al. 2023), to define ACiSD in the empirical study (Sass et al. 2023a, b), and in studies concerning the action-oriented content of school textbooks (Biström and Lundström 2021a, b).

12.2.3 Methods of Studying AC

Both qualitative and quantitative approaches are used in the empirical studies (Appendix 1). Two studies are qualitative, using interviews (Vesterinen et al. 2016; Ottander and Simon 2021) and four studies use content analysis to study secondary school textbooks (Biström and Lundström 2021a, b; Biström 2022; Mongar 2022). Two studies use both strategies (Ojala 2008; Mongar et al. 2023). Five studies are quantitative, using surveys. Solís-Espallargas and Morón-Monge (2020) use descriptive statistics in their analysis, but in the other studies, modelling is commonly used to analyse the survey data. Weber et al. (2020) use confirmatory factor analysis and apply the theory of planned behaviour, and Ratinen and Linnanen (2022) use principal component analysis to construct factors for environmental awareness and anticipatory competence. Sass et al. (2023a, b) analyse their quantitative data survey with multiple linear modelling.

12.2.4 The Focus of AC in Science Education Research

Most empirical studies consider AC in the context of school science education, but the focus is more on EE or in ESD than on subject-specific questions (Appendix 1). Two studies investigate prospective teachers' SC (Solís-Espallargas and Morón-Monge 2020; Weber et al. 2020). Other studies focus on participants' AC, perceptions, awareness, attitudes or emotions towards socio-scientific or sustainability issues. Ojala (2008) studied young people's coping with emotions to climate change. Ratinen and Linnanen (2022) studied the environmental awareness of different-aged people and varying educational background in relation to their anticipatory competence for making a sustainable future through sustainability actions, for instance, in biodiversity conservation or climate change prevention.

Ottander and Simon (2021) and Sass et al. (2023a, b) studied secondary school or university students' AC in various contexts. The study of Vesterinen et al. (2016) considers citizenship education and AC in an international science camp with participants aged 15–19 who answered to the question "How to make the world a better place". In general, the responses of the interviewees show that they felt themselves competent to take sustainability actions in their personal lives. Three types of actions were identified: personally responsible actions, participatory actions and preparations for the future. Ottander and Simon (2021) studied upper secondary school students' democratic participation and found that tasks built around ideological dilemmas can create learning opportunities about democratic participation and awareness about discourses and subject positions that inform SSI discussions. Knowledge in and about science was used in a multifaceted way in students' discussions. In the learning situation, the students were engaged to discuss the controversial topic of "Do we have enough food?" The results revealed many conceptions of sustainability and had the potential to uncover norms and values in relation to the issues discussed. Ottander and Simon (2021) state that dilemmas can lead students to ask more questions and identify their need for scientific knowledge to further their ability to evaluate and scrutinise information about socio-scientific issues, thus taking one step forward in critical thinking.

With their empirical multilevel models based on large-scale survey data, Sass et al. (2023a) predict that school students' perceived action-orientedness in ESD will facilitate students' AC, which means sustainability knowledge, willingness to act and self-efficacy in sustainability actions. In another multilevel study (Sass et al. 2023b) found that at the classroom level, teaching had impact on students' AC during a study year. Mongar et al. (2023) studied Bhutanian secondary school students' environmental actions in school and the perceptions of students, teachers and headmasters on sustainability issues on environmental science lessons. They found that all participants valued the idea of taking environmental action to solve ecological issues and promote intergenerational equity. However, the development of AC remained a minor issue in teaching and learning. In this situation Mongar et al. (2023) emphasise the professional learning and development of various stakeholders and the development of teacher education.

12.2.5 AC in Biology Education

To find out how subject education contributes to sustainability or AC research, studies focusing on biology education were also collected for the literature review. Seven studies deal with some aspect of AC in the context of biology education. In their studies on EE, Jensen and Schnack (1997) and Breiting and Wickenberg (2010) criticise the role of biology as school subject because of its "hegemony" in EE, by teaching mostly about zoology, ecology and environmental problems but lacking the AC viewpoint. Weber et al. (2020) investigated biology student teachers' inclination towards sustainable nutrition, which in turn was supposed to be important for professional AC when teaching the issue. Weber et al. (2020) found that positive attitudes, high subjective norm and high perceived behavioural control explain teachers' intention to eat sustainably.

There are four studies on the quality of textbooks used in secondary school education to promote students' AC. Biström and Lundström (2021a) studied biology and geography textbooks for sustainable development, Biström (2022) biology and religion textbooks in sustainable sexuality, and Biström and Lundström (2021b) gender equality in biology, civics, home and consumer studies textbooks. They found that AC was insufficiently addressed in the textbooks. Mongar (2022) studied the quality of secondary school environmental science textbooks, the examinations and curriculum framework to promote action-oriented teaching. He found that the textbooks and examinations lacked action-oriented contents to promote students' AC.

12.3 Discussion

The literature review showed the main trends in the research focusing on competence to act for sustainability over a period of almost three decades of science education research. The results indicated that most studies were framed with EE or ESD and there were not many studies considering sustainability issues and the competence to act for sustainability in the context of science education. For biology education, only few studies concerning AC were found. The Vision II and III types of approaches, as defined by Sjöström et al. (2017) and Sjöström and Eilks (2018, 2020), could be found in the studies; however, they lack discussion of connections between scientific competence and AC or SC. The focus of the studies remains on a general level, emphasising the goals of sustainability education. However, the science education research would also need the viewpoints of subject didactics and subject-specific powerful knowledge in sustainability education (c.f. Gericke 2022; Uitto et al. 2024; Young and Muller 2013).

A group of empirical studies address participants' awareness, perceptions or emotions around sustainability issues, for instance, on climate change and biodiversity loss. In these types of studies, AC mostly serves the general frame of the study, without a specific concern for the definition of the competence. These studies do not focus on science literacy or scientific literary as a competence, either. However, all these studies focus on participants' future-oriented perceptions on how to cope with sustainability challenges. The results suggest that different conceptual frames of science and sustainability education have influenced the research approach. In general, according to Shephard et al. (2019), it has been a challenge to define the concept of competence in sustainability education research.

12.3.1 AC and SC

The theoretical studies on AC found in the literature review are part of the debate carried out in the Nordic countries on the environmental movement, environmental education and sustainability education. Respectively, science education research focusing on the competence issues used the concept of actions competence. The results are in line with the literature review by Chen and Liu (2020), who found that the concept of AC was used in several studies concerning interdisciplinary projects and several subjects in varying levels of education. Despite the concept of sustainability competence being widely used in educational research (e.g. Corres et al. 2020; Redman and Wiek 2021), it is hardly used in science education research.

In general, the terms of SC and AC are used, implying different meanings. Mogensen and Schnack (2010) define AC as an educational philosophical ideal, referring to the North-European concept of *Bildung*, in which the core of AC is in “action” and the other features such as sustainability knowledge and values are components that can foster students’ AC. This definition differs from the theoretical definitions of SC (c.f. Wiek et al. 2011; Redman and Wiek 2021), the educational GreenComp framework (Bianchi et al. 2022) and the competences and learning objectives for achieving the Sustainable Development Goals (UNESCO 2017).

The varying definitions reflect the challenges in understanding the meaning of competence within ESD (Shephard et al. 2019). It is unclear how to define the goal of AC and how to teach, learn and evaluate the competence. Definitions are crucial, because in some studies, AC is understood as *the goal of ESD*, not only as a competency among other competencies and learning objectives (c.f. Mogensen and Schnack 2010; Sjöström and Eilks 2018).

Most of the studies in the literature review concern AC in authentic teaching and learning situations, but conclude that there are uncertainties in the outcomes. However, Ottander and Simon (2021) show that it is possible to study and evaluate the characteristics of action-oriented teaching and learning in specific classroom situations. The study is in accordance with the delineation by Mogensen and Schnack (2010) that evaluation should focus *on the learning conditions* that are expected to develop and qualify AC, instead of measuring the achievement of specific SC per se. On the other hand, Sass et al. (2020) and Sass et al. (2023a, b) position their quantitative research by using AC as *the goal of action-oriented ESD* and define their constructs by using psychological terms to study and assess AC on individual and class levels. They suggest that only the perceived action-oriented education have a positive impact on students’ self-reported ACiSD.

The results presented in the previous paragraph are partially in line with the empirical model based on the analysis of a large-scale survey on Finnish grade 9 students’ self-reported perceptions and behaviours (Uitto et al. 2015). According to that empirical model, sustainability-oriented school education enhances students’ inclination towards sustainability actions in their own life outside the school. The model predicts that participation in various ecological sustainability activities was not directly connected to students’ nature values, sustainability self-efficacy beliefs, or sustainability actions outside the school. However, if the teaching also enabled students' *sustainability agency experiences* such as being a competent actor in the school's sustainability actions, the experiences were linked to students' positive nature values and sustainability self-efficacy, which in turn predicted students' sustainability actions outside the school. In the study of Uitto et al. (2015), the agency experiences in the school can be interpreted as action-oriented learning experiences.

The present study suggests that theoretical definition and operationalisation of the concepts of action-orientedness, AC and actions are crucial for the research and educational implications of ESD/SE research. Some studies are interested in methodologies to evaluate AC as the ESD learning outcome. However, according to the theoretical *Bildung* perspective, the idea is to evaluate the *learning conditions*, which are expected to develop and qualify the educational ideal of AC (Mogensen and Schnack 2010).

However, the basic question is: What is really happening in the school teaching and learning situations? This would direct research towards the questions of critical-constructive *Didaktik*, and how it is implemented, for instance, in the interactions of the teaching-studying-learning (TSL) process (e.g. Kansanen and Meri 1999). Although TSL is a theoretical model on interactions, it can be a tool for analysing educational situations in the school systematically. Analysis carried out with the TSL model could reveal the types of the most and least frequently occurring didactical interactions in teaching and learning the issue of sustainability and AC. Recently, the model has been applied when studying, for instance, subject didactics in research literature on science education (Kinnunen et al. 2016; Lampiselkä et al. 2019).

12.3.2 Subject-Specific AC: The Case of Biology Education

The present study found that in biology education, AC has been used only when analysing textbook content. The results indicate that subject-specific education has not been the focus of AC research. However, it is known that science education provides conceptual tools to consider and value sustainability issues (Laherto et al. 2023; Uitto et al. 2024). As the European Commission / EACEA / Eurydice (2024) report reveals, sustainability as a theme is included in existing subjects in all countries, most frequently in the natural sciences, geography, citizenship education, economics and the social sciences, history and technology. The role of biology education in ESD is remarkable because it provides concepts for ESD (ecology, biodiversity, ecosystem services; the human body, health, well-being, etc.) that are crucial in systems thinking (Assaraf and Orion; Gilissen et al. 2020). Systematic, cumulative and transferable knowledge (Young and Muller 2013) provides well-defined concepts to deal with large conceptual systems that require systems thinking competence. The concern of subject-specificity is important, because each subject has its unique contribution to sustainability teaching and learning (Borg et al. 2012; Uitto and Saloranta 2017).

The key themes for biology education are defined, for instance, by Campbell et al. (2020). Hence, the "educational path" in biology lessons towards AC can use the ideas of the scientific literature's Visions I-III by Sjöström and Eilks (2018, 2020). Teaching and learning, for instance, on biodiversity and ecosystem services can start with a discussion of students' experiences in forest environments and SSI connected to forestry and forest protection (Visions II–III) and continue with learning subject-specific knowledge of forest ecosystem and practice to investigate the ecosystem (Vision I, Sjöström and Eilks 2018) and discuss meaning-making on biodiversity loss and protection and the controversial concept of forest ecosystem services. The learning can be action-oriented (scientific literacy in Vision III, Sjöström and Eilks 2020) by practising argumentation and decision-making skills when pondering, for instance, well-grounded and constructive actions to protect or utilise local forest biodiversity (c.f. Uitto et al. 2024). The learning can include out-of-school activities, such as forest ecosystem enquiry. Participation in sustainability projects by collaborating with informal learning organisers can facilitate students' learning across boundaries between school and out-of-school contexts as recommended by Bronkhorst and Akkerman (2016) and Mogensen and Schnack (2010). In the assessment, students' scientific and AC, such as argumentation skills can be assessed, but the evaluation of teaching methods and learning experience is also important.

12.4 Conclusions

The results of the present study show that to this end there are few empirical studies concerning AC for sustainable development in science education, and particularly on specific science subjects, such as biology. Scientific literacy (c.f. Vision II, Sjöström and Eilks 2018) is the focus of only two empirical studies concerning AC and socio-scientific issues (Vesterinen et al. 2016; Ottander and Simon 2021). The situation may reflect difficulties in crossing epistemological boundaries between teaching and learning practices and research on ESD and science education. Multidisciplinary and interdisciplinary projects with other school subjects are important, but they need subject-specific key concepts and competence to build knowledge on sustainability issues and act for sustainability. By using subject-specific sustainability key concepts, teachers have the possibility to link the "pipeline science" of Vision I with learning on socio-scientific and sustainability issues in their teaching. Thus, to develop action-oriented science education, more research concerning the links between scientific literacy and action competence with the *Bildung* orientation (e.g. Sjöström and Eilks 2018, 2020) is needed in subject-specific science education.

According to the results of the literature review, it can be suggested that AC as a main goal of ESD is a valid theoretical definition to study authentic teaching and learning situations in the school. As stated by Bianchi et al. (2022), GreenComp provides an agreed definition of what SC entails and how it can be used to develop and evaluate curricula at different levels of education. For instance, the objective of the Finnish secondary school biology curriculum, "The students get the skills to influence and participate in the development of their own neighbourhood and to keep it vital", is transformative and action-oriented (FNAE 2016, p. 380). The objective is related to AC, but depending on the definition, its role may vary in the research focusing on sustainability issues in science education. The results of the literature review suggest that in educational research, the meaning of AC should be more clearly defined, as well as the goal of AC and ESD (c.f. Shephard et al. 2019) and whether AC is the goal of ESD or one competency among other sustainability competencies.

Appendix 1. Results of the literature review

Authors	Research type	Definition of competence	Focus of the competence study	Connection to competence
Jensen and Schnack (1997)	Theor	AC	Definition of EE and AC, Ao	EE and SEd in school education, criticism of SEd
Breiting and Wickenberg (2010)	Theor	AC	Definition and development of EE, ESD and AC, Ao	History and policy of environmental protection, EE and ESD in school education, criticism of SEd
Bishop and Scott	Theor	AC	Relation between EE and SEd, Ao	Arguments for SEd in EE, emphasising the importance of scientific literacy in EE
Ashley	Theor	ACs, *scientific action competence of EE*	Formal and informal education, EE	Arguments for SEd in EE emphasising the importance of SL in EE
Rauch (2002)	Theor	AC	School reform, definition of EE and ESD	AC in EE and ESD frameworks
Sass et al. (2020)	Theor	ACs, *action competence in SD (ACiSD)*	ACiSD, ESD, Ao	AC as a generic psychological concept, competence of people
Lampert et al. (2023)	Theor & Meth	ACs, *Action competence for insect conservation (ACIC)*	Creating an education framework for ACiC, Ao, EE, ESE	ACIC framework to identify and develop an effective intervention approach to support insect biodiversity
Gericke (2022)	Theory	*ACs, action competence for sustainability*	Subject-specific teaching, Bildung perspective, ESD	The feasibility of future 1, 2 and 3 scenarios in subject-specific teaching in ESD

Authors	Research type	Definition of competence	Focus of the competence study	Connection to competence
Ojala (2008)	Emp Quant Qual Survey	AC	Young people's sustainability emotions, worry, hope, coping strategies, ESD	Coping with emotions related to climate change in everyday life to try to contribute to the solution of the problem
Vesterinen et al. (2016)	Emp Qual interview	AC	Young people's perceptions, science camp, citizenship science education, EE, ESD, SSI	Students' actions to make the world a better place, students see citizenship as a process of continuous self-improvement
Solís-Espallargas and Morón-Monge (2020)	Emp Quant survey	SC	Teacher education, self-perception of the participants, ESD	Teacher training in promoting prospective primary school teachers' sustainability competences in ESD
Weber et al. (2020)	Emp Quant survey	ACs *professional action competence* & (SC)	Teacher education, attitudes, norms, perceived behavioural control, concerns; ESD	Factors behind pre-service biology teachers' intention to eat sustainably
Biström and Lundström (2021a)	Emp Qual thematic textbook analysis	AC	Lower secondary school biology and geography textbooks, Ao, ESD, SE	Quality of textbooks in promoting students' AC for SD
Biström (2022)	EmpQual thematic Textbook analysis	ACs, *action competence for sustainable sexuality*	Lower secondary school biology and religion education textbooks, ESD, SE	Quality of textbooks in promoting students' AC in sustainable sexuality
Biström and Lundström (2021b)	Emp Qual thematic textbook analysis	AC	Lower secondary school biology, civics, home and consumer studies textbooks, Ao, ESD	Quality of textbooks in promoting gender equality
Ottander and Simon (2021)	Emp SSI Qual interview	*Competence for scientific literacy*	Secondary school students, SSI, SL, interpretative repertoires, ESE	Learning of democratic participation, argumentation, SL and SSI in Sed
Mongar (2022)	Emp Qual content analysis	AC	Secondary school environmental science text books, examinations and curriculum framework, Ao, ESD	Textbooks and examinations lack action-oriented contents in promoting AC

Authors	Research type	Definition of competence	Focus of the competence study	Connection to competence
Ratinen and Linnanen (2022)	Emp Quant Web survey	ACs *anticipatory competence*	Environmental awareness, anticipatory competence, climate change, biodiversity, ESD, SE	Connections between environmental awareness and anticipatory competence
Mongar et al. (2023)	Emp Quant Qual	AC	Secondary school students, teachers, school leaders' perception on AC, Ao, EE, ESD	Action-oriented teaching in environmental science, teachers' and stakeholders' professional learning, development of education
Sass et al. (2023a)	Emp Quant survey multilevel analysis	ACs, *action competence in SD (ACiSD)*	Factors behind secondary school students' self-reported ACiSD, Ao EfS, ESD	Students' ACiSD and action-oriented teaching and learning, instructional practices, effectiveness of ESD
Sass et al. (2023b)	Emp Quant survey multilevel analysis	ACs, *action competence in SD (ACiSD)*	Factors behind secondary school students' learning on ACiSD, Ao ESD	ACiSD in students' sustainable behaviour, class-group differences

Abbreviations: Theor: theoretical study or position paper, *Meth:* methodology, *Emp:* empirical study, *Quant:* quantitative or *Qual:* qualitative research method, *AC:* action competence, *ACs:* specified action competence, *Ao:* action-orientedness, *SC:* sustainability competence, *EE:* environmental education, *ESE:* environmental and sustainability education, *ESD:* education for sustainable development, *SEd:* science education, *SL:* scientific literacy, *SSI:* socio-scientific issues, *SD:* sustainable development

References

Aivelo T, Uitto A (2021) Factors explaining students' attitudes towards learning genetics and belief in genetic determinism. Int J Sci Educ 43(9):1408–1425. https://doi.org/10.1080/09500693.2021.1917789

Aivelo T, Neffling E-M, Karala M (2022) Representation for whom? Transformation of sex/gender discussion from stereotypes to silence in Finnish biology textbooks from 20th to 21th century. J Biol Educ:1–15. https://doi.org/10.1080/00219266.2022.2047099

Bianchi G, Pisiotis U, Cabrera Giraldez M (2022) *GreenComp* – the European sustainability competence framework. In: Puni Y, Bacigalupo M (eds) EUR 30955 EN. Publications Office of the European Union. https://doi.org/10.2760/13286

Borg C, Gericke N, Höglund H-O, Bergman E (2012) The barriers encountered by teachers implementing education for sustainable development: discipline bound differences and teaching traditions. Res Sci Technol Educ 30(2):185–207. https://doi.org/10.1080/02635143.2012.699891

Biström B, Lundström R (2021a) Textbooks and action competence for sustainable development: an analysis of Swedish lower secondary level textbooks in geography and biology. Environ Educ Res *27*(2):279–294. https://doi.org/10.1080/13504622.2013.833584
Biström B, Lundström R (2021b) Action competence for gender equality as sustainable development: analyzing Swedish lower secondary level. Textbooks in biology, civics, and home and consumer studies. Comp Educ Rev *65*(3):513–533. https://doi.org/10.1086/714607
Biström E (2022) Action competence for sustainable sexuality: an analysis of Swedish lower secondary level textbooks in biology and religious education. Sex Educ *22*(5):538–551. https://doi.org/10.1080/14681811.2021.1966408
Breiting S, Wickenberg P (2010) The progressive development OF environmental education in Sweden and Denmark. Environ Educ Res *16*(1):9–37. https://doi.org/10.1080/13504620903533221
Bronkhorst LH, Akkerman SF (2016) At the boundary of school: continuity and discontinuity in learning across contexts. Educ Res Rev 19:18–35. https://doi.org/10.1016/j.edurev.2016.04.001
Campbell N, Urry L, Cain M, Wasserman S, Minorsky P, Reece J, Orr R (2020) Biology and its themes. In: Biology: a global approach, global edition. Pearson Education
Chen S-Y, Liu S-Y (2020) Developing students' action competence for a sustainable future: a review of educational research. Sustainability 12(4):1374. https://doi.org/10.3390/su12041374
Chen KH, Hsu LP (2020) Visioning the future: evaluating learning outcomes and impacts of futures-oriented education. J Futures Stud 24(4):103–116. https://doi.org/10.6531/JFS.202006_24(4).0011
Corres A, Rieckmann M, Espasa A, Ruiz-Mallén I (2020) Educator competences in sustainability education: a systematic review of frameworks. Sustainability 12(23):9858. https://doi.org/10.3390/su12239858
Dempster ER (2023) What is 'powerful knowledge' in school biology? J Biol Educ 57(2):245–247. https://doi.org/10.1080/00219266.2023.2190269
European Commission / EACEA / Eurydice (2024) Learning for sustainability in Europe: building competences and supporting teachers and schools. Eurydice report Publications Office of the European Union. https://data.europa.eu/doi/10.2797/81397
Feinstein NW, Allen S, Jenkins E (2013) Outside the pipeline: reimagining science education for nonscientists. Science 340(6130):314–317. https://doi.org/10.1126/science.1230855
FNAE (Finnish National Agency for Education) (2016) National core curriculum for basic education 2014. Finnish National Agency for Education
FNAE (Finnish National Agency for Education) (2020) National core curriculum for general upper secondary education 2019. Finnish National Agency for Education
Gericke N (2022) Teaching for the Anthropocene: Bildung-oriented education for sustainable development in a subject-specific curriculum. In: Krogh E, Qvartrup A, Ting Graf S (eds) Bildung, knowledge, and global challenges in education: Didaktik and curriculum in the Anthropocene era, 1st edn. Routledge, pp 49–66. https://doi.org/10.4324/9781003279365
Gilissen MGR, Knippels M-CPJ, van Joolingen WR (2020) Bringing systems thinking into the classroom. Int J Sci Educ 42(8):1253–1280. https://doi.org/10.1080/09500693.2020.1755741
Jensen BB, Schnack K (1997) The action competence approach in environmental education. Environ Educ Res 3(2):163–178. https://doi.org/10.1080/1350462970030205
Jensen BB (2002) Knowledge, action and proenvironmental behaviour. Environ Educ Res vol 8(3), pp 325–334. https://doi.org/10.1080/13504620600943053
Kansanen P, Meri M (1999) The didactic relation in the teaching-studying-learning process. In B. Hudson, F. Buchberger, P. Kansanen & H. Steel, (Eds.), Didaktik/Fachdidaktik as science (-s) of the teaching profession? Thematic network on teacher education in Europe. TNTEE Publicat 2(1), 107–116
Kinnunen P, Lampiselkä J, Meisalo V, Malmi L (2016) Research on teaching and learning in physics and chemistry in NorDiNa papers. Nordic Stud Sci Educ 12(1):3–20. https://doi.org/10.5617/nordina.907

Klafki W (2000) Didaktik analysis as the core of preparation of instruction. In: Westbury I, Hopmann S, Riquarts K (eds) Teaching as a reflective practice: the German didaktik tradition. Routledge, pp 139–159. https://doi.org/10.4324/9780203357781

Laherto A, Rasa T, Miani L, Levrini O, Erduran S, S. (2023) Future-oriented science education building sustainability competences: an approach to the European GreenComp framework. Springer EBooks, pp 83–105. https://doi.org/10.1007/978-3-031-37391-6_5

Lampert P, Goulson D, Olsson D, Piccolo J, Gericke N (2023) Sustaining insect biodiversity through action competence: an educational framework for transformational change. Biol Conserv *283*(2):110,094. https://doi.org/10.1016/j.biocon.2023.110094

Lampiselkä J, Kaasinen A, Kinnunen P, Malmi L (2019) Didactic focus areas in science education research. Educ Sci 9(4):294. https://doi.org/10.3390/educsci9040294

Laurie R, Nonoyama-Tarumi Y, McKeown R, Hopkins C (2016) Contributions of education for sustainable development (ESD) to quality education: a synthesis of research. J Educ Sustain Dev 10(2):226–242. https://doi.org/10.1177/0973408216661442

Lederman NG, Zeidler DL, Lederman JS (2023) Handbook of research on science education. Routledge. https://doi.org/10.4324/9780367855758

Mezirow J (1997) Transformative learning: theory to practice. In New directions for adult and continuing education 74:5–12

Mogensen M, Schnack K (2010) The action competence approach and the 'new' discourses of education for sustainable development, competence and quality criteria. Environ Educ Res 16(1):59–74. https://doi.org/10.1080/13504620903504032

Moher D, Liberati A, Tetzlaff J, & Altman DG (2009) Preferred reporting items for systematic reviews and meta-analyses: The PRISMA statement. *PLoS Medicine 6*(7), e1000097. https://doi.org/10.1371/journal.pmed.1000097

Moher D, Shamseer L, Clarke M et al (2015) Preferred reporting items for systematic review and meta-analysis protocols (PRISMA-P) 2015 statement. Syst Rev 4:1. https://doi.org/10.1186/2046-4053-4-1

Mongar K (2022) Alignment of the environmental science textbooks, examinations, and curriculum framework to achieve the teaching objectives. J Turkish Sci Educ *19*(1):52–70. https://doi.org/10.36681/tused.2022.109

Mongar K, Quinn F, Elliott S (2023) Action-oriented approaches to teaching environmental science in Bhutanese secondary schools: Stakeholder perceptions. J Environ Educ *54*(2):132–147. https://doi.org/10.1080/00958964.2022.2157368

Ojala M (2008) Regulating worry, promoting hope: how do children, adolescents, and young adults cope with climate change? Int J Environ Sci Educ *3*(3):537–561

Ottander K, Simon S (2021) Learning democratic participation? Meaning-making in discussion of socioscientific issues in science education. Int J Sci Educ *43*(12):1895–1925. https://doi.org/10.1080/09500693.2021.1946200

Pedaste M, Mäeots M, Siiman L, de Jong T, van Riesen S, Kamp E, Manoli C, Zacharia Z, Tsourlidaki E (2015) Phases of inquiry-based learning: definitions and the inquiry cycle. Educ Res Rev 14(14):47–61. https://doi.org/10.1016/j.edurev.2015.02.003

Perignat E, Katz-Buonincontro J (2019) STEAM in practice and research: an integrative literature review. Think Skills Creat 31:31–43. https://doi.org/10.1016/j.tsc.2018.10.002

Rauch F (2002) The potential of education for sustainable development for reform in schools. Environ Educ Res *8*(1):43–51. https://doi.org/10.1080/13504620120109646

Ratinen I, Linnanen L (2022) The connection of Finns' environmental awareness to their anticipatory competence. Front Sci Educ *7*(838):005. https://doi.org/10.3389/feduc.2022.838005

Redman A, Wiek A (2021) Competencies for advancing transformations towards sustainability. Front Educ 6:785163. https://doi.org/10.3389/feduc.2021.785163

Roberts DA (2007) Scientific literacy/science literacy. In: Abell SK, Lederman NG (eds) Handbook of research on science education. Lawrence Erlbaum Associates, pp 729–780

Roberts DA (2011) Competing visions of scientific literacy: the influence of a science curriculum policy image. In: Linder C, Östman L, Roberts DA, Wickman P-O, Erickson G, MacKinnon A (eds) Exploring the landscape of scientific literacy. Routledge, pp 11–27

Roberts DA, Bybee RW (2014) Scientific literacy, science literacy, and science education. In: Lederman NG, Abel SK (eds) Handbook of research on science education. Routledge, pp 545–558

Sadler TD, Zeidler DL (2005) Patterns of informal reasoning in the context of socioscientific decision making. J Res Sci Teach 42(1):112–138. https://doi.org/10.1002/tea.20042

Sass W, Boeve-de Pauw J, Olsson D, Gericke N, De Maeyer S, Van Petegem P (2020) Redefining action competence: the case of sustainable development. J Environ Educ *51*(4):292–305

Sass W, De Maeyer S, Boeve-de Pauw J, Van Petegem P (2023a) Effectiveness of education for sustainability: the importance of an action-oriented approach. Environ Educ Res. https://doi.org/10.1080/13504622.2023.2229543

Sass W, De Maeyer S, Boeve-de Pauw J, Van Petegem P (2023b) Honing action competence in sustainable development: what happens in classrooms matters. Environ Develop Sustainab *25*:3649–3670. https://doi.org/10.1007/s10668-022-02195-9

Schwartz RS, Ledermann NG, Crawford BA (2004) Developing views of nature of science in an authentic context: an explicit approach to bridging the gap between nature of science and scientific inquiry. Sci Educ 88(4):610–645. https://doi.org/10.1002/sce.10128

Shephard K, Rieckmann M, Barth M (2019) Seeking sustainability competence and capability in the ESD and HESD literature: an international philosophical hermeneutic analysis. Environ Educ Res 25(4):532–547. https://doi.org/10.1080/13504622.2018.1490947

Sjöström J, Eilks I (2018) Reconsidering different visions of scientific literacy and science education based on the concept of Bildung. In: Dori YJ, Mevarech ZR, Baker DR (eds) Cognition, metacognition, and culture in STEM education. Innovations in science education and technology, vol 24. Springer, pp 65–88. https://doi.org/10.1007/978-3-319-66659-4_4

Sjöström J, Eilks I (2020) The Bildung theory: from von Humboldt to Klafki and beyond. In: Akpan B, Kennedy TJ (eds) Science education in theory and practice. Springer texts in education. Springer, pp 55–67. https://doi.org/10.1007/978-3-030-43620-9_5

Sjöström J, Frerichs N, Zuin V, Eilks I (2017) Use of the concept of Bildung in the international science education literature, its potential, and implications for teaching and learning. Stud Sci Educ 53(2):165–192. https://doi.org/10.1080/03057267.2017.1384649

Solís-Espallargas C, Morón-Monge H (2020) How to improve sustainability competences of teacher training? Inquiring the prior knowledge on climate change in primary school students. Sustainability *12*:6486. https://doi.org/10.3390/su12166486

Sterling S (2011) Transformative learning and sustainability: sketching the conceptual ground. Learn Teach High Educ 5:17–33

Tonnetti B, Lentillon-Kaestner V (2023) Teaching interdisciplinarity in secondary school: a systematic review. Cogent Educ *10*:2216038. https://doi.org/10.1080/2331186X.2023.221603

Uitto A, Saloranta S (2017) Subject teachers as educators for sustainability: a survey study. Educ Sci *7*(1):8. https://doi.org/10.3390/educsci7010008

Uitto A, Boeve-de Pauw J, Saloranta S (2015) Participatory school experiences as facilitators for adolescents' ecological behavior. J Environ Psychol 43:55–65. https://doi.org/10.1016/j.jenvp.2015.05.007

Uitto A, Jeronen E, Yli-Panula E (eds) (2024) Biologian opetus ja tiedekasvatus kestävän tulevaisuuden edistäjinä (Biology and science education for a sustainable future). Aineidaktisia tutkimuksia 27. Publications of the Finnish Research Association for Subject Didactics DOI: https://doi.org/10.23988/sats.1066

UNESCO (United Nations Educational, Scientific and Cultural Organization) (2017) Education for Sustainable Development Goals: Learning objectives. https://unesdoc.unesco.org/ark:/48223/pf0000247444

Vesterinen V-M, Tolppanen S, Aksela M (2016) Toward citizenship science education: what students do to make the world a better place? Int J Sci Educ *38*(1):30–50. https://doi.org/10.1080/09500693.2015.1125035

Vesterinen M, Ratinen I (2023) Sustainability competences in primary school education: a systematic literature review. Environ Educ Res 30(1):56–67. https://doi.org/10.1080/13504622.2023.2170984

Weber A, Büssing AG, Jarzyna R, Fiebelkorn F (2020) Do German student biology teachers intend to eat sustainably? Extending the theory of planned behavior with nature relatedness and environmental concern. Sustainability *12*(12):4909. https://doi.org/10.3390/su12124909

Wiek A, Withycombe L, Redman CL (2011) Key competencies in sustainability: a reference framework for academic program development. Sustain Sci 6(2):203–218. https://doi.org/10.1007/s11625-011-0132-6

Young M, Muller J (2013) On the powers of powerful knowledge. Rev Educ 1(3):229–250. https://doi.org/10.1002/rev3.3017

Young M (2015) Powerful knowledge as curriculum principle. In Young M, Lambert D, Roberts CR, Roberts MD (Eds), Knowledge and the future school: curriculum and social justice (2nd ed, pp 65–88). Bloomsbury Academic

Anna Uitto is an Emerita Professor in biology education at the University of Helsinki, Finland. She also holds a title of docent in hydrobiology in UH, which means an academic expertise above the doctoral level. In education, her research interests include the nature of science, conceptual understanding, curriculum studies, inquiry-based science education, out-of-school learning, and sustainability education. Also, she is interested in educational challenges the schools meet and solutions they produce when combining sustainability issues with subject-specific school science education. Besides publishing numerous research articles, she has edited publications on biology and sustainability education. She has supervised doctoral theses and published research-based university level learning materials for teacher education. She has led or collaborated on several research projects, both nationally and internationally.

Chapter 13
Science and Health Education Through Hands-On Learning: Bacteria Cultivation in Collaboration with a Scientist

Lili-Ann Wolff, Sari Vuorenpää, and Pia Sjöblom

Abstract This chapter presents a two-step science education study, on the topic of bacteria cultivation and hygiene. Among the many purposes of science education is to develop an understanding of health-related issues, of which hygiene is a central theme. Hygiene had been in focus for years in countries with poor water and sanitary systems, when it ended up in the spotlight around the whole world because of the outbreak of the COVID-19 pandemic in late 2019. Before the pandemic, in early 2018, we observed a Swedish fifth-grade science classroom when focusing on health in collaboration with a scientist specialised in the studies of bacteria (biochemist). The students' immediate engagement in the bacteria topic was obvious, and it was apparent that the biochemist played a relevant role. Therefore, we decided to interview the teacher and a few of the students at the end of 2021 to find out how they reflected on the bacteria cultivation experience four years later. The aim of the study was to investigate the meaningfulness for the students and the teacher involved in a school project on bacteria cultivation in collaboration with a visiting scientist. The results showed that, according to both the teacher and the students, the project had influenced their thoughts and actions. They stated that the learning had been important and enjoyable, and they especially remembered the useful help and inspiration they received through the biochemist, who opened the hands-on part of the project.

Keywords Collaboration with scientists · Health education · Health literacy · Science education · Scientific literacy · Sustainability education

L.-A. Wolff (✉)
Faculty of Educational Sciences, University of Helsinki, Helsinki, Finland
e-mail: lili-ann.wolff@helsinki.fi

S. Vuorenpää
Department of Special Education, Stockholm University, Stockholm, Sweden
e-mail: sari.vuorenpaa@specped.su.se

P. Sjöblom
Faculty of Human and Social Sciences, Åbo Akademi University, Vaasa, Finland
e-mail: pia.sjoblom@abo.fi

F. Skarstein, L.-A. Wolff (eds.), *Sustainability and Science Education*, Environmental Challenges and Solutions,
https://doi.org/10.1007/978-3-032-24408-6_13

13.1 Introduction

Health and well-being are essential for sustainable development, and health is the third of Agenda 2030's sustainable development goals (UN 2015). As with the other sustainable development goals, schools are important settings for promoting health (Mannix-McNamara and Simovska 2015). A few years after Agenda 2030 had been agreed on and published, the world faced the COVID-19 pandemic, which caused a global health crisis, and especially called for purposeful health education (Mody and Gontjes 2023). Two overlapping concepts that relate to health education at school and other contexts are *scientific literacy* (also called science literacy) and *health literacy*. Scientific knowledge and thinking are parts of the democratic decision-making of contemporary societies.

"Developing science literacy means learning how to communicate and interpret science through texts and spoken words. Science literacy includes actions and dialogue that allow individuals to synthesise and defend ideas, formulate arguments, and reframe concepts to accommodate new information" (Croce and Firestone 2020, p. 1).

Health literacy is strongly related to scientific literacy, since scientifically relevant knowledge is a base for understanding health. In addition, health is multidimensional, and context related, and is linked with economic, social, and other complex daily life factors. Based on a review focusing on the many definitions of health literacy, Sørensen (2019) finds common elements in the various interpretations of health literacy. They are that health literacy aims for the development of lifelong capacities to search for, understand, employ, and evaluate knowledge that could improve people's health and, thus, their quality of life. Sørensen also emphasises that health literacy is not only self-centred, but also aims at interaction and participation to promote health in society.

The role of *health education* relates to geographical, political, economic (Nutbeam 2019; Sharma 2021), and other circumstances. In low- and middle-income countries, the lack of clean drinking water and appropriate sanitation, and the capacity to maintain adequate hygiene levels have a considerably negative impact on children's health (Yamauchi et al. 2022). Knowledge about hygiene is not self-evident in richer countries either, and definitively not health in its broader sense (Shah et al. 2019). The COVID-19 pandemic showed that this is also the case in Europe (Paakkari and Okan 2020). In addition, the severity of the pandemic showed that health literacy relates to social responsibility and solidarity (see also, Resnik 2007). Therefore, students everywhere need to learn how bacteria flourish, the benefits that arise from adapting appropriate hand hygienic practice, and how this knowledge will promote health (Song 2014). They also need to learn how health is linked to global issues like sustainability (Mannix-McNamara and Simovska 2015). Hygiene and health literacy are about knowing why and how to act, but also about acting wisely and taking responsibility by caring about others both near and far (see Paakkari and Okan 2020).

Health education is an often-neglected part of science education, and therefore, it is important to develop and spread knowledge of best practices (Zeyer 2012; Zeyer and Dillon 2012). With this chapter, we strive to do so by describing a science education project taking place in a Swedish school class. The students experimented with bacteria cultivation in the classroom as a part of a sequence focusing on the topic "body and health".

Like in many other countries, the Swedish curriculum for compulsory school (grades 0 to 9; ages 6 to 16) includes science (biology, physics, and chemistry; see Swedish National Agency for Education 2018). Health education,[1] including hygiene, belongs to the key science content "body and health", which trains the students to apply biological knowledge, communicate, and take a stand in questions related to health. In addition, the students receive training in systematic biological investigations, to use biological concepts, models, and theories and to describe the relationship between the human body, nature, and society (Swedish National Agency for Education 2018). According to the curriculum, Swedish students in grades four to six learn about common diseases, and how to avoid and treat them, whereas the curriculum for grades seven to nine specifically mentions study content about viruses, bacteria, infections, and the spread of infections (Swedish National Agency for Education 2018).

As educational researchers, two of us had the opportunity to observe a grade five class for two weeks in early 2018, when their studies focused on science, but also to meet a few of the students and their teacher for a retrospective focus group discussion in late 2021, almost four years later. The intention with the observation section was to have a cross-disciplinary view of what took place in the classroom, since as researchers, we represented both the language and science perspectives on teaching. During the observation period, an overarching teaching topic happened to be health and bacteria cultivation involving a biochemist. Due to the pandemic, this topic became particularly relevant afterwards. Therefore, we decided to arrange a second sub-study four years later, to follow up the outcome of the first one.

The aim of this two-step study was to investigate the meaningfulness for the students and the teacher involved in a school project on bacteria cultivation in collaboration with a visiting scientist. The two research questions were:

RQ1: What created meaning during the lesson with the visiting biochemist?
RQ2: How did a group of students and their teacher reflect on the meaning of the bacteria cultivation project when they met four years later?

The theoretical approach is hermeneutics, and the empirical investigations are influenced by ethnography, hermeneutics, and dialogism. The interactions are in focus, especially how the biochemist interacted with the teacher and the students.

[1] A part of health education is also included in the subject 'physical education and health'.

13.2 Literature Review

As mentioned in the introduction, health is a part of the science content in the Swedish curriculum for grades four to six, but not especially microbiology. However, there are many opportunities for younger students to explore and learn about bacteria and viruses in informal contexts, such as social and traditional media, and at home from their parents. In this section, we summarise previous research on students' conceptions about bacteria and viruses and appropriate learning methods when teaching about them.

13.2.1 Students' Conceptions of Bacteria and Viruses

A lot of knowledge about bacteria and viruses[2] among students is based on informal educational experiences, for example, from families. This may be a problem, since these experiences are not necessarily scientifically grounded. Therefore, they often result in conceptions which are resistant to conceptual change and can thereby be obstacles to learning (Carvalho et al. 2004). The COVID-19 pandemic remarkably increased the amount of microbiology-related information in the media. Naturally, the emphasis in this information has been on the harmful effects of the virus, rather than on the more positive role of bacteria and viruses in ecosystems and the vital role of microbes for human health (Simard 2023). The many roles that bacteria and viruses have in the ecosystem are seldom clear to young students. Due to biased information, they see them as mainly something harmful (Jones and Rua 2006). Overall, in Jones and Rua's study, an anthropocentric (human-focused) perspective of microorganisms was noticeable among both students and adults, for example, they thought that microorganisms can be mostly found at sites which are often touched by humans, such as computer keyboards, door handles, and desktops.

The respondents also had an anthropocentric view of microorganisms as creatures intending to spread diseases among humans. Anthropocentric and anthropomorphic (human-like) perspectives have also been reported in other studies (Byrne et al. 2009; Byrne 2011). Simard (2023) summarises research from the last few decades about students' conceptions of microorganisms and states that the prevailing conception is that microorganisms are dangerous, cause diseases, and live in dirty environments. Nevertheless, the positive aspects connected to microorganisms seemed to be gaining ground before the COVID-19 pandemic, but the negative discourse has been reinforced during the pandemic (Simard 2023). Simard

[2] While bacteria are classified as microorganisms (also called microbes), viruses are not. Bacteria are single-cell organisms, producing their own energy and reproducing on their own. Viruses grow and reproduce in the cells of a host. This difference is often unclear, and in some contexts, viruses are also included in microorganisms, like in some of the references we use. Microorganisms include bacteria, protozoa, algae, and fungi. We will mainly use the concept microorganism, if the author has used it non-specifically.

problematises this and argues that teaching about microorganisms after the pandemic will be challenging. In this situation, schools will play important roles in promoting scientific understanding and health literacy.

Students' conceptions about bacteria and viruses might be different now in the 2020s, due to the pandemic. For example, Rönner et al. (2023a) found that 10–12-year-old students did not include anthropomorphic or animal-like characteristics in their drawings of bacteria and viruses but draw them as cell-like. Yet, there were still many misconceptions among the students in this population regarding size, if bacteria and viruses are living or not, and about the similarity of bacteria and viruses. Rönner et al. (2023b) conclude that the COVID-19 pandemic also has influenced students' understanding of infectious diseases. In addition, they found that the 10–12-year-old students had a receiver's perspective rather than a provider's perspective on the spread of diseases. The students talked about how they are infected by ingesting or inhaling, but not what they spread themselves. Another theme that appeared in their answers was the overrepresentation of bacteria in discussions on causes of infections, even when it came to COVID-19 and colds. Progression in conceptual understanding about microorganisms is most obvious between the ages of 7 and 11 years and decreases after the age of 11 (Byrne 2011). Hence, children could learn about microorganisms earlier than the curriculum in many countries stipulates (Byrne 2011; Rönner et al. 2023a), particularly when it comes to contagions and other relevant microorganisms (Rönner et al. 2023a).

13.2.2 Teaching and Learning Science Particularly About Microorganisms

What, then, are the learning methods in science education that support the development of scientific literacy and health literacy among students, and hence, also are suitable for learning about microorganisms? Researchers have emphasised hands-on learning methods (Fitzgerald 2012; Satterthwait 2010), such as experiments (Satterthwait 2010), inquiry-based learning (Crawford 2012; Pedaste et al. 2015), and fieldwork (Winks 2018) as being crucial in promoting students' understanding and interest when learning science. These methods have also been highlighted in curricula for decades. Despite that, in many science education contexts, these methods do neither occur as frequently as they could, nor as often as students would prefer (Murphy et al. 2012; Uitto et al. 2013). Kontra et al. (2015) have further called for more physical experience in science learning, and Niebert et al. (2012) for embodied learning, which will require more active teaching approaches. Hands-on learning is also important for enhancing interest and learning in science among students of various ages (Satterthwait 2010; Jeronen et al. 2017). According to Kalemkus et al. (2021), laboratory experiments and argumentation increased fourth-grade students' metacognitive awareness and attitudes about science. Nevertheless, hands-on activities in themselves will not automatically contribute to students' interest and learning; the hands-on activities must be qualitative (Holstermann et al. 2010). To ensure that science education is qualitative and equal, Vuorenpää et al. (2020) recommend a focus on language, the use of tangible representations, and encouragement of student initiatives. Hands-on learning is tangible and needs time, and at its best, it can involve student initiative (Wolff et al. 2018).

Jeronen et al. (2017) searched for features of useful teaching methods for promoting sustainability goals in the context of biology and sustainability education, and concluded that students' activity, participation, and interactivity are features that often appeared in the literature. However, they point out that the teaching methods must always be related to the current context and subject. Therefore, it is difficult to rank the methods on a general level. Nevertheless, the features mentioned can be regarded as guidelines when planning activities for fruitful sustainability learning in the context of biology and sustainability education.

When it comes to teaching about microorganisms, bacteria cultivation is a possible hands-on activity. An intervention study comparing theoretical and practical teaching about microorganisms, in which the practical intervention included bacteria cultivation, did not reveal any significant differences in knowledge after the teaching sequences (Ruis-Gallardo and Paños 2018). However, the group that had participated in the practical work used scientific language at a higher level when discussing the theme. Furthermore, quite few studies point out the added value by using bacteria cultivation. Song (2014) used hands-on activities in the form of bacteria cultivation to teach university students about how infections are spread. The course also included writing tasks and encouraged the students' critical thinking and creativeness when employing scientific methods. In addition, the course strengthened their scientific knowledge, and trained them in collaborative learning. Mafra et al. (2014) found that practical activities are effective when young students (9–10 years old) learn about microorganisms. The activities they describe included examination of dental plaque under a microscope and observing microbial growth of dental plaque before and after teeth brushing.

When Denessen et al. (2015) studied students at various ages, they found that they were more eager to learn science if the teacher had an enthusiastic interest in the topic. Another notion these authors made was that the younger students were more eager to learn science than the older students. If the students' enthusiasm started to fade, it could even relate to the teacher's enthusiasm for the topic, according to Denessen et al. Taber (2015) also stresses the role of the teachers and sees their guidance as important from both a cognitive and an affective perspective. The teachers are the ones who both challenge and inspire the students and create learning activities that make sense to them. On the other hand, Jaber and Hammer (2016) emphasise that epistemic affect—"the feelings and emotions experienced within science, such as the excitement of having a new idea or irritation at an inconsistency" (p. 189)—is a central part of learning science through inquiry. They also mention "feelings of uncertainty, anticipation, fascination, curiosity, and frustration" (p. 213), as well as "confusion, or the anticipated joy of discovery" (p. 213). Similarly, Winks (2018) means that feelings of uncertainty and discomfort can disrupt the human–environment narratives, and in such situations the students are triggered to ask questions. Thus, the teaching is fruitful, even if it can border on being discomfort.

Hands-on activities are not the only way of learning science. Hodson (2014) distinguishes between "learning science", "learning about science", and "doing science activity". Even if hands-on *learning* is crucial in science education, *teaching* is also needed. It is impossible to implement studies of complex topics like bacteria merely by students' self-directed learning. Both before, after, and during the hands-on part, the teaching plays a big role. It is the teacher who introduces the topic, leads the activities, and encourages the students. Hands-on experiences develop in a critical and supportive learning environment (Hodson 2014). The teacher's role is also to teach about how scientific knowledge is theoretically and empirically proved and produced, and to raise discussions. Furthermore, it is the teacher who ensures that students comprise meanings (in line with the curricula) from the teaching (Hodson 2014). However, teaching is not only about providing "everyday relevance", but also about feeding "epistemic hunger", according to Taber. However, people other than professional schoolteachers can also teach in a classroom, as we will discuss in the next section.

13.2.3 Collaboration with Scientists

It has become usual to invite real or digital guests into classrooms and higher education settings. A combination of classroom teaching and hands-on fieldwork in collaboration between scientists, teachers, and students may result in improved knowledge and attitudes in science (Houseal et al. 2014). Taylor et al. (2022) confirm that these partnerships can affect student learning positively. Furthermore, their research shows that the students' attitudes about scientists were influenced positively, thanks to the partnership. Another study indicated that students gained richer learning experiences when scientists offered them the opportunity to have access to the scientific community (Fadzil et al. 2019). Students also felt that the hands-on work resulted in better procedural skills and that they could discover current topics in science through the partnership. At the same time, they became familiar with careers in the science fields.

Jones and Rua (2006) found that the elementary and middle school teachers' conceptions of microorganisms were only marginally more elaborate than their students' conceptions. Teachers may also have poor experience of how to arrange science activities (Hodson 2014). Therefore, when involving scientists in the classrooms, not only students, but also teachers learn from them.

Casper and Balgopal (2020) analysed the common characteristics which made guest speakers "memorable" for university students, in comparison with other less "memorable" guest speakers. When the speakers' lectures about socio-scientific issues were analysed from a student perspective, the guest speakers who left a positive impression had also shared personal content in their presentations. They did not start their presentations with theory, but rather with a case, or embedded the case in the theory. These guests also used emotive, engaged, and explicit storytelling. Metaphors and analogies are powerful tools in storytelling, and they are common in the language of science and science education, in which they are often used as synonyms (e.g. Aubusson et al. 2006). According to Aubusson et al., using a metaphor means that A is said *to be* B, but in an analogy, A *is like* B. Thus, analogy is about "spotting a parallel" (Fischer 2019). In biology, terms like "invasive" species, food "chains" and ecological "footprints" are common (Taylor and Dewsbury 2018). In the literature, these linguistic tools are alternately called analogies, metaphors, and models, and they are crucial in science communication and teaching. However, there is a risk that they can constrain scientific understanding and contribute to misconceptions (Taylor and Dewsbury 2018). Using representations, analogies, or the like is not always enough to enhance learning, but they are still useful tools (Srivastava and Ramadas 2012; see also Niebert et al. 2012).

Yet, in addition to saying, doing is crucial. The participation in real investigations gives teachers and students insight into how scientific research is conducted, and the researcher learns to communicate science popularly. Authentic activities can improve both content knowledge and attitudes about science among students (Houseal et al. 2014). Monroe et al. (2019) confirmed this in their meta-analysis relying on 49 studies in fields related to and overlapping with science education. They found that the most effective teaching strategies in environmental education use content, which is personally meaningful and relevant to the students, as well as teaching methods through which students are activated and engaged. In climate change education per se, interacting with scientists was a visible element of effective teaching (Monroe et al. 2019).

13.3 Material and Methods

This section starts with a description of the theoretical approach of the study, and the qualitative data collection used in Sub-studies 1 and 2. It continues with a report on data analysis. Finally, it discusses the ethical considerations of the entire study.

13.3.1 Ethnography, Hermeneutics, and Dialogism

This study is influenced by ethnography, hermeneutics, and dialogism, and the philosophical background is hermeneutics. Consequently, we do not think it is possible to make observations and objectively capture what somebody experiences. Instead, our result is subjective, and influenced by many elements.

Even if there are many variations of hermeneutics, most researchers agree on is that it is about interpretation of meaning (Bleicher 2017; Føllesdal 2001). Traditionally, hermeneutics focuses on the interpretation and understanding of texts and symbols. However, even in hermeneutics, text includes much more than written data (Nåden 2010). In hermeneutics, the concept of *meaning* is central (Føllesdal 2001). The researchers search for meaning by examining the parts and the whole in a circle-like process, which becomes an interplay between the researchers and the data (Rasmussen 2004). Even the researchers' pre-understanding needs to steadily be recalled (From and Holmgren 2000), which in our case included various prejudices and general experiences, but also our experiences and pre-understanding of teaching contexts, classroom observations, bacteria cultivation, linguistic perspectives, and science education. The hermeneutic circle altered our pre-understanding, reading the literature, teaching artefacts, observations, as well as the focus group data. When the researchers' beliefs and attitudes meet the data, the meeting promotes reflections that might change the researchers' views. Thus, the data and the researchers' views encounters in a "horizon" (Føllesdal 2001), and even "meld".

An ethnographic case study is characterised by depth rather than width. It strives to answer questions about *what* occurs and *how* it occurs, and thereafter problematises the occurrence—*why* something occurs (e.g. Yin 2003). According to a popular metaphor, ethnography is called "a way of seeing" (Wolcott 2008). Classroom ethnography refers to the investigation of activities, interaction, and discourses in various educational settings (Watson-Gegeo 1997). Participating in a specific school context for a specific time span gives the researchers opportunities to "see", and listen, and thus to obtain a deeper understanding of daily school life.

Since the language of science at school is filled with special concepts, the linguistic aspect needs attention. Yet, what is counted as language is much more than words (Nåden 2010). In Fairfield's words (Fairfield 2011, p. 1): "There is more to education than what can be spoken of within an empirical, technical, and utilitarian vocabulary, and perhaps a great deal more". While hermeneutics traditionally studies texts, and our study has many kinds of interaction, we are also inspired by dialogism in our interpretation. From a dialogical view of interaction, life is dialogic (see Linell 2011). From this view, not only the verbal communication, but also the context and all kinds of interaction become crucial parts of the analysis. To provide a deep understanding of the case, we initially draw the background including the observation procedure, and then present the data from Sub-study 1 and the focus group data from Sub-study 2.

13.3.2 Observations and Focus Group Discussion

This case study is a part of a larger ethnographically inspired research project on science and literacy conducted between 2018 and 2021 in Finland and Sweden (see Vuorenpää et al. 2020; Wolff et al. 2018). The specific case took place in a primary school in a municipality close to a large city in Sweden. At that time, the school had about 500 students from preschool class (6–7 years old) to grade five (11–12 years old).

The empirical research is based on video observations during most of the lessons in a fifth-grade classroom in two weeks in 2018, because the class dealt mostly with science during this period. When applying video observation, there are many decisions to consider, both before and during the observation, such as where cameras are placed, which camera to use, film angles, and when the camera should be running or not (Derry et al. 2010). The camera provides a lens for studying teaching and describing who does what and how, since video-recorded data provides the conditions to divide the observed teaching into smaller units. Thus, the researcher can return to the material several times during the analyses.

At the time of the observation, the 26 students in the class ran a bacteria project accompanied by a scientist who worked as a biochemist at a Swedish university. The teacher contacted the biochemist and asked if the class could borrow material for the bacteria cultivation. The biochemist answered that they do not lend that type of material without monitoring the use themselves. Instead, the biochemist was willing to participate in the project.

Two of the authors of this chapter visited the school and made observations together, while the third author took part by using the data gathered in transcribed form. This approach provided double views of the gathering of information in the school and in the focus group discussion. By the engagement of a third author, the view became more reliable. Overall, the data consist of about 30 hours of film and audios. In addition to video recordings, the data include notes, interviews with the teacher and students, and photographs of items in the classroom. These additional data were necessary to describe and understand the context of both sub-studies.

To analyse the collaboration and creation of meaning, we used various material. The recordings and the still images were captured with iPads. We also captured audio recordings of students' small group discussions when they were working independently with the bacteria cultivation tasks. For RQ 1, we used 3.5 hours to build up the background and analysed 1.5 hours of the 5-hour video and audio material about the bacteria cultivation project in detail.

The dialogue arose from both the conversation between the participants and from the artefacts they used as teaching material (the textbook extract, notebooks, laboratory equipment, etc.). The researchers' roles during this research process varied. We did not participate as teachers in the lessons, although the teacher and the students occasionally asked us questions. However, we came to know the students, and they invited us back when we left the school. Even if we did not have any plan to continue our study then, it made it easier to contact them again a few years later.

In December 2021, two years after the COVID-19 outbreak, the field researchers arranged a retrospective focus group discussion with the teacher and three of the students from the class in focus during 2018. The students were now in ninth grade, but they were now in several classes and schools. The teacher helped to connect the students from the earlier class, and they also contacted their guardians. Yet, it was difficult to reach them all. Five students said that they were willing to participate in a digital discussion, but on that occasion, only three of them turned up.

Sub-study 2 was methodically inspired by the focus groups' reference point, in which the fundamental in the informants' interaction is the mutual construction of meaning (see, Puchta and Potter 2004). Conversations in focus groups are a specific form of conversation between individuals who share common experiences and discuss a given topic (Dahlin-Ivanoff and Holmgren 2017). In the focus group, R1 and

R2 participated mainly as discussion leaders, but also as intrigued discussion partners.

The core of a focus group discussion is what the participants achieve together through their interaction. In a focus group, the participants, who are the experts in the topic in question, meet with the researchers to share common experiences. In this case, the students and the teacher were the experts, while R1 and R2 occupied the role as curious participants. This procedure transfers the prerogative, and therefore the power, from the researchers to the interlocutors (Puchta and Potter 2004). In a focus group discussion, focus means that the topic of the conversation is framed within a specific area, in this case, the earlier bacteria experiment and the collaboration with the biochemist. The recorded focus group discussion lasted one hour. The discussion built on a semi-structured interview guide, asking questions in relation to the bacteria-cultivating project, as well as how the project had affected the students' and teachers' view of hygiene afterwards, especially during the pandemic.

13.3.3 The Data Analysis

It is always challenging to try to capture the complexity from a dialogue by transcriptions, and at the same time, the transcriptions outline what to analyse (see Ochs 1979). Both the teaching and the focus group discussion took place in the Swedish language, and were transcribed close to the spoken language, using a few adaptions to the written language to make the transcriptions more readable. Even when the transcriptions were translated into English, the ambition was to maintain the spoken language character of the teaching situation, as well as of the focus group discussion.

The empirical data consist of video observations and audio recordings. The analyses are based on the following principles: what the participants *say*, *do,* and *mean* through a participant's perspective (see further in Linell 2011). Thus, the observations also include bodily expressions in addition to verbal expressions, while the focus group discussion was only analysed verbally.

13.3.4 Ethical Considerations

This study follows the ethical guidelines in Sweden (Swedish Research Council 2017), where it was reviewed. The researchers informed the participants about the study's purpose, and they voluntarily participated in the project. All empirical data have been de-identified and only the researchers have access to the original material. The guardians approved the students' involvement in 2018 in writing, and both the guardians and the students gave consent to the students' participation in the focus group discussion in 2021. The students were now 15 years old and were ninth-grade students. Even if they were old enough to decide about their research participation themselves, according to the Swedish research rules, we chose to contact the guardians again.

13.4 Results

This section first describes what happened during the lesson in 2018. Second, it gives a view of the focus group discussion in 2021, which, on the one hand, relates to the earlier-described lesson, but on the other hand, the entire bacteria cultivation project. In the following text, the biochemist is called B, the teacher T, the students S1 to S6, and the two field researchers (author 1 and 2) R1 and R2.

13.4.1 A Science Lesson in 2018

Sub-study 1 relates to RQ1: What created meaning during the lesson with the visiting biochemist? In this section, we describe the lesson with an extract of the analysis of interaction data from the video observation from the day in which the biochemist participated. Observation according to hermeneutics is an interpretation practice (From and Holmgren 2000). Therefore, we both describe and interpret the situation.

The lesson in which B participated, started with T introducing the topic. T reconnected to an earlier lesson about bad and good bacteria and viruses. T also intimated that they had not been prepared enough for the lesson the day before. After that, T distributed a textbook extract with a short instructive story about viruses and bacteria. This short and easy-to-read text dealt with the differences between bacteria and viruses, and how they spread. In addition, the text stated that bacteria are both good and bad, humans cannot live without bacteria, and a lot of vital bacteria are in the intestines. Some of the students and T read the text aloud to the class. Afterwards, T explained and asked questions in relation to the text and its illustration. T also introduced the bacteria cultivation task, and asked about where bacteria might thrive. Then the guest, B, stepped in without any extra wording about themselves, and directly continued with the topic T had started. B first asked if the students had any questions, and then B thoroughly explained basic facts about bacteria, beginning with their size, by drawing on the white board and asking questions again. B encouraged the students to try to draw a dot in their own notebooks, and the following dialogue took place between B, T and students S1 and S2:

B: have you drawn a dot?
T: do you see the dot?
B: a little bit / is it hard to see? Ok, for the bacteria you see here (points to a piece of paper) they are a hundred times smaller than (writes 100 on the board) your dot
S1: Oh, so a hundred
S2: So, bacteria like that, if you put a row with a hundred of them, it's like the dot

B: If you do / so / how can I explain / (takes out the marker pen) they are / they say that two micrometres if you put five hundred in a row in a row it becomes a millimetre / do you know how big a millimetre is?

Several students: yeah

B: So, it means that in a centimetre / how big is a centimetre? / yeah so (shows a centimetre between the fingers) that means you need 5000 in a row / so they are quite small / they are quite important to us because we have lots of them / we talked about how many / your own body / and I thought I could explain to you / how many bacteria we have in the body (blurs on the board)

The example above demonstrates how, with simple representational means, such as a dot on the board, B described and illustrated the size of a bacterium step by step. Thus, the students could understand that a dot is 100 times larger than a bacterium. By doing this, B related to the students' knowledge in mathematics. B also described how bacteria multiply and how they are structured. Like T, B also described that there are both good and bad bacteria and viruses, and that they affect human health.

After the theoretical part, B started to introduce the practical part of the project. B distributed the petri dishes that came from a laboratory and asked the students about ideas on how to catch the bacteria, and where to find beneficial bacteria and where to find harmful ones. Both students and T gave examples, like the whiteboard, the mobile phones, the furniture, the clothes, and lip balms. T talked about why it is good to wash hands and proposed that the students can do a test before and after washing hands. B explained the nutrition in the petri dishes, and every student was invited to smell a dish. This made them eager, since some found the smell bad, and others good. B also showed the pins they would use to catch the bacteria and explained how all the equipment had been in a sterilisation oven for three hours.

B demonstrated the basic laboratory equipment that included material like petri dishes with agar solution, tape, long cotton swabs, water, and felt-tip pens. After that, B chose a student for a demonstration, instructed them to put on gloves, and showed how to take a bacteria sample after watching how B did it. The student took the task seriously, but was a little unsure. B also showed how to mark the dish and put the sample in the dish. In addition, B raised the issue of hygiene related to the task, and how to close the petri dish lid to make the sample safe. B forbade the students to open the lid in any case. When B asked if anybody else wanted to try, all the students raised their hands. T asked B: "What did you expect", and all laughed.

After the theoretical instruction, the demonstration, and the handing out of equipment, the groups started their work. Each group chose two items from which to take a sample for their bacteria cultivation. The students planned to take samples from several surfaces in the classroom, the corridor, and the school canteen, and to cultivate bacteria from these samples. When writing, the students followed T's instructions, and they used these headings: experiment, material, hypothesis, result. This showed that they followed a similar inquiry-based approach as used by scientists. When the groups started planning their experiments, B was still in the room, and walked around like the teacher, helping the students and answering questions, if needed. The atmosphere was engaging and positive. Before B left, T asked how many students wanted to become biochemists, and in reply, all the students raised their hands. Their eagerness during the entire lesson could not be mistaken, and neither during the afternoon, when the students worked in groups. The observations after B's visit and the audio recording of one of the groups' work show how committed the students were to their tasks.

B's presence in the classroom was in general interactive. B described and told them about the theoretical aspects, but this happened in dialogue with T, and B also asked the students questions to see that they understood and were actively engaged. Both T and B asked the students to draw and write in their notebooks, and they both regularly referred to the textbook extract, which the students had in front of them. Beside the notebooks and the white board, this artefact united new knowledge with old, and the scientific expertise with daily school life.

Both T and B tried to flesh out what they said by storytelling, representations, gestures, use of analogies, and comparing with examples from the students' lives (anthropomorphic comparisons). For example, B used analogies when comparing the cell membrane with a jacket, said that "bacteria were eating", soon "built up a big town", and that good bacteria have bad cousins. B also often used their fingers and arms to better express the topic (gestures). The students were engaged, and followed B with their gaze during the lesson, and they also carefully followed B's instructions, after B had left.

After B's visit, and parallel with other schoolwork during the coming two weeks, the students worked with the practical bacteria cultivation topic and continuously followed up the test results. T contacted B when problems with the cultivation occurred, or questions arose.

Many important aspects emerge when interpreting the creation of meaning during the lesson involving the biochemist in the light of Sub-study 2 and in relation to what took place in the classroom before and after that lesson. Together with B, T had planned a science education section, on the topic of body and health, and allocated two weeks for the entire project, of which this lesson was an important part. Theoretical lectures in collaboration with a scientist were altered with hands-on learning activities. As a scientist, B had expertise knowledge, laboratory work experience, and brought laboratory equipment to the classroom. The education included both teacher-centred (both T and B) theoretical parts, and student-centred hands-on parts, which obviously engaged the students cognitively, bodily and emotionally. T initiated the bacteria cultivating project by placing it in a larger health content. Meaning was created through the entirety, as well as the details.

13.4.2 Retrospection in 2021

Sub-study 2 relates to RQ2: How did a group of students and their teacher reflect on the meaning of the bacteria cultivation project when they met four years later? The discussion begins by R1 and R2 asking T and the students about what they recalled from the bacteria cultivation in fifth grade. T remembered having taught about the human body and health to this group of students. Since the topic was illnesses, bacteria, and viruses,

T had contacted B and asked if they could assist with petri dishes and pipettes. The conversation between them resulted in B visiting the school in person, and started the bacteria project with the class, along with T. T remembered B bringing a large trolley with equipment, and starting the lesson by presenting facts about bacteria and explaining how they would run the experiment. S3 stepped in and summed up the reasoning regarding the swabbing with "and then we put it into some goo in the bowl". S4 followed, stating that they even swabbed mobile phone screens. The conversation then continued with what happened after the samples. S4 remembered them following the development of the bacteria. The samples had to be placed next to the classroom heaters. Then a dialogue followed about the results of the experiment. Below, we present two selected and representative parts from the focus group discussion. R1 began with a question: "What was the result?" It was silent a couple of seconds after the researcher's question, and then the following conversation sequence started:

Conversation sequence 1:

S4: I don't remember.
T: Do you remember which one was the most?
S5: The handle or the seat or something else?
T: Do you remember that we touched that thing you use to wipe of the counter with?
S4: Yeah, I remember that.
S5: Right, dishcloths!
T: That was actually kind of gross.
R1: Remember? How did that feel?
S5: I think it was interesting. I don't remember being exactly grossed out by the bacteria, it was more like, cool, to see how it developed.
R2: Do you, S4, want to say something about how you felt?
S4: Like, the same. You kind of felt like a real scientist at that moment.
R1: How did you feel about doing this, S6?
S6: Quite interesting, not like an ordinary school day. In the fifth grade you have your subjects and your schedule, and this felt a bit special, like something important. In that sense, it was a lot of fun.
R1: And other science experiments you had?
S6: This was probably the most interesting experiment I've conducted at school so far.
R1: Did you get any new ideas after this?
S6: When I get a test about it, I know more than the majority in the class. I notice the same in my two classmates who also were part of that bacteria experiment. I particularly remember this one class, I don't remember what year, but when we discussed bacteria, we had our hands up in the air all the time in class, because we knew all that, we had learned that stuff.

A bit further ahead in the discussion, R1 asked the following question:

Conversation sequence 2:

R1: Now you are ninth graders and this is the third year at lower secondary school, and you've surely performed chemistry experiments and biology experiments. Which areas are your students most committed to? When would you say that the teachers are most successful in engaging their students in various things?
S6: The topic has to either be interesting, which this was, so that everyone feels that they want to be a part of it, or that swabbing the toilet might be fun at that age, in fifth grade, to see what happens. And then I think that when a teacher starts grading, that's when the students get more committed.
R1: You weren't graded in fifth grade?
S6: No.

Further in the conversation, R1 asked the students and T about their experience in the collaboration with B and the fact that B was managing the lesson, the students unanimously answered that it was the visit in itself that made them feel more committed. It was a unique visit, and it engaged the students. They tried to squeeze out what they could, since "the scientist isn't going to come back too often" (S4).

The students expressed gratitude for what they had experienced as being highly significant. The bacteria cultivation was not a learning experience that happened every day at the school. In addition, B was specifically trained and was a specialist in this field. T emphasised that B excelled in explaining things in a way that everyone understood. "For instance, B drew a dot on the blackboard and explained how many bacteria would fit on a pin. To think that so incredibly much would fit on that tiny little dot that B drew on the whiteboard".

When R1 and R2 asked the students and the teacher if the bacteria project had influenced their way of acting, both the students and the teacher said that it had had an effect. They all said that they had learnt that many places and items in the school environment carried numerous bacteria, and therefore, they started washing their hands more carefully when they had touched items which they knew contained bacteria. The project provided the students with more knowledge about bacteria than most of the ninth-grade students had, according to one of the students. In addition, the teacher explained about having learnt a lot. The teacher remembered how surprised they all were when the result did not correspond with the expectations according to the students' hypotheses. However, they also evaluated the results and discussed the surfaces and material on which it is more likely to find bacteria.

R1 and R2 discussed with the students and teacher whether the learning project in 2018 affected how they behaved when the pandemic started, and people were supposed to be careful not to receive the COVID-19 virus infection. They explained how the bacteria project had made them think more about the need to wash hands to avoid becoming infected by unwanted microorganisms. When they had seen with their own eyes that there are many bacteria in a room, they understood that even if COVID-19 is a viral disease that is an airborne infection, viruses and other bacteria, not only stomach infections bacteria, can also spread via surfaces being touched by unclean hands. S4 argued: "We became more careful about washing our hands, yes through the experiment it became more important to use soap and to disinfect our hands, otherwise we might have been satisfied with just washing with soap". T further explained that: "yes, we compared the results of handwashing when we washed with just soap or just hand sanitiser or both". S5 added that "I always wash my hands as soon as I get home, especially now during Corona, I wash every time I change environment". The students summarised that they do not feel uncomfortable when they know more about bacteria, because knowledge is "easy to carry". R2 repeated the question that the teacher asked the class at the end of the lesson in grade five: "How many of you would be interested in becoming a biochemist? Did you think about it afterwards?" When the teacher first asked the question (2018), almost everyone in the classroom raised their hands. However, the huge enthusiasm the students had expressed for the chemistry profession after the classroom observation seemed to have cooled down, or the future prospect might have had more influences.

13.5 Discussion

The aim of this two-step study was to investigate the meaningfulness for the students and the teacher involved in a school project on bacteria cultivation in collaboration with a visiting scientist. This section discusses the results based on RQ1 and RQ2 (see 1, 4.1 and 4.2), and the relevant literature. We will discuss the answers to the questions in conjunction, and finally consider the limitations of our research.

13.5.1 The Meaningfulness of the Project

When observing the students during Sub-study 1, we noticed that they were all engaged and followed the lesson with excitement. They were also eager to start the bacteria cultivation when the scientist had left the school. In Sub-study 2, we discovered that the students used many descriptive words when they talked about the bacteria project. They found it to be authentic—they had dealt with something real and relevant to them. Simultaneously, the bacteria cultivation had been something special and beyond the ordinary schedule. The students also talked about the project as being interesting and even important. According to Stuckey et al. (2013), this could imply that the project was relevant from the students' view. It is interesting that the students thought that taking bacteria samples from a toilet might be more exciting for students in fifth grade than students later in the adolescence. The reason might be that younger students are more open to methods of this kind, and curious in a more child-like way. Nonetheless, it is in line with Denessen et al. (2015), that younger students are more enthusiastic about laboratory tasks. In Sub-study 2, one of the students also said that a project is more interesting if it is graded, which might be more the case with older students. In addition, this claim is from one student and would need more voices and more investigations to further discuss in this context.

Emotions are a crucial part of science education (Jaber and Hammer 2016; Winks 2018). During the bacteria cultivation, the learning experience became both a shared responsibility and enjoyment. Smelling a petri dish and talking about dishcloths and other items that could contain harmful bacteria was exciting and emotional. Finding the dishcloth to be "gross" seems to have triggered both the students' and the teacher's learning, since they made a joke of it.

Like the students, the teacher enjoyed participating in the project. The teacher both enjoyed watching the students' eagerness and experienced the activities as being authentic. It is obvious that both the students and the teacher found that the project dealt with an issue that was close to their daily life reality (Fitzgerald 2012). Also, other authors have emphasised that science education must be authentic (Crawford 2012; Jeronen et al. 2017; Rule 2006). Authentic learning is built on the idea that learning is efficient when the learning situation resembles relevant and real-world contexts (Crawford 2012), and includes hands-on experiences (Jeronen et al. 2017). The bacteria project can be regarded as being authentic in many ways.

The students took the samples from their own school environment, and the project related to their real life, the learning was social, they used real scientific equipment and methods and, thus, imitated the work of a scientist, and the inquiry was open ended and triggered their thinking. According to Niebert et al. (2012), thoughts are embodied, which means that basic conceptions originate in bodily experiences, and depend on perception, bodily engagement, and experiences from both the physical and social environment. The bacteria cultivation project was embodied, activating, and acknowledged the actual environment.

13.5.2 The Role of the Visiting Biochemist

B played a major role in the bacteria project. B's visit (see Sub-study 1) made an impact on the students, but also on T. Obviously. B brought expertise, but B did not act as though they were a superior expert, rather like a well-prepared co-teacher. B continued directly where T had ended teaching, and was in constant dialogue with T. This made B into a natural part of the lesson without any unnecessary interruption. B used the same textbook extract as T had done, and thus the textbook extract had a role as connector between B, T, and the students, but also between previous and present knowledge. It levelled the hierarchical difference between school and scientific knowledge. Similarly, B happened to have pedagogical talent, and managed to describe things to the students in a way that appealed to them and was easy to understand. We experienced this when we watched and listened to the students during and directly after the biochemist's lesson, but also in the retrospective focus group discussion, in which T emphasised that B managed to explain things in a way that the students seemed to understand. The students also explained how they appreciated having a scientist explaining the bacteria cultivating process. B used representations and gestures when describing the size of a bacterium. Representations have the potential to help student's mentally visualise a complex issue like cell biology (Srivastava and Ramadas 2012). B also used analogies to make the students better understand topics like what a cell membrane is. Using analogies is a linguistic method that can enhance science teaching and learning, since it promotes higher level thinking (Aubusson et al. 2006).

As Holstermann et al. (2010) argue, hands-on activities are not beneficial as such; they must be qualitative. According to Vuorenpää et al. (2020) qualitative hands-on activities focus on adequate language and tangible representations. In addition, quality needs time, and so does clarifying (Wolff et al. 2018). In the bacteria cultivation project, the communication had an important role, since there was time enough for explanations, questioning, and discussions. The class also had time for the practical hands-on activities, and many tangible representations were included. With a visiting scientist who also is a good pedagogue, who can communicate with the students and engage them in both knowledge and activities, the quality increases (Casper and Balgopal 2020).

The focus group discussion showed that the teacher learnt new things during the bacteria cultivation project. This result correlates with Houseal et al. (2014). When B was already waiting on the doorstep of the classroom and ready to teach (Sub-study 1), T admitted not having theoretically prepared the previous lesson very well. Inviting scientists into school classes can fill the scientific knowledge gaps among teachers. Then, the teachers can learn new things themselves, co-teach with the guests, and co-learn with the students (Houseal et al. 2014). However, inviting guest scientists is not a single long-term solution to the problem; the teachers should also have opportunities to improve their own content knowledge in science by, for example, participating in in-service courses and visiting research laboratories. Not all teachers invite guests, and not all scientists could suddenly fit in a school environment. Similarly, not all teachers invite scientists, but more could do so. However, it demands teachers with open and curious attitudes, and scientists who can teach about their expertise topics on a level appropriate for the actual grade.

13.5.3 Limitations and Future Studies

As already stated, there is much general information about bacteria cultivation in classrooms, but we found few academic studies. An interesting point in our study was that we had the opportunity to follow up the bacteria project almost four years later. This was because of collaboration with a teacher who trusted us as researchers and helped us to realise Sub-study 2. Without this teacher, it would have been impossible.

Like all studies, this one also has its limitations. Since there were few participants in Sub-study 2, it is impossible to draw any distinct conclusions based on merely the result. However, previous research supports our results (Fitzgerald 2012; Houseal et al. 2014; Rule 2006; Vuorenpää et al. 2020; Wolff et al. 2018).

To draw conclusions about the benefit of bacteria cultivation activities in science learning, more studies are needed. Because of the COVID-19 pandemic situation in 2021, we did not discuss the benefit of bacteria at all in Sub-study 2. This could have been another interesting topic—to hear how the teacher and the pupils thought about that issue. We could also have involved the scientist in the discussion. To continue this poorly explored topic of bacteria cultivation in the classroom, we suggest in-depth longitudinal studies with teachers and students representing a range of school grade levels.

As a practical application of our research, we recommend close collaboration between educational researchers, teachers, and scientists in the development of science education and health education. During the focus group discussion in 2021, the students clearly showed their previous teacher that they were grateful that the teacher had run the bacteria project with them and invited the scientist. Teachers receiving valuable feedback from previous students is uncommon in the hectic daily school routine. Feedback could perhaps increase the teachers' enthusiasm and have a positive influence on future students. At least it could increase the teachers' confidence, and thus have an effect similar to more knowledge and skills (Denessen et al. 2015). Interaction and collaboration between students, teachers and scientists adds positive value to science education by contributing to more authenticity and valuable experiences for students.

13.6 Conclusion

Even if the topic of goal no 3 in Agenda 2030 is health-related, there are also other goals in the Agenda. Yet, health education is an important part of sustainability education. Bacteria cultivation can be an engaging part of health education, a part that may attract both teachers and students because of its realistic nature. Even if health education is urgent for many reasons, the COVID-19 pandemic has updated the need for health education, especially with a focus on hygiene, virus infections, but also bacterial infections. COVID-19 has clearly shown that health is a joint global issue. Health promotion policy and action is urgent in every part of the world. If the method and the framework are chosen carefully, education can be inspiring, and trigger the students' interest. Our results show that the students appreciated the project very much, and one of the students said to the teacher: "So, if it is possible, do this with the others [future students], too!"

To cultivate bacteria not only demands proper equipment (Song 2014), but also knowledge and experience about how to do it. In this respect, the school in our study was well-equipped, because of the collaboration with the biochemist who offered both knowledge and provided the class with basic equipment. Without expertise, experiments of this kind could even be catastrophic, since bacteria cultivation demands caution. Treated without enough understanding, harmful bacteria (and viruses) could be transmitted into the school environment. Nevertheless, in collaboration with experts and multiple stakeholders, schools can develop their health-promoting knowledge and practice, and contribute to health literacy and sustainability among children and adults (Mannix-McNamara and Simovska 2015). Well-planned bacteria cultivation in the classroom can make a difference and become meaningful for all involved.

Well-planned science lessons have the potential to improve students' scientific literacy (and health literacy), make them familiar with possible careers in the field of science (Fadzil et al. 2019; see also Houseal et al. 2014), and affect the students practically in their daily life (see Sub-study 2). Having a guest scientist serves all these goals, but the role of a committed teacher cannot be underestimated. However, the teachers also learn from having expert guests (Houseal et al. 2014), which indicates that well-planned and well-performed science education can become meaningful in the long run for both students and teachers.

References

Aubusson PJ, Harrison AG, Ritchie SM (2006) Metaphor and analogy: serious thought in science education. In: Aubusson PJ, Harrison AG, Ritchie SM (eds) Metaphor and analogy in science education. Springer Netherlands, pp 1–9

Bleicher J (2017) Contemporary hermeneutics: hermeneutics as method, philosophy and critique, vol 2. Routledge

Byrne J (2011) Models of micro-organisms. Children's knowledge and understanding of micro-organisms from 7 to 14 years old. Int J Sci Educ 33(14):1927–1961. https://doi.org/10.1080/09500693.2010.536999

Byrne J, Grace M, Hanley P (2009) Children's anthropomorphic and anthropocentric ideas about micro-organisms. J Biol Educ 44(1):37–43. https://doi.org/10.1080/00219266.2009.9656190

Carvalho GS, Silva R, Lima N, Coquet E, Clément P (2004) Portuguese primary school children's conceptions about digestion: identification of learning obstacles. Int J Sci Educ 26(9):1111–1130. https://doi.org/10.1080/0950069042000177235

Casper AMA, Balgopal MM (2020) How guest experts tell stories about environmental socio-scientific issues in an undergraduate class. Int J Sci Educ 42(9):1568–1584. https://doi.org/10.1080/09500693.2020.1772519

Crawford BA (2012) Moving the essence of inquiry into the classroom: engaging teachers and students in authentic science. In: Tan KCD, Kim M (eds) Issues and challenges in science education research. Springer, pp 25–42. https://doi.org/10.1007/978-94-007-3980-2

Croce KA, Firestone JB (2020) Defining what it means to develop science literacy. In: Croce KA, Firestone JB (eds) Developing science literacy in the 21st century. Information Age Publishing, pp 1–4

Dahlin-Ivanoff S, Holmgren K (2017) *Fokusgrupper* [Focus groups]. Studentlitteratur

Denessen E, Vos N, Hasselman F, Louws M (2015) The relationship between primary school teacher and student attitudes towards science and technology. Educ Res Int:534690. https://doi.org/10.1155/2015/534690

Derry SJ, Pea RD, Barron B, Engle RA, Erickson F, Goldman R, Hall R, Koschmann T, Lemke JL, Sherin MG, Sherin BL (2010) Conducting video research in the learning sciences: guidance on selection, analysis, technology, and rthics. J Learn Sci 19(1):3–53. https://doi.org/10.1080/10508400903452884

Fadzil HM, Saat RM, Awang K, Adli D (2019) Students' perception of learning stem-related subjects through scientist-teacher-student partnership (STSP). J Balt Sci Educ 18(4):534–548. https://doi.org/10.33225/jbse/19.18.537

Fairfield P (2011) Introduction. In: Fairfield P (ed) Education, dialogue and hermeneutics. A & C Black, pp 1–4

Fischer O (2019) Analogy in language and linguistics. Oxford University Press

Fitzgerald A (2012) Science in primary schools: examining the practices of effective teachers. Sense Publishers

Føllesdal D (2001) Hermeneutics. Int J Psychoanal 82(2):375–379

From J, Holmgren C (2000) Hermeneutik och pedagogik. Nordisk Pedagogik 20(4):219–229

Hodson D (2014) Learning science, learning about science, doing science: different goals demand different learning methods. Int J Sci Educ 36(15):2534–2553. https://doi.org/10.1080/09500693.2014.899722

Holstermann N, Grube D, Bögeholz S (2010) Hands-on activities and their influence on students' interest. Res Sci Educ 40:743–757. https://doi.org/10.1007/s11165-009-9142-0

Houseal AK, Abd-El-Khalick F, Destefano L (2014) Impact of a student–teacher–scientist partnership on students' and teachers' content knowledge, attitudes toward science, and pedagogical practices. J Res Sci Teach 51(1):84–115. https://doi.org/10.1002/tea.21126

Jaber LZ, Hammer D (2016) Learning to feel like a scientist. Sci Educ 100(2):189–220. https://doi.org/10.1002/sce.21202

Jeronen E, Palmberg I, Yli-Panula E (2017) Teaching methods in biology education and sustainability education including outdoor education for promoting sustainability: a literature review. Educ Sci 7(1):1. https://doi.org/10.3390/educsci7010001

Jones MG, Rua MJ (2006) Conceptions of germs: expert to novice understandings of microorganisms. Electr J Res Sci Math Educ 10(3):1–40. https://ejrsme.icrsme.com/article/view/7741

Kalemkus J, Bayraktar S, Çiftçi S (2021) Comparative effects of argumentation and laboratory experiments on metacognition, attitudes, and science process skills of primary school children. J Sci Learn 4(2):113–122. https://doi.org/10.17509/jsl.v4i2.27825

Kontra C, Lyos DL, Fischer SM, Beilock SL (2015) Physical experience enhances science learning. Psychol Sci 26(6):737–749. https://doi.org/10.1177/0956797615569355

Linell P (2011) Samtalskulturer. Kommunikativa verksamhetstyper i samhället. Vol. 1 och 2. [Conversation cultures: communicative activity types in society. Volume 1 and 2]. Linköping University

Mafra P, Lima N, Carvalho GS (2014) Experimental activities in primary school to learn about microbes in an oral health education context. J Biol Educ 49(2):190–203. https://doi.org/10.1080/00219266.2014.923485

Mannix-McNamara P, Simovska V (2015) Schools for health and sustainability: insights from the past, present and for the future. In: Simovska V, McNamara PM (eds) Schools for health and sustainability. Springer, pp 3–17

Mody M, Gontjes KJ (2023) Health education after COVID-19: a time to revisit, revamp, and revitalize? Am J Infect Control 51(4):466–468. https://doi.org/10.1016/j.ajic.2022.11.018

Monroe MC, Plate RR, Oxarart A, Bowers A, Chaves WA (2019) Identifying effective climate change education strategies: a systematic review of the research. Environ Educ Res 25(6):791–812. https://doi.org/10.1080/13504622.2017.1360842

Murphy C, Varley J, Veale Ó (2012) I'd rather they did experiments with us…. than just talking: Irish children's views of primary school science. Res Sci Educ 42:415–438. https://doi.org/10.1007/s11165-010-9204-3

Nåden D (2010) Hermeneutics and observation: a discussion. Nurs Inq 17(1):75–81

Niebert K, Riemeier T, Gropengießer H (2012) The hidden hand that shapes conceptual understanding: choosing effective representations for teaching cell division and climate change. In: Treagust DF, Tsui C-Y (eds) Multiple representations in biological education. Springer Netherlands, pp 293–310

Nutbeam D (2019) Health education and health promotion revisited. Health Educ J 78(6):705–709. https://doi.org/10.1177/0017896918770215

Ochs E (1979) Transcription as theory. In: Ochs E, Schieffelin B (eds) Developmental pragmatics. Academic Press, pp 43–72

Paakkari L, Okan O (2020) COVID-19: health literacy is an underestimated problem. Lancet Public Health 5(5):e249–e250. https://doi.org/10.1016/S2468-2667(20)30086-4

Pedaste M, Mäeots M, Siiman LA, De Jong T, Van Riesen SA, Kamp ET et al (2015) Phases of inquiry-based learning: definitions and the inquiry cycle. Educ Res Rev 14:47–61. https://doi.org/10.1016/j.edurev.2015.02.003

Puchta C, Potter J (2004) Focus group practice. SAGE. https://doi.org/10.4135/9781849209168

Rasmussen J (2004) Textual interpretation and complexity. Radical hermeneutics. Nordic Stud Educ 24(3):177–193. https://doi.org/10.18261/ISSN1891-5949-2004-03-01

Resnik DB (2007) Responsibility for health: personal, social, and environmental. J Med Ethics 33(8):444–445. https://doi.org/10.1136/jme.2006.017574

Rönner A-C, Jakobsson A, Gericke NM (2023a) "Bacteria are not viruses; viruses are more malicious": young pupils' understanding of bacteria and viruses in the aftermath of COVID-19. J Biol Educ. https://doi.org/10.1080/00219266.2023.2247409

Rönner A-C, Jakobsson A, Gericke NM (2023b) Cough, sneeze, pass it on: pupils' understanding of infectious diseases in the aftermath of COVID-19. J Biol Educ. https://doi.org/10.1080/00219266.2022.2159492

Ruis-Gallardo J, Paños E (2018) Primary school students' conceptions about microorganisms. Influence of theoretical and practical methodologies on learning. Res Sci Technol Educ 36(2):165–184. https://doi.org/10.1080/02635143.2017.1386646

Rule AC (2006) Editorial: the components of authentic learning. J Authen Lear 3:1–10

Satterthwait D (2010) Why are "hands-on" science activities so effective for student learning? Teach Sci 56:7–10

Shah R, Hagell A, Cheung R (2019) International comparisons of health and wellbeing in adolescence and early adulthood. Nuffield Trust. https://www.nuffieldtrust.org.uk/research/international-comparisons-of-health-and-wellbeing-in-adolescence-and-early-adulthood

Sharma M (2021) Theoretical foundations of health education and health promotion. Jones & Bartlett Learning

Simard C (2023) Microorganism education: misconceptions and obstacles. J Biol Educ 57(2):308–316. https://doi.org/10.1080/00219266.2021.1909636

Song P (2014) A handful of bacteria: a simple activity that engages students to think and write like a scientist. J Technol Sci Educ 4(1):3–11. https://doi.org/10.3926/jotse.89

Sørensen K (2019) Defining health literacy: exploring differences and commonalities. In: Okan O, Bauer U, Levin-Zamir D, Pinheiro P, Sørensen K (eds) International handbook of health literacy: research, practice and policy across the lifespan. Policy Press, pp 5–20

Srivastava A, Ramadas J (2012) Analogy and gesture for mental visualization of DNA structure. In: Treagust DF, Tsui C-Y (eds) Multiple representations in biological education. Springer Netherlands, pp 311–329

Stuckey M, Hofstein A, Mamlok-Naaman R, Eilks I (2013) The meaning of "relevance" in science education and its implications for the science curriculum. Stud Sci Educ 49(1):1–34. https://doi.org/10.1080/03057267.2013.802463

Swedish National Agency for Education (2018) Curriculum for the compulsory school, preschool class and school-age educare. The Swedish National Agency for Education. https://www.skolverket.se/getFile?file=3984

Swedish Research Council (2017) Good Research Practice – Swedish Research Council (vr.se)

Taber KS (2015) Meeting educational objectives in the affective and cognitive domains: personal and social constructivist perspectives on enjoyment, motivation and learning chemistry. In: Kahveci M, Orgill M (eds) Affective dimensions in chemistry education. Springer, pp 3–27. https://doi.org/10.1007/978-3-662-45085-7_1

Taylor C, Dewsbury BM (2018) On the problem and promise of metaphor use in science and science communication. J Microbiol Biol Educ 19(1):1–5. https://doi.org/10.1128/jmbe.v19i1.1538

Taylor JA, Adams CT, Westbrook AL, Creasap Gee J, Spybrook JK, Kowalski SM, Gardner AL, Bloom M (2022) The effect of a student–teacher–scientist partnership program on high school students' science achievement and attitudes about scientists. J Res Sci Teach 59(3):423–457. https://doi.org/10.1002/tea.21733

Uitto A, Kärnä P, Hakonen R (2013) Työ- ja toimintatapojen yhteys biologian osaamiseen ja biologiasta pitämiseen peruskoulussa [Relationship between the working methods and lower secondary school students' performance and attitudes towards biology]. LUMAT 1:263–278. https://doi.org/10.1080/13504622.2017.1360842

UN (United Nations) (2015) Transforming our World: the 2030 Agenda for Sustainable Development; 21 October 2015 A/RES/70/1. https://sdgs.un.org/2030agenda

Vuorenpää S, Wolff L-A, Sjöblom P (2020) Social hållbarhet och scientific literacy i flerspråkiga naturvetenskapsklassrum [Social sustainability and scientific literacy in multilingual science classrooms]. Ainedidaktiikka 4(3):23–40. https://doi.org/10.23988/ad.97377

Watson-Gegeo KA (1997) Classroom ethnography. In: Hornberger NH, Corson D (eds) Encyclopedia of language and education, vol 8. Springer. https://doi.org/10.1007/978-94-011-4535-0_13

Winks L (2018) Discomfort in the field: the performance of nonhuman nature in fieldwork in South Devon. J Environ Educ 49(5):390–399. https://doi.org/10.1080/00958964.2017.1417219
Wolcott HF (2008) Ethnography: a way of seeing. Altamira Press
Wolff L-A, Vuorenpää S, Sjöblom P (2018) Chicken raising in a diverse Finnish classroom: multidimensional sustainability learning. Sustainability 10:3886. https://doi.org/10.3390/su10113886
Yamauchi T, Otsuka Y, Agestika L (2022) Influence of water, sanitation, and hygiene (WASH) on children's health in an urban slum in Indonesia. In: Yamauchi T, Nakao S, Harada H (eds) The sanitation triangle: socio-culture, health and materials. Springer Nature, pp 101–129
Yin RK (2003) Case study research: design and methods. Sage
Zeyer A (2012) A win-win situation for health and science education: seeing through the lens of a new framework model of health literacy. In: Zeyer A, Kyburz-Graber R (eds) Science | environment | health. Springer, pp 147–173
Zeyer A, Dillon J (2012) Science, environment, health: towards a reconceptualization of three critical and inter-linked areas of education. Int J Sci Educ 34(2):327–328. https://doi.org/10.1080/09500693.2011.647111

Lili-Ann Wolff is an Associate Professor in environmental education at the University of Helsinki in Finland. In addition to education, she has degrees in science and philosophy. Her research interest includes sustainability education, and science education from transdisciplinary and critical views, combining education (theory, policy, empirical studies, teacher education, etc.) and philosophy (classical philosophy, philosophy of education, ethics, philosophy of science, Bildung theories, etc.). She is an active editor and author with many articles and book chapters in international publications. She has a long experience of both national and international networking and collaboration.

Sari Vuorenpää is a Senior Lecturer at the Department of Special Education at Stockholm University, Sweden. Her research interests revolve around the language and knowledge development of students and include interdisciplinary studies in multilingual schools based on students' diverse needs. An important starting point in her research is therefore the different types of literacy environments offered to children and young people in school. The goal in her research is new knowledge about learning in various learning environments. The focus is on the role of language and communication in shaping belonging and further a profound understanding of the world.

Pia Sjöblom is a University Lecturer in biology and geography education at the Faculty of Human and Social Sciences at Åbo Akademi University in Finland. She holds a PhD in environmental education and works as a teacher educator and researcher. Her research interests are science education, biology education, science teacher education, and outdoor education, as well as environmental and sustainability education from several perspectives. Her published research includes articles based on both empirical and theoretical studies.

Concluding Reflections on Science Education and Sustainability

Frode Skarstein, Lili-Ann Wolff, and Kerry Shephard

This volume attempts to draw together insights from the traditional but rapidly evolving domain of science education, and the more modern and urgent domain of sustainability education. But the task has not been easy. Both domains are highly contested historically and currently. It is challenging to find generally accepted definitions of the basic concept of science and of sustainability, even within these pages, and attempts to discuss how to educate for, or about, science and for or about sustainability add additional levels of uncertainty to the task. The combined knowledge domains of science and sustainability encompass a wide diversity of epistemological and ontological ways of knowing and ways of understanding the world. The authors who have contributed so conscientiously to the challenges of reflecting on sustainability through the lens of science education, in this volume, themselves present a wide diversity of perspective on the core differences between instrumental, normative, approaches to education and more emancipatory, hopeful, approaches. The problems inherent to science education and to sustainability education are not necessarily diminished by reflecting on one through a lens provided by the other, but perhaps they are helpfully brought into focus together.

To reflect on sustainability through the lens of science education is to reflect on some of the most fundamental questions of our time. How should people live together on a finite planet? What forms of knowledge and reasoning can guide people's choices? And how can education prepare people not just to survive but to

F. Skarstein
Department of Education and Sports Science, Faculty of Arts and Education, University of Stavanger, Stavanger, Norway

L.-A. Wolff
Faculty of Educational Sciences, University of Helsinki, Helsinki, Finland

K. Shephard
Higher Education Development Centre, University of Otago, Dunedin, New Zealand

F. Skarstein, L.-A. Wolff (eds.), *Sustainability and Science Education*, Environmental Challenges and Solutions,
https://doi.org/10.1007/978-3-032-24408-6

flourish in conditions of uncertainty, diversity, and constraint? These questions are not answered by science alone, nor by values alone, but by the interplay between them. Science provides insights into how the world works and the boundaries within which human societies must operate, while values shape what kind of world people aspire to create. Education, situated at the intersection of knowledge and value, is therefore both a deeply intellectual and a profoundly ethical endeavour.

Contributions to this volume remind the readers that progress in sustainability is rarely stalled because of missing facts. More often, it is shaped by conflicts of interest, competing priorities, and divergent worldviews. Energy policies, biodiversity strategies, and public health measures are contested not because people lack evidence, but because evidence can be or must be interpreted within frameworks of meaning and purpose. For science education, and for science educators, this recognition is pivotal. It challenges the assumption that more knowledge and understanding will lead to wiser decisions, and it calls instead for an education that equips learners with tools to connect knowledge and understanding with values, and to deliberate across differences.

At the same time, the chapters in this book insist that some versions of science, and so of science education, remain indispensable to the future of humanity on this planet. Science does not go away because somebody chooses not to teach it, or to combine it with social science or the humanities, or to market it as science communication. Reliable knowledge and understanding of ecosystems, climate systems, and health processes are prerequisite for any serious conversation about sustainability. Without such knowledge and understanding, debates risk becoming ideological, detached from the material realities of the Earth, the solar system, and what people know about the universe. In the context of sustainability, education thus has a dual task: to foster understanding of scientific concepts and methods, and to situate them within the social, political, and ethical contexts where sustainability decisions are made. This duality runs through this volume, giving it both intellectual depth and practical urgency, but does not define in which discipline this task resides, and it often leaves open the question of what limits could or should be placed on the responsibilities of science disciplines, of science education, and of science educators.

To this end, this volume demonstrates the necessity of interdisciplinarity. No single discipline can capture the full complexity of sustainability challenges. Learners must draw on economics, ethics, history, politics, and the arts, as well as on science. To be sure, science has a distinctive role in this wider conversation. It provides ways of testing claims against evidence, of distinguishing stronger explanations from weaker, and of offering shared reference points for dialogue. When students understand both the reach and the limits of scientific inquiry, they become better equipped to navigate contested debates and to participate in the collective task of shaping sustainable futures.

And so, in addition to interdisciplinarity, a focus on agency is also crucial. Learners are not simply receivers of knowledge but active constructors of meaning. The contributors to this volume illustrate how inquiry-based learning, interdisciplinary projects, and critical reflection can foster students' capacity to act. Agency here cannot necessarily be directly equated with purposefully trying to change other people's behaviour but may rather require an education that triggers others to develop the ability to question assumptions, to deliberate about alternatives, and to imagine alternative futures. Education that emphasises sustainability must therefore cultivate not only scientific literacy but also civic competency and imaginative capacity.

Such an education may, in the future, have at its centre an evolving core of science education. It will certainly be increasingly interdisciplinary and focused on agency. But, at present, it is incompletely conceptualised.

The contributions of this volume enable the authors of this chapter to imagine ways in which science education could evolve and connect more deeply with other forms of education. We start by considering what science education currently has to offer a future interdisciplinary- and agency-focused education in the context of sustainability. We note that science education recognises complexity, embraces uncertainty, and fosters resilience in the face of misinformation. In a world where public debate is shaped by social media, algorithmic echo chambers, and deliberate attempts to mislead, science education enables learners to distinguish between empirically grounded knowledge and manipulation.

Looking beyond what is probably currently achieved, science education promotes critical thinking and so potentially opens pathways to media literacy reflexivity about how science itself is produced and validated and to openness to plurality, acknowledging that sustainability is about reconciling different values, not arriving at a single truth. Our vision for science learning is much more than an intellectual undertaking; it is also something emotional, social, aesthetical, and ethical. Our vision of science education needs to engage the entire individual and its mutual engagement in social contexts, but how much of this will be achieved by science educators and how much in partnership with other educators much remain to be imagined. Such an education leans on goals that extend far beyond measurable results accrued by individual learners, but it does not negate the necessity for educators to engage in ongoing and comprehensive evaluation of the extent to which the contemporary educational mission is progressing.

In conclusion, this anthology shows that science education has a vital role to play in the search for sustainability, but not as a provider of readymade solutions. Its role is to preserve and transmit scientific knowledge and understanding, to teach how that knowledge and understanding is generated and tested, and to foster critical engagement with how it interacts with values and politics. Science cannot prescribe the futures anyone chooses, but without scientific contextual understanding, those choices risk being blind. Science remains at the core of human capacity to understand planetary limits and to imagine how to live justly and safely within them.

The task ahead is therefore clear but demanding. Education built around a core of science education must empower learners to navigate value conflicts with openness, to resist the distortions of misinformation with critical insight, and to ground collective action in empirically tested knowledge of the natural world. If it succeeds, it can help build societies that are not only informed but also wise, societies capable of deliberating across differences and organising their lives within the safe and just boundaries of planet Earth. To fail would be to leave future generations without the intellectual tools they need to face the defining challenge of the century. To succeed is to make science education a cornerstone of humanity's shared work towards a sustainable future.

MIX
Papier aus verantwortungsvollen Quellen
Paper from responsible sources
FSC® C105338

If you have any concerns about our products,
you can contact us on
ProductSafety@springernature.com

In case Publisher is established outside the EU,
the EU authorized representative is:
Springer Nature Customer Service Center GmbH
Europaplatz 3, 69115 Heidelberg, Germany

Printed by Libri Plureos GmbH
in Hamburg, Germany